Planet of the Grapes

PLANET OF THE GRAPES

A Geography of Wine

Robert Sechrist

An Imprint of ABC-CLIO, LLC
Santa Barbara, California • Denver, Colorado

All illustrations in the text are by Ryan Brown.

Library of Congress Cataloging-in-Publication Data

Names: Sechrist, Robert P.
Title: Planet of the grapes : a geography of wine / Robert Sechrist.
Description: Santa Barbara, California : Praeger, an imprint of ABC-CLIO, LLC, [2017] | Includes bibliographical references and index.
Identifiers: LCCN 2016058638 (print) | LCCN 2016059489 (ebook) | ISBN 9781440854385 (alk. paper) | ISBN 9781440854392 (ebook)
Subjects: LCSH: Wine and wine making. | Wine districts. | Wine industry.
Classification: LCC TP548 .S4655 2017 (print) | LCC TP548 (ebook) | DDC 663/.2—dc23
LC record available at https://lccn.loc.gov/2016058638

ISBN: 978-1-4408-5438-5
EISBN: 978-1-4408-5439-2

21 20 19 18 17 1 2 3 4 5

This book is also available as an eBook.

Praeger
An Imprint of ABC-CLIO, LLC

ABC-CLIO, LLC
130 Cremona Drive, P.O. Box 1911
Santa Barbara, California 93116-1911
www.abc-clio.com

This book is printed on acid-free paper ♾

Manufactured in the United States of America

Contents

Preface

All night I drink, and study hard all day: Bacchus and Phoebus hold divided sway.[1]

—Martial

I wrote this book to help others understand the fundamental role wine played in past civilizations, in the shaping of the modern world, and the pivotal role wine plays in modern geography, society, and economy. This role is expressed in the landscapes, trade patterns, and economic structures of many nations. Adored and abhorred, wine is unique among all global commodities. The cultural, historical, economic, social, religious, and medical significance of wine is not given its due in the American educational system, a wrong I hope to partially correct with this book. My title, *Planet of the Grapes*, refers to the pervasive importance that fermented grape juice had and continues to have in world affairs. Pierre Boulle, I apologize.

Geography is the science that studies patterns, flows, and distributions in and on the earth's surface and beyond. Geographers identify what makes places unique, what makes them similar, what is possible at a location (climatically, biologically, agriculturally, technologically, or socially), and what is not possible, the ebbs and flows of phenomena across the earth's surface, and the interaction of phenomena in space. Many geographic principles, theories, and methods are exemplified here by studying the spatial distribution, patterns, and place interactions resulting from the human love affair with wine. The models discussed in the following pages apply to wine's discovery, its spread, its transport, its role in globalization, its trade, its character, its spatial variability, its role in society, its manufacture, its horticultural practices, its flavors and aromas, its lovers, and its detractors.

Vitis vinifera, the wine grape, is endlessly variable; even clones show changes. Each seed carries a unique vine, a new variety. Successful seeds have characteristics suited to the local environment. Each seedling emerges in a unique location and experiences unique conditions. Prehistoric peoples over many generations went from eating ripe grapes once a year, to nurturing particular plants, to creating vineyards and wineries. How grapes from those vines became wine (the symbol and the drink) results from the cultural and geographic heritage of Western civilization.

The varietal distribution of wines is an interpretation of physical, cultural, and economic geography. Economic geography evaluates the shape and structure of the global flow of wine from source (vine) to destination (mouth). Among those who consume wine, knowledge of geography improves their ability to select wine based on the character of regional variations in wine production methods. Tastes and aromas trigger memory. Knowledge of the local environmental characteristics and tradition enhance the wine drinking experience by tying those pleasant (and even unpleasant) memories to the specific flavors and fragrances of a specific wine.

Reflect on the quotations that introduce each chapter and you begin to appreciate the depth of the role of wine in Western society, from the transformative act of domesticating a wild man with wine (as portrayed in the Epic of Gilgamesh and the Greek rise from barbarism) to modern political battles regarding public access to alcoholic beverages.

One indication of the importance of wine in Western civilization is reflected in the number and location of places named for wine or for the vine. Vienna, Austria and Venice, Italy are the two most prominent European cities whose name translates as wine. In the United States many towns have a Vine Street, Concord is a common town name. Many regional names have also become synonymous with the types of wines produced there. Champagne, Port, Burgundy, Tokay, and Marsala are just a few wine place names protected by international agreement.

To maximize your experience, I suggest readers do two things. First, visit your local wine merchant. Examine the bottles, varieties, and styles that correspond to the portion of this text you are reading. Second, take the words of the Roman poet Martial (above) to heart and accompany the reading of each section with a glass of the appropriate wine.

Indiana, Pennsylvania

A Note on Numbers and Maps

Standard conventions are followed when recording and reporting numbers. Dates are presented in years BCE and CE. BCE ended about 2015 years ago. Whenever possible statistics are presented in their native, published, measurement units. Measures of area are generally presented in acres, thousands of acres, or hectares (2.5 acres). Volume is either in gallons or hectoliters (100 liters or 26.4 gallons), depending on the source. Tonne(s) refers to a weight of 1,000 kg and tons refers to 2,000 lbs. Descriptions of bottle and barrel sizes, capacities, and additional measures can be found in chapter 11.

The maps accompanying the text are simple and general references to major wine regions. There are two sources I recommend for accessing high-quality detailed maps of wine regions. The first is the Society of Wine Educators' online wine maps at: http://winewitandwisdomswe.com/wine-spirits-maps/wine-maps-2015-csw-study-guide/. Produced by Bill Lembeck, they show all the world's wine regions in a consistent format. The second source is the *World Atlas of Wine* by Hugh Johnson and Jancis Robinson. This volume contains large-scale topographic maps of wine regions and vineyard locations. A third set of sources are the regional and national websites promoting each region's wine. However, while often quite informative individually, these often lack consistency and availability across regions.

Introduction: The World of Wine

When it comes to wine, there is no ingredient more important than location. The land, air, water, and weather where grapes are grown are what make each wine unique.[1]

—Origins.wine

The world of wine was much simpler just a few years ago. At the beginning of the millennium New Zealand's Sauvignon Blanc swept the planet as *the* newly discovered source for this already famous international variety. Since 2003, a succession of places and grape varieties have become media darlings. In 2005, thanks to the movie *Sideways*, Pinot Noir popularity spiked and Merlot sales plummeted. Argentinean wines made from Malbec and Torrontes were introduced to North America with great success. Calls for Riesling revivals echoed in the corners of wine shops and magazines. Many U.S. consumers got headaches trying to distinguish between Pinot Gris, Grauburgunder, and Pinot Grigio (all the same variety but from France, Germany, and Italy, respectively).

The establishment and subsequent discovery of new wine regions are a global trend, with wine consumers the winners. Wine, and the search for new places where grapes could grow, was a driving factor in European colonialism. China, which had very limited vineyards under the strict communist regime of the past, is now number six in wine grape acreage. Virginia is becoming known for its Petit Verdot and the Finger Lakes region of New York for its Riesling, which tastes entirely different from Washington State's Riesling. The countries of Georgia and Turkey have reemerged as net exporters. The hunt is on for new wine lands, often in unlikely places. Researchers are developing varieties capable of succeeding where no grape has grown before. What or where will be next?

Wine has been a major component of the global economy for centuries. Carefully stored and transported on wooden ships, wine was a widely traded liquid commodity. Proven by millennia of consumers, wine is a tasty, healthy food, consumed around the world by tens of millions daily. Producing the world's wine supply employs millions across six continents. The distribution and sales network delivering wine to consumers, including food service personnel, employs many millions more.

Grapes have the amazing capacity to produce a high-value crop in soils and locales where other fruits, vegetables, and even grains wither. The vine makes otherwise barren land green and productive, resulting in significantly higher population densities than otherwise possible. The potential for great financial returns makes grape growing one of the most studied agricultural fields. The first books on farming dealt solely with grape growing. In the regions where they grow, grapes are a keystone species, a defining element of the cultural landscape. In these places (and others) the grape symbolizes wealth, plenty, sophistication, and civilization.

Portland state geographer Teresa Bulman supported this view in her 2003 presidential address to the Association of Pacific Coast Geographers when she said, "Many subfields of geography intersect with the world of wine. . . . [T]raining in geography enables geographers to see and understand numerous aspects of the physical and human dimensions of . . . wine."[2] Or, as geographers John Dickenson and John Salt expressed it in their 1982 analysis of the relationship between wine and geography, wine

> may be studied from a variety of perspectives and encompasses the influence of the physical environment, historical diffusion of the vine and viticulture, economic geographies of cultivation and marketing, political influences on trade and production, and cultural perceptions of landscape, products, and people.[3]

The evolution of the global network of wine producers, shippers, distributors, retailers, and consumers is rife with opportunities to explain geographic theories, concepts, and models regarding economics and global trade. Geography, and the related concept of *terroir*, can go a long way toward explaining why each wine is as individualistic as a fingerprint. Geographer Warren Moran explained terroir from the geographic perspective this way: "That the physical environment influences the vine and its production is incontestable. The vine has its roots in the soil and leaves and fruit in the atmosphere. Its juice, musts, and wines inevitably reflect these conditions."[4]

Despite the acknowledged connection between place and grape, wine has a murky and ill-studied geography. In the United States nativism and religious fervor led to ostracism of those who consumed alcohol. For decades the topic was shunned by social scientists because of the long-standing social stigma associated with it. Research dollars flowed from state and federal agencies, whose sole existence is the demonizing of alcoholic beverages, to those who "discovered" health and social problems caused by drinking. The open mind of the researcher is questioned when government officials like U.K. chief medical officer Sally Davies said on British TV, "There is no safe level of drinking."[5] To be fair, other state and federal agricultural agencies promote the production of wine grapes and alcoholic beverages. Revenue and treasury agencies promote drinking as a source of tax revenue.

Drinking fermented products with premeditated intoxication in mind began more than 100,000 years ago. Initially consumption was fortuitous: eating overripe and fermenting fruit or honey mixed with enough water to permit fermentation. Later it became more regularized, as people camped near the fruit groves and saved honey for subsequent fermentation at ceremonial events. Drinking fermented grape juice (wine) began at least 10,000 years ago in the foothills of the Caucasus Mountains. In these places and others, the grape symbolizes wealth, plenty, sophistication, and civilization. Diffusionary studies tell us of wine's spread from the Caucasian foothills to Mesopotamia, the Mediterranean, and the world.

For most of Mediterranean history wine was a given; everyone knew the process (although not why or how grape juice transformed into wine) as well as anyone today in that same region. Numerous Egyptian tomb paintings depict wine making and consumption. Cato's *De Agri Cultura,* written in 134 BCE, is the oldest extant Latin text. Its topic is grape production and wine making.

Where and when people could not obtain wine they drank less potent alcoholic beverages (grain beer, fermented milk, banana beer, and fruit wines) while still craving the grape wine enjoyed by royalty. Since its discovery wine has always been the leading high-value liquid commodity. As a liquid commodity it required specialized containers and transport.

Wine, as identified by cultural anthropologist Donald Horton, satisfies four basic human needs: (1) hunger and thirst, (2) medicine, (3) religious ecstasy, and (4) social jollification.[6] Together these characteristics make wine an important factor in human physical and mental health.

Ancients considered wine to be the "food with two faces." The comedy and drama masks derive directly from the worship of Dionysus, the Greek god of wine. In moderation wine is good for both mind and body. Wine was used to kill microorganisms in water millennia before the microscope first allowed us to see them. Unlike beer and liquor, wine contains many trace nutrients. People throughout the centuries have consumed wine to escape their daily worries and physical ailments. Wine, a natural tranquilizer and depressant, has the capacity to dull mental and physical pain. For centuries it was practically the only medication available to most of the world's population.

The ancients also drank wine for its nutritional value. Red wine, through the compound resveratrol, scavenges free radicals. Charred foods have free radicals; physiologically red wine is, therefore, an appropriate counterbalance to grilled (charred) food. Radiation also creates free radicals. Medical researchers believe red wine is especially useful for patients undergoing chemotherapy. This capacity of wine was well understood as early as the 1950s, when it was rumored that Soviet nuclear submarine crews were issued red wine to combat exposure. The advice "In Case of Nuclear Attack, Drink Red Wine," long an urban myth, was verified by University of Pittsburgh researchers in 2008.[7]

Consume too much, however, and the alcohol begins to have a poisonous effect on internal organs. Consume way too much (in one sitting) and your body will violently reject it. Before you reach that point, however, you will probably be physically discordant and your mental processes will be attenuated. Worst of all, you will reach a state of boorishness that can only be tolerated by others in your condition.

The newly (re)discovered health advantages of wine drinking encouraged many Americans to start enjoying it in the 1990s. During the Prohibition era medical texts were purged of any mention of the positive effects of wine. For almost fifty years after Prohibition's repeal, this information was suppressed by government health professionals. In the 1970s medical research discovered that moderate wine consumption reduced coronary disease, but the government was uninterested in promoting wine as having medicinal benefits.

The concepts of regional geography and terroir are closely related. They both consist of studying and classifying a location's geology, topography, soil, weather, and climate. Geographers commonly specialize in a region, cataloging, inventorying, and tracking some or all of the characteristics of that region with the goal of deriving explanations for the patterns observed. Grape growers rely on the products of geography, including climatological, meteorological, geomorphological, and pedological (soil) data, to select vineyard sites and grape varieties. Someone evaluating a location's terroir does so with the goal of identifying the potential of the land to produce a grape crop and ultimately revenue.

Finally, wine is politically and socially controversial. The conflicting messages associated with wine are expressed and visible on the American landscape. Social wine geography in the United States is greatly influenced by paternalistic governmental control, which in turn is controlled by vocal, often fear-mongering, minorities whose primary arguments are replete with pseudo-science, opinion expressed as fact, and outright lies. All wine labels require federal approval. When *60 Minutes* aired the story "The French Paradox" in 1991, the government's anti-alcohol stance came to light. Since then, grudging officials have acknowledged that moderate daily wine consumption is good for the average adult. Political kickback from the "concerned" resulted in the use of scary terms to describe alcohol consumption and reduced legal alcohol limits for vehicle operation.

Governmental suppression of freedom of speech regarding wine ended in 1996 in the Supreme Court case *44 Liquormart, Inc. v. Rhode Island.* A Rhode Island law forbade the public advertising of wine prices in flyers, newspapers, and/or posters in store windows. The Court decided the Rhode Island law was an unconstitutional suppression of free speech, as the store's statements were true. The Court's action simultaneously struck down several other state laws and federal policies.

In international geo-econo-politics wine is among the most intensely debated commodities. Historically, wine exports and imports were controlled by national policy either to assist friendly nations or to hurt enemies. For the last 1,000 years British wine imports fluctuated with changing Continental relationships. Following the Iranian revolution in 1979, Iranians smashed wine bottles found in the cellars of Western hotels. In 1991, Americans smashed French wine bottles when the French government objected to U.S. actions in Iraq. In local socio-geo-econo-politics the locations and rules under which wine may be produced, possessed, and consumed demonstrate the complexity and intensity of opinions regarding this simple natural product. Wine is seen as a source of government revenue, a cultural trait depicting both elegance and ruin, wholesome, and a danger to our youth.

Look around! Are there grape and vine symbols incorporated into the architecture of the buildings? Is there any artwork that contains grapes, wine, vines, or wine gods in sight? How about advertising? How many stores retail wine in your community? They exist all around you and you will see them when you look.

Chapter 1

The Historical Geography of Wine

The strange power of intoxicants to release the human spirit from the control of the mind led to their being regarded with superstitious awe and, seized upon by shamans, witchdoctors, and priests, they became early and everywhere instruments of religious experience. Their use became a religious right, and this was the case of wine.[1]

—Edward Hyams

There is no other liquid that flows more intimately and incessantly through the labyrinth of symbols we have conceived to mark our status as human beings from the rudest peasant festival to the mystery of the Eucharist. To take wine into your mouth is to savor a droplet of the river of human history.[2]

—Clifton Fadiman

I then discovered wine's power: its historical significance; its use as a civilizing drink, both as a food and a social beverage; its political influence; its investment opportunities; and as a subject for lengthy and interesting discussions.[3]

—Walter J. Clore

WINE IN PREHISTORY

Watching the National Geographic special on the Okavango Delta in Botswana we are treated to images of wildlife eating fermented fruits until they stagger and pass out. Many species of birds and mammals seek altered states of reality in this way. Zebras, elephants, warthogs, baboons, and birds cavort in their inebriated joy before falling unconscious after eating fermented fruits. You can see this video on YouTube; simply search for "drunken animals." This proclivity, practically a biological urge, is not, obviously, uniquely human.[4]

Wine was certainly not the first alcoholic beverage consumed by humanity. Annual visits to groves of trees whose fruit spontaneously ferments would have been the first intermittent source of alcohol. The first regular supply of a fermented beverage probably came from honey. A mixture of one part honey to three parts water will readily ferment from natural yeasts present in the honey.[5] Evidence from Mesopotamia and pre-Columbian Mexico indicates that people made alcoholic beverages from a variety of plants.[6] Our ancestors certainly consumed these products greedily.

Archaeologists discovered ancient grape seeds and residue with other artifacts in the vicinity of Nice, France, dating from about 350,000 BCE. Chronologically, the next grape discovery was in a Greek cave, where seeds dating to 12,000 BCE were discovered. From about 8000 BCE onward, grape seeds are plentiful at archaeological sites across the Mediterranean and Middle East. There is no proof that pre-agricultural ancient people made wine at these sites, though; perhaps they simply ate the grapes.

The earliest physical evidence of any fermented beverage comes from Henan Province, China, from a site called Jiahu. The pottery at this Yellow River valley site contained the mixed residue of grapes, hawthorn, rice, and honey.[7] People living in the southern foothills of the Caucasus Mountains domesticated the grape between 10,000 and 8000 BCE. At first, wine was only available shortly after harvest, due to its perishable nature. Eventually people learned to make pottery, and pottery allowed storing wine for year-round consumption. The hypothesis that these people became sedentary to protect their wine supply is as plausible as the common, less controversial belief that they settled to protect the grain crop for bread making. The insistence that grains were grown for bread, not beer, is a product of the American educational system and is refuted by University of Pennsylvania archaeologist Patrick McGovern:

> Fermented beverages have been preferred over water throughout the ages: they are safer, provide psychotropic effects, and are more nutritious. Some have even said alcohol was the primary agent for the development of Western civilization, since more healthy individuals (even if inebriated much of the time) lived longer and had greater reproductive success.[8]

Producing, protecting, and trading the valuable "holy" fluid must have been much more lucrative than growing wheat. The earliest definitive archaeological evidence of wine making comes from the Neolithic village of

Hajji Firuz Tepe (in modern Iran) where six ceramic pots dated from 5400 BCE to 5000 BCE were found. The second archaeological appearance of wine comes from Godin Tepe, Iran (about 300 miles south of Hajji Firuz Tepe) and dates from between 3300 and 3000 BCE.

The most ancient known grape-pressing facility, discovered in 2007, is in modern-day Armenia. The wine press and fermentation jars date to around 4000 BCE. The sophistication of the site's content and organization suggest a well-understood process, indicating that wine-making technology is significantly older.[9]

The ancients did not know how grape juice transformed itself into wine. For thousands of years, no one knew. Without an answer, our ancient ancestors invoked the supernatural, and the biochemical transformation took on mystical and religious symbolism. The symbolism extends to the sources of the mystic fluid, the grape, the vine, and even to wine's material culture. In multiple religious ceremonies wine came to symbolize blood, replacing the need to spill actual blood in sacrifice. Greek theater and plays evolved from the symbolism of Dionysian religious rites.

The ancients used wine to both deaden pain, physical and mental, and to clean wounds. Drinking fermented beverages was essential when clean water was unavailable, as was usually the case. Wine dissolved medicinal chemicals that water could not. Wine converted to vinegar was heated and used to crack and dissolve limestone in construction projects.

ANCIENT WINE: ORIGIN AND DIFFUSION

Vitis vinifera was discovered growing wild in the Caucasus foothills of modern Georgia, Azerbaijan, and Armenia perhaps 100,000 years ago. At first, people could not store wine and thus consumed it only seasonally. The invention of pottery allowed year-round storage at least 9,000 years ago.[10] From these foothills grape growing and wine making originated and diffused. Noted wine authority Hugh Johnson traced wine origins linguistically and reports, "The Georgian name *gvino* spread to the whole of Europe and the world from Georgia and was established in a lot of languages in the same or changed form: vino, vin, wein, wine."[11] It can be argued that the ancient decision to domesticate the grape and make wine launched the agricultural revolution every bit as much as the decision to domesticate wheat.

From the beginning, and continuing today, wine is a commodity that inspires trade. Wine trade descended the Tigris and Euphrates rivers from the Caucasus region. About the same time, grape growing expanded southward onto the slopes of the nearby Persian plateau. The Mesopotamian centers of civilization were located in the riverine lowlands near their irrigated fields. Wine, made in the hills upstream, was transported down the Tigris and Euphrates to the rulers of these cities.

Wines imported from the highlands were expensive. Among Mesopotamians wine was a royal luxury obtained by import from the mountains to the north over a secure transportation network. The commoners drank what Dr. Patrick McGovern has termed "grog." He describes it as a fermented mixture of available grain, fruit, and honey whose flavor and character changed with the seasons.[12] Grog was consumed quickly before it soured and new grog was started each day. Grog was routinely consumed through a straw to help strain out the bigger chunks in the mixture. One of the earliest female occupations was making and selling ale and date wine from jars kept outside the front door. European women provided this service from their back porches well into the Middle Ages.

There was clearly a beer-wine boundary in 4000 BCE as archaeologist Marvin Powell states:

> The points of equilibrium where barley and wine—beer and grape—culture met were probably, on the Euphrates, around the Syrian–Iraqi border and, on the Tigris, around modern Tikrit. True wine cultures seem to have begun still further up the river valleys and at slightly higher altitudes, in other words, within the perimeter of natural rainfall agriculture.[13]

Meanwhile, wine imported from the highlands and floated down the Tigris and Euphrates rivers, as Jeffery Munsie describes, satisfied the nobility and priesthood. Commoners drank McGovern's grog:

> The Code of Hammurabi from the Sumerian empire around 1700 BCE made it clear that beer was the most important drink, but indicated that wine was consumed by the ruling elite and was essential for religious ceremonies. Paintings, statues, and papyri also illustrate the social and religious functions that wine played during Egyptian times (2000 to 1000 BCE), even though beer was still more popular. The Assyrian empire then left its mark in the form of carved reliefs depicting wine and tablets from 800 BCE that provided a list of those who were entitled to a daily ration of wine due to their employment in the king's service.[14]

As transportation became smoother and the trade routes longer and safer, wine culture expanded beyond the Mesopotamian river trade. Wine culture diffused to the Mediterranean by following the arc of the Fertile Crescent westward. The people of the Fertile Crescent joined the wine trade and expanded it to Turkey (then ruled by Hittites) and then to coastal peoples of Phoenicia (Canaan) (in the area of modern-day Lebanon). Wine came to Egypt as gifts from Mesopotamian nobles, at first. The establishment of wine culture in Phoenicia and Palestine led to a rapid expansion of wine consumption in and around the Mediterranean Basin.

ANATOLIA (MODERN TURKEY)

The Hittites had a greater influence in diffusing wine than the Egyptians. The Hittites (living in what is today Turkey, Lebanon, and Syria) produced wine adequate for their entire population and for export. The Hittite Empire flourished between 2000 and 1200 BCE when they were the region's only master of iron. The Hittites deified wine, providing the foundation for the Greek god Dionysus. The Hittites overwhelmed and occupied Troy about 1800 BCE.

The Trojan War occurred near the end of the Hittite Empire. Most modern accounts ignore the many jars of wine accompanying the wooden Trojan horse in Homer's *Odyssey*. No wonder the Trojans were overwhelmed by the coupled joy of apparent victory and the intoxication of free wine.

After the demise of the Hittites the Phrygians came to dominate Anatolia. They introduced wine to the Greeks on the Anatolian coast. It was the Phrygians who introduced the Muscat grape to Europe. With the collapse of the Hittite Empire, the Phoenicians emerged as the principal movers and shakers in the Levant.

THE LEVANT (LEBANON, ISRAEL, SYRIA, AND JORDAN)

No single culture had more influence on the spread of the vine across the Mediterranean Basin than the Phoenicians. In *Food: A Culinary History*, Flandrin and Montana report that "Egyptian annals refer to 'wine running like water in the presses' at Ullaza. The fact that winemaking was highly developed during the second millennium B.C.E. [2000–1000 BCE] is shown by the texts from the cities of Ugarit and Alalakh, both sited north of Phoenicia."[15] Ullaza was a coastal Phoenician city near Biblos.

The Phoenicians embraced and spread the vine throughout the ancient world. Wine featured prominently in their religion as observed in this passage from their religious text, the *Rapiuma*.

> Day long they pour the wine, . . . must-wine, fit for rulers.
> Wine, sweet and abundant, Select wine . . .
> The choice wine of Lebanon, most nurtured by El.
> Eat, O Gods, and drink, drink wine till you are sated.[16]

The Phoenician society was based in trade. Beyond wine they bought and sold wood, spices, gold, purple dye, and other commodities. They routinely traded across the Mediterranean and Black Sea. Their trading took them to Britain for tin and Senegal for ivory, gold, and gems. The original Phoenician cities were Tyre, Byblos, Caesarea, and Sidon. The colonies they established include some of Europe and North Africa's most famous cities: Carthage,

Marseilles, Malaga, Tangier, Utica, Genoa, Venice, Palermo, and Cadiz. Referring to what was probably a Muscat wine, Samuel Morewood wrote in 1838, "The wine of Byblos, in Phoenicia, was much esteemed for the strength of the perfumes with which it was impregnated."[17] Phoenician historian George Rawlinson also attests to the quality of the region's wines: "Syria of Damascus gives the 'wine of Helbon'—that exquisite liquor which was the only sort that the Persian kings would condescend to drink."[18]

Phoenicians regularly sailed as far as Cornwall, England, to acquire tin (necessary to make bronze) and Senegal to obtain gold. By 1000 BCE Phoenicians had diffused wine culture through most of the ancient world. Phoenician sailors were also the first to circumnavigate Africa, about 600 BCE. They were secretive about their ways and their sources for trade goods. The full nature of their culture remains somewhat of a mystery to us, as a result of their need for secrecy.

By the time of the Phoenicians the pattern for trade between strangers was established. The trader sailed into a Mediterranean coastal village. His first act would be to pay tribute (by making gifts) to the local ruler. Wine was foremost among these gifts as were wine-related trinkets (goblets, *kottabus* poles, figurines, mixing bowls, kraters, painted pottery, etc.). Upon meeting, wine might be distributed. The wine loosened trade and heightened sales of all goods. The wine itself was a valuable trade item. The wine trinkets helped cement wine culture: staring at an empty wine goblet would invoke memories of the last time it was full and the owner happy. Edward Hyams suggests this led to a desire to profit from producing one's own wine:

> The steps in the introduction of the vine would have been much as they were in Thrace: first wine is sold; the natives have become very fond of it and, perhaps addicted to it, want more than they can buy; as soon as they begin to settle and farm—that is, as civilization spreads to them—they plant the vine. More often, a trading-post develops into a small, true colony, with wine plantation. The vine is one of the very first plants to be established, because wine is not only easily made with very little gear, but it is an immensely profitable object of trade. From the colonial vineyard the plants make their way in due course into the still-independent native country.[19]

The trade between the ancient Israelites and their Phoenician neighbors was brisk. The ancient Hebrews called the Phoenicians Canaanites, Hebrew for "merchants." The Phoenicians provided much of the material, labor, and expertise for constructing the Hebrew temple in Jerusalem.[20] In compensation, Phoenician historian George Rawlinson asserts,

> the Israelite monarch on his part made a return in corn [not maize], wine, and oil, supplying Tyre, while the contract lasted, with 20,000

cors of wheat, the same quantity of barley, 20,000 baths of wine, and the same number of oil, annually.[21]

Obviously, the ancient Hebrews produced much of the wine traded around the Mediterranean by the Phoenicians.

The influence of the traveling Phoenicians appears to have parented a secondary domestication event. Genetic research conducted on Iberian grapes suggests that, after bringing the idea (and some samples) of wine making to the people of Iberia, the Iberians themselves began making wine from local grape varieties instead of (or in addition to) the varieties of the eastern Mediterranean. The genetic mapping and subsequent analysis by Arroyo-Garcia and colleagues revealed, "The western Mediterranean region . . . gave rise to many of the current Western European cultivars. Indeed, over 70% of the Iberian Peninsula cultivars . . . are . . . derived from western *Sylvestris* populations."[22]

A Carthaginian religious organization known as the *marzeah*, "linked to the cult of certain divinities and the dead, made considerable use of wine during its rites."[23] This tradition probably originated in the Phoenician homelands and might relate back to rituals associated with the Phoenician deity El.

EGYPT

Wine arrived in Egypt before 3500 BCE. The god Osiris was, among other things, the god of wine and much akin to Dionysus. The earliest evidence for wine in Egypt comes from the tomb of Pharaoh Scorpion I, circa 3100 BCE. Hundreds of wine-filled amphorae lined his tomb. The amphorae and presumably the wine were made in Palestine and transported hundreds of miles.

For the Egyptians wine was a luxury reserved for nobles and priests who had large wine cellars. The Egyptians were never able to produce enough wine to satisfy demand. Many nobles and temples had private irrigated vineyards, which they personally tended, to provide their owners with grapes and wine. The common people drank beer, and, according to food historian Thomas Pellechia, "[l]aborers on the Pyramids were among the few Egyptians commoners permitted to taste wine, which their masters treated them to during festivals or on certain prescribed occasions."[24] The Egyptians kept good records of the wines they made and imported. Many amphorae are stamped with both vintage and vineyard of origin information.

In the western portion of the Nile delta larger wine estates were established. Over the centuries, Egyptians learned to grow several varieties of grapes using multiple techniques to produce multiple styles of wine. Egyptologist Leonard Lesko's translation of hieroglyphs on wine jars stored in King Tut's tomb carried messages "such as 'wine for offering', 'wine for merrymaking', [and] 'wine for a happy return'."[25] By 2500 BCE the Egyptians had appointed an official to taste test wines and determine the tax rate.[26]

Numerous tomb paintings depict trellising, training, pruning, harvesting, pressing, storing, and serving activities. Several tomb paintings show Egyptians using a wringer method to gently squeeze the juice from the grapes, others show peasants mashing grapes with their feet. Apparently, drinking to excess was common among the nobility. Women, never men, unable to hold their wine are depicted in multiple tomb paintings.

CRETE

Wine making found its way to Crete from Phoenicia some time before 4000 BCE. According to Sanford Holst, the Minoan civilization of Crete probably was a colony of the Phoenicians. They had a fascination with bull dancing (performing acrobatic feats over and around bulls), an early form of bull fighting. Among other innovations of their culture was indoor plumbing. They built and used large clay vessels to press and ferment wine. Archaeologically, artifact amphorae abound on Crete and in the surrounding sea. The Minoans were almost certainly worshipers of the Earth Mother goddess and her consort. The Minoans/Phoenicians almost certainly brought wine and the wine god to the Peloponnesus (southern mainland Greece).[27]

The Minoan culture was destroyed by the eruption of a volcano on the nearby island of Thera around 1600 BCE. The caldera (blown-out volcanic cone) of ancient Thera is the modern-day crescent-shaped island of Santorini.

GREECE

As elsewhere, the Phoenicians, the great traders and colonizers of the ancient Mediterranean, brought wine to Greece. With wine, they brought wine culture and artifacts. The earliest known evidence of wine in Europe comes from Dikili Tash in eastern Macedonia and is dated at 4200 BCE. Yet wine and its god, Dionysus, were foreign to the Greeks. Theater historian John Donaldson relates the following myth about the integration of wine into their culture and Dionysus into their pantheon:

> There cannot be the slightest doubt that the worship of Dionysus or Bacchus was of oriental origin, and that it was introduced into Greece by the Phoenicians, who, together with the priceless gift of the Semitic alphabet, imparted to the Pelasgian inhabitants of the Mediterranean coasts a knowledge of those forms of elementary worship which were more or less common to the natives of Canaan and Egypt. The

> mythical founder of Thebes, the Phoenician Cadmus, is connected with both these innovations. For while he directly teaches the use of letters, it is his daughter Semele, who, according to tradition, in B.C. 1544 gave birth to Dionysus, the Theban wine-god.[28]

The cult of Dionysus diffused to northern Greece from Anatolia (Turkey) by way of Thrace, as Mediterranean historical geographer Daniel Stanislawski elaborates:

> In certain places and at certain times the . . . practice of his cult was terrifying, running the gamut from benevolence to violence and, at times, to temporary madness. Commonly, dances were a part of the worship, a tarantism provoked by contagious religious fervor. This frenzied, orgiastic savagery has been attributed to Anatolia, particularly to Phrygians and Lydians, from whom such traits were acquired by Greeks.
>
> In Thrace, where great excesses were performed in the ceremonies . . . a great fervor grew, at least as early as the ninth century BCE. The newly designed cult moved down both sides of the Aegean Sea, especially to Lesbos on the east; and western from Thrace the cult spread into Macedonia and down the western side into Boeotia, where it probably met the earlier Cretan version.[29]

The first region of Greece to adopt wine culture was Thrace (northwest Greece). As Hyams tells us, "Myth and poetry point to Thrace as the first wine country of Greece, so does the immense antiquity of that Maronean wine which Ulysses used to make Polyphemus blind-drunk. . . . And if it is certain that if Thrace was first, then Macedonia must have been second."[30]

Wine production technology matured by the time of its arrival in Greece. It was perhaps the ancient Greeks who invented the technique for enriching wines by partially drying the grapes. Samuel Morewood explained:

> The method of making wine among the Greeks was nearly as follows:—About the end of September, or early October, when the fruit was deemed sufficiently ripe, the grapes were collected, and usually exposed for ten days to the sun and the coolness of the night, in order that they might become more luscious and juicy. With many it was a practice to make three gatherings of the fruit during the vintage, for the purpose of producing wine of different qualities, while other means were resorted to for improving strength, taste and flavor.[31]

The drying process is still used today in the production of strong and dessert wines across the Mediterranean.

Hugh Johnson tells us the Greeks enjoyed drinking games. Their favorite was *Kottabus*. Some dedicated whole rooms of their houses to the play of this game.[32] According to Johnson,

> On the top of a high stand, something like a candelabrum, is balanced rather delicately a little saucer of brass. The players stand at a considerable distance with cups of wine. The game is to toss a small quantity of wine into the balanced saucer so smartly as to make the brass give out a clear ringing sound, and to tilt upon its side. Much shouting, merriment, and a little wagering ensues.[33]

Traditional Greek wine, Retsina, tastes of sap from the terebinth tree. The terebinth is a member of the cashew family. It is more famous as the original source for turpentine; a terebinth plug and resin sealed many ancient amphorae. Greeks, used to the flavor, still enjoy it today even though it is no longer essential for sealing bottles.

ROME

Prior to Roman conquest many Mediterranean regions were producing wine. The Phoenicians brought grape cultivation and wine making to Iberia long before 1500 BCE. An independent producer of wine for many centuries, Spain produced much of the wine traded by Carthage and became an important wine-producing region for Rome after 100 BCE. Wine production in Portugal was highly limited in ancient times because of its aridity, thin soil, and peripheral location.

A large krater (think of it as the biggest punch bowl you ever saw) capable of holding forty amphorae of wine was found in central France. This vessel predates the Roman conquest of Gaul by nearly 400 years. Its Greek characteristics suggest it was transported there from Greece before 600 BCE. It no doubt helped indoctrinate the ancient Gauls in wine culture. Southern France was the scene of much ancient wine production. Production moved northward as former Roman soldiers settled throughout Gaul. Vines reached the Bordeaux region by 100 CE.

When Rome gained control of southern Italy in the early third century BCE it gained control of the numerous Greek colonies established in southern Italy in the seventh through fourth centuries BCE. Before Rome expanded by conquering southern Italy, Morewood tells us, "wines were so scarce, sacrifices, the libations to the gods were ordered to be made only with milk."[34] The new Roman masters adopted many of the Greeks' cultural traits, including the worship of Dionysus. The Romans renamed him Bacchus. Among the Greeks, the rituals associated with Dionysus included secret ceremonies and orgies. The leaders of the cult of Bacchus, like modern cult leaders, took control of the lives and fortunes of the average followers. When drunk, debauched, and sated Romans readily offered up their family's wealth. When the leaders claimed the properties given them, the Roman Senate began to worry. This excerpt from Livy, written about 1 CE, partially relates the extent of the problems Rome had with Bacchus:

> When the wine had enflamed their minds and the dark night and the intermingling of men and women, young and old, had smothered every feeling of modesty, depravities of every kind began to take place because each person had ready access to whatever perversion his mind was inclined. There is not just one kind of immorality; not just promiscuous and deviant sex between freeborn men and women, but false witness and forged seals, wills, and documents of evidence also issued forth.[35]

Stuart Fleming, in his book *Vinum: The Story of Roman Wine*, tells us what happened in 186 BCE when the appalled Roman Senate outlawed the cult:

> About 7000 people from all over Italy were accused of conspiracy against the State . . . most of these were put to death for their beliefs. . . . Savage though this and several later persecutions were, the Bacchic cult not only survived but flourished off-and-on through the years. By mid-1st century BCE, it had shaken free of its roots as a faith for the common man and was finding adherents amongst Rome's wealthy and powerful. The latter pressured Julius Caesar to lift the ban on Bacchic festivals. For centuries thereafter their festivals were one of the many annual street celebrations which entertained and amused the city's citizenry with the somewhat bizarre antics and unabashed bent towards pleasure.[36]

Despite the trouble brought by the worship of Dionysus (Bacchus), the consumption of wine continued apace. Wine bars in Roman cities were as common as coffee shops today, and just like coffee shops today, the wine bar carried a variety of flavoring agents including sea water, peppercorns, and opium. Wine came to the bar in amphorae. The wine was served in glass decanters and drunk from glasses. The customers often stood in the street as they drank, although some bars had tables. Many bars had brothels attached.

Cato, who probably plagiarized from Magon, wrote *De Agri Cultura* in 134 BCE. Cato's book is the earliest surviving Latin prose text. Pliny the Elder catalogued grape varieties from around the empire in his *Natural History*, and "His list included 91 varieties of wine, 50 kinds of quality wine, and 38 varieties of foreign wines, and a range of other salted, sweet, and artificial wines."[37]

Most famous and treasured by Roman emperors was Falernian wine. This wine was made from Armenian grapes. The vintage of 121 BCE was regarded as the best in Roman history. Roman emperors are reported to have served this wine 150 years after harvest. According to Edward Hyams, author of *Dionysus: A Social History of the Wine Vine*, the Romans had access to many wines, and "there were other famous growths: Pompeii and Cumae had each about a dozen vineyards known for their wines. . . . The most popular wine with all classes was the Setian. . . . the Romans clearly liked the southern syrupy wines, not the thinner ones of the colder north."[38] One way the rulers of Rome maintained control of the populace was by distributing bread and wine to the masses in a policy known as "bread and circuses."

The Romans took the vine with them as they penetrated into northern Europe. Retired Roman soldiers were largely responsible for establishing the wine-growing regions of northern France and Germany. The expansion of grape production in Roman Gaul was slow and surreptitious before 270 CE because wine production in Gaul was forbidden by Roman law. This law served three purposes: it provided a market for Italian wine, it forced the Gauls to grow grain to feed the residents of Italy, and it provided a constant stream of slaves to satisfy the labor needs of Rome. In his travelogue, Didore, a Sicilian visitor to Gaul in 50 BCE, states:

> The Italian merchants exploit the Gallic passion for wine. On the boats which follow the waterways or by wagons which roll across the plain, they transport wine, from which they make fantastic profits, going as far as trading one amphora for one slave, in such manner that the buyer brings his servant to pay for the drink.[39]

Comparatively, this is a rather mild example of the wealth and power of Rome and wine. During the imperial period emperors, heroes, and those seeking political advantage satisfied crowds with gladiatorial extravaganzas, bread, and wine. Bread and circuses, it has been said, kept the empire going. At these events tens of thousands of amphorae of wine were handed out. To one feast Ptolemy Philadelphus, son of Cleopatra, delivered wine to the crowd in a bag made of panther skins sewn together. This skin bag is reputed to have held more than 20,000 gallons of wine.[40]

Long before Rome collapsed in the West in 476 CE, everyone who could have a vineyard, did. Wealthy Romans spread throughout the empire creating villas. Villas were managed and fed by a host of slave labor; their organization tended toward the feudal. As the Roman Empire shrank, the owners retreated, leaving their villas to the slaves or invaders. Some villas were taken over by Christian monks who continued wine production.

From the collapse of Rome in the early Christian era forward to the time of Napoleon, generations of monks experimented with growing wine grapes and selecting locations and varieties to maximize wine quality. Meanwhile, in the winery new methods of fermentation, storing, blending, and aging were developed. In Western Europe varieties adapted to successively cooler climates were developed and new vineyards in new places such as Champagne and the Rhine valley resulted. After 500 CE changes in the northerly wine–beer boundary stabilized along the old Roman–Teutonic border, which followed the Rhine and Danube rivers.

Bigger changes happened in the Eastern Mediterranean. By 800 CE Islam had a firm hold on the Arabian Peninsula, North Africa, and Southwest Asia. Muhammad's prohibition against alcohol meant social changes for the peoples of these regions. Under new social pressure, the grape became an unpopular crop in much of its homeland.

THE INVENTION OF DISTILLATION

The Neolithic era, the period of stone tools, ended when someone realized that the mixture of elements in a rock could be separated by heat. Smelting ores from rocks is like extracting oils and alcohols from water, except that a solid is transformed into a liquid, instead of transforming a liquid into a gas. Appropriately shaped ceramic vessels provided the boiling pot, lid, and extraction conduit for perfumes and distilled products other than alcohol for many years.[41]

Water freezes at a higher temperature (32°F) than alcohol (–173.2°F). Anyone living where a winter was cold enough, and the wine lasted long enough, would quickly discover this phenomenon and identify the principle of freeze distillation. The first firm evidence comes from Mongolia and dates to about 800 BCE, but in reality freeze distillation must have been discovered much earlier. Consumers of freeze-distilled products suffer greatly from hangovers because freeze distillation retains and concentrates harmful organic chemicals and alcohols (such as methanol). Heat distillation, when properly performed, separates out these toxins.

A number of Asian drinks perhaps had their earliest expression through the happy discovery by the first steppe dweller who consumed the liquid that did not freeze on a January morning. The drinks and their potential Asian steppe low-alcohol sources for distillation include:

- Skhou: In the Caucasus, from kefir (mare's milk)
- Sochou/Shochu: In Japan, from sake (rice)
- Saut/Sautchoo: In China, from tehoo (rice, millet)
- Arrack: In India, from toddy (rice with molasses or palm sap)
- Arika: In Mongolia, from koumiss (mare's milk)

The Norse and the Normans learned to freeze-concentrate mead between 500 BCE and 500 CE.

Circa 100 CE heat distillation was invented. The inventor probably found a new use for an existing principle and modified an existing device. Perhaps the still was invented by someone reading old texts. The inventor probably lived in or around Alexandria, Egypt. The early still was used to separate tar from oils.

The Persian alchemist Abu Musa Jabir Ibn Hayyan (721–815 CE) (aka Geber) invented the alembic still to distill rosewater and other perfumes about 800 CE. He was a scientific genius on par with Archimedes, Leonardo da Vinci, Louis Pasteur, and Thomas Edison. His distillation apparatus, originally made from pottery, included a boiling chamber, lid, and condenser cone. The condenser cone greatly increased purity and quantity over older style stills where the collection took place in the boiling chamber. Modern improvements to the alembic still include placing a water jacket around the condenser tube to accelerate condensation.

Archaeologists working in pre-Columbian Mexico discovered multiple stills in the Philippine design dating to the ninth century. At the most complete site is a hollowed log used to store fermented liquid and shallow bowls used as receiver and drinking vessels. The still itself was fashioned from clay.[42]

Distillation technology came to Europe before 1100 CE via Sicily. Sicily was conquered and partially held by North African Moors from 832 to 1091 CE. The Moors were a group of Islamic peoples of multiple ethnicities who lived in North Africa and Iberia from circa 711 CE to 1492 CE. The Moors spread Ibn Hayyan's invention throughout the Islamic world. The Moors used distillation to extract essences from plants to create perfumes. They also learned how to distill alcohol from grapes for use as lamp fuel and as a disinfectant. They taught the Sicilians how to distill essences. The wine-drinking Europeans quickly learned to drink the strong lamp liquor,[43] which they called the "Water of Life."

By 1200, the Spanish were making a distillate called *aqua vini*. This and similar terms, such as *aqua vitae* (Latin), *acqua della vita* (Italian), *uskiah* (Irish), and *eau du vie* (France), translate to "the water of life." Apothecaries took over distillation of alcohol and produced it on a large scale, with distilling becoming "more or less an industry, first in Italy, where we find a burgher of Modena producing large quantities of alcohol for sale as early as 1320."[44] Brandy was inexpensive and was widely consumed by the time of the Black Death in the 1340s. Cognac made its appearance in the 1500s, rendering the thin weak wines of the Loire into an exquisite, highly valued product:

> In *The Newe Iewell of Health* by Conrad Gesner, and dated 1576, is "the fourth Booke of Dyſtillations, conteyning many ſingular ſecrete Remedies " The first chapter is about "Of the distilling of Aqua vitae, or as some name it, burning water, and of the properties of the same. "

> It goes on to say . . . "that the water which is distilled out of wine, is named by some the water of life, in that it recovers and maintains life, yes and slays old age. But this may rightly be named the water of death, if it shall not be rightly and Artly prepared."[45]

The big push for distillates came from the Dutch in the early 1600s. Wanting to reduce transport costs, Dutch entrepreneurs built stills in western France to remove the useless water from the wine. In 1624, the Dutch built the first large-scale commercial alembic still near the mouth of the Loire. This still was quickly joined by others, establishing Cognac as the first region devoted to brandy production.

Rod Phillips in his *Short History of Wine* sums up the advantages the Dutch saw in brandy over wine:

> Brandy had several commercial advantages over wine. First, distilling wine increases the amount of alcohol. It took about five or six units of wine to make one of brandy, but brandy had up to eight times the alcohol content of wine by volume. In short, grapes produced more alcohol as distilled wine than as ordinary wine. There were also immense advantages in terms of transportation because the costs of shipping brandy were much lower, per unit of alcohol, than shipping wine. For consumers, brandy provided a different but palatable benefit: an immediate feeling of warmth that wine and beer lacked. . . . Brandy was quickly adopted as the alcohol of choice on merchant and naval ships, for it occupied less space and traveled far better than wine.[46]

Not long after, Europeans began to make distilled beverages from non-grape sugar sources (including fermented molasses to make rum and fermented grains to make whiskey). Early distillers experimented with adding herbs to their distilled spirits. In this manner gin, Chartreuse, and Absinthe were created.

As cheap distilled spirits became available throughout Europe, Europeans developed a serious drinking problem. Rampant, continuous drunkenness was everywhere. The common saying was "Drunk for a penny, dead drunk for two." Stupor was common. People stopped working. Social control and organization broke down. Workmen sold their tools for gin. Artist William Hogarth's Gin Lane and Beer Street lithographs parody the scale of the mayhem on the streets once gin became available.

The English call the period between 1730 and 1750 the Gin Craze. At that time English per capita annual gin consumption exceeded ten liters. The limited health and sanitation systems collapsed because people stopped caring. People became argumentative. Assaults, murder, and robbery became common. Disease spread across the land. The Gin Craze ended, but never really went away. A combination of harsh penalties, taxes, and forced migration (to the American colonies) rid England of the most seriously addicted and made it expensive for the remainder.

Distillation is important for the wine industry. It absorbs a great deal of excess wine, not only for brandy, but for industrial purposes as well. Europeans currently distill substantial amounts of low-quality wine into ethanol for fuel.

It is legal to make and own a still in the United States. It is, however, illegal to operate it without government licensing and supervision. A still is a simple device to construct, however, and this prohibition is often ignored. In the recent past, the operation of legal stills was much less common because state laws prevented it. In 2006, however, the federal government made obtaining artisan-scale distilling permits easier. Most states have now followed, creating opportunities for new distilleries. Relatively low-cost and readily available tax-paid products have displaced the need, but not the desire, for clandestine distillation.

GLOBAL DIFFUSION OF WINE

The year 1500 is a convenient marker for the beginning of the modern era and the beginning of wine's global diffusion. Printing, thanks to Johannes Gutenberg, was a recent invention. In 1453 Constantinople fell to the Muslim Ottomans, cutting off European access to the luxury goods of the Orient. Access was reestablished by the Portuguese. Meanwhile, in 1492, Columbus failed to reach China but opened the New World. In 1494, Spain and Portugal agreed to split the world, with the papal blessing. Martin Luther set the Reformation in motion in 1517. Spain, in response, launched an era of anti-Protestantism and anti-Semitism that lasted for centuries. Together the events of 500 years ago provide the basis for ethnic, religious, and political European wars as each nation sought to establish its own version of mercantilism.

In the year 1500 wine was the drink of choice throughout the Christian world. At that time the global distribution of wine production correlated more highly with the distribution of Christianity than any other crop or cultural trait. Southern Europeans drank wine every day. Northern Europeans drank wine when they could get it and beer when they could not. Beer soup was the typical breakfast of northern Europe. Distillation was known, but not yet widely in use.

The early years of the modern era are known as the Age of Discovery. Discovery was a global search for treasure, be it gold, spices, knowledge, or other groups of humans. Wherever Europeans went they hoped to discover lands suitable for grapes. For some viticulture was a religious imperative; for others it was a desire to civilize the discovered lands. Most hoped to make money.

When the modern era began in 1500, southern Europeans made wine for their own pleasure and to trade with the peoples of northern Europe. Within fifty years Spain and Portugal controlled vast empires. While searching for gold and glory Spanish conquistadors spent their energies sending natives to God. They did not have time to await vineyard maturation. Wine shipments from Spain could not keep up with demand. To combat the wine shortage Mexican governor Hernán Cortés ordered the planting of grapes in 1525. No matter the success of a colony's vineyards, the demand was only partially satisfied. Colonists still needed European wines to augment the local products.

Mexican wine making was successful enough to scare Spanish winemakers into successfully petitioning the Spanish king to ban further plantings in 1595. Lobbyists for jealous Spanish winemakers kept the ban in place for 150 years, although it was widely violated. Satisfying church requirements for sacramental wine alone led to successful long-term viticulture in the Spanish colonies. Exceptions were granted for the church to make communion wine in remote locations like California, Peru, and Chile. These wines were made by monks directing Indian labor. The Mission grape is still grown in many of these same locations today.

As Europeans explored further, grapes were among the first crops attempted in these new lands. Far-flung places like South Africa, Madeira, the Canaries, Chile, and Argentina entered the wine trade, but not eastern North America.

Every group of North American colonists attempted to grow grapes. French colonists attempted viticulture from Canada to Florida. Dutch colonists tried in New York. Swedish colonists tried in Connecticut and Delaware. English colonists tried everywhere else. Making a profit from wine is not easy. The grape is a labor-intensive crop, as is vinification. Vines take several years to mature. Most colonists planting vineyards usually abandoned them in favor of annual crops like tobacco. This topic is expanded further in chapter 9.

While the Portuguese discovered and regularly visited South Africa on their way to the Indian Ocean, the Dutch colonized it in 1653. The first Dutch immigrants brought vine cuttings, which thrived in the new environment. Before long, Cape Town was a thriving way station providing naval stores, fresh foods, and wine to the ships traveling between Europe and the Orient. The wine called Constantia was especially favored for its delicious orange smokiness. Later shipgoing migrants took cuttings from the South African vines to Australia.

In Australia and New Zealand the first colonists (convict transportees) attempted to grow grapes. Australia quickly developed a wine industry. Australia, first colonized in 1788, was exporting significant quantities of wine to other English colonies before 1860. Initial failure in northernmost New Zealand, around 1815, led to a long hiatus before commercial wine production was successful further south near Gisbourne.

WINE SCIENCE

The first significant discovery in wine science was made by Louis Pasteur. Less than 200 years ago he discovered yeast's role in converting juice to wine. Before his discovery of yeast, the process by which juice became wine was a (religious) mystery.

Shortly thereafter, wine science diverted into finding a cure for the vine-killing insect phylloxera. Germany established the Geisenheim Grape Breeding Institute in 1872, which became the prototype for wine research. In 1880 the California legislature followed, directing the establishment of a program for instruction and research in viticulture and enology. Scientific inquiry for the next fifty years largely concentrated on developing new and hybrid varieties with good taste and disease resistance.

Events since then have accelerated and we now have a greater understanding of the relationship between grape varieties, the causes of spoilage, microbial influences, and how to make pure wines. The modern science of wine got a great initial boost when University of California, Davis faculty published *General Viticulture* in 1974. Since 1975 the number of technical wine books has exploded as new genetic, agricultural, and vinification techniques have emerged thanks to scientific inquiry.

SUMMARY

The wild vine was domesticated in the foothills of the Caucasus Mountains 12,000 years ago by the earliest horticulturalists. The domesticated vine diffused across ancient Mesopotamia (modern Iran, Iraq, Jordan, and Syria). The Phoenicians actively spread wine throughout the Mediterranean basin. The Phoenicians introduced wine and its material culture to the Greeks. The Greeks revered the wine god, spreading his message to the Romans. The Romans expanded wine territory north of the Alps. The Muslims curtailed wine production in the Middle East and North Africa. Yet, throughout the 1600s the demand for wine increased. The single biggest factor in the increased demand was the discovery of distillation. In midcentury the Dutch built the first commercial distilleries in France's Loire Valley. Brandy fueled the African slave and American fur trades before cheaper rum production, which began about 1650, replaced it.

The Europeans spread wine and its culture around the world through colonization. With the exception of eastern North America, colonists succeeded in growing grapes in most temperate locations. Today, wine remains a primary global commodity. In volume its transport is second only to petroleum. Appendix B distills key events in North American wine history.

Chapter 2

The Biogeography of the Vine

The evolutionary descent, some would argue ascent, from an aboriginal species has resulted in grapes as varied and diverse as the regions they inhabit. Some species produce dwarf shrubs 2–6 feet high in harsh terrain where roots must find interstices in rock to gain foothold. Others produce enormous trunks supporting great canopies of branch and vine that produce fruit for two or more centuries. Between these extremes are a score of species that vary greatly and yet are close enough alike that identification is sometimes a difficult task. Their leaves, in particular, with few exceptions, are uncannily similar.[1]

—Jack Keller

That the vineyard, when properly planted and brought to perfection, was the most valuable part of the farm, seems to have been an undoubted maxim in the ancient agriculture as it is in the modern through all the wine countries.[2]

—Adam Smith

North America has one of the richest troves of wild grape species in the world, rivaled only by the valleys of central China, west of Shanghai.[3]

—George Gale

NATURAL HISTORY OF THE VINE

In the wake of the K-T mass extinction, 65 million years ago, many plant and animal species evolved to fill the void left by the dinosaurs and their associated plant communities. After the impact, the ancestor of the grape, a species of woody berry-producing shrub, erupted into a number of species over the next few million years, filling the void left by pre-impact plants. Evidence suggests that the proto-grape ancestor appeared about 60 million years ago on the great northern continent of Pangaea, before North America and Eurasia drifted apart. The proto-grape made its original home in the foothills of a large mountain chain with a humid continental climate. Remnant parts of that ancient mountain chain include the Caucasus Mountains of Asia, the Atlas Mountains of Morocco, the Scottish highlands, and the Appalachian Mountains of the eastern United States. Genetically diverse and easily mutated, the proto-grape responded to changing environmental conditions and geographic isolation brought on by continental drift. The isolated populations experienced genetic drift leading to speciation.

Perhaps one hundred species of grape currently exist, and, while different, they have much in common. They are woody climbers with tendrils. They root deeply. In the wild each vine is either male or female. Genus members can cross-reproduce sexually. When cultivated they produce self-pollinating flowers.

Two subgenera of the genus *Vitis* are identified: *Euvitis* and *Muscadinia*. All members of the subgenus *Euvitis* have 38 chromosomes. The *Muscadinia* have 40 chromosomes. *Euvitis* species are plentiful and widespread across the Northern Hemisphere. There are two major branches of the *Euvitis* line, the North American and the Asian. There are more than sixty *Euvitis* species native to North America and thirty in Eurasia. North American species lack acidity, producing unbalanced, grape-tasting, and often sweet wine. Most species have little economic value. *Muscadinia* evolved in and is native to North America. The Muscadine, also known as the Scuppernog and Bullace, resulted from a mutation occurring after North America separated from Eurasia 65 million years ago.

Two American species have exceptional value, *Vitis vinifera* and *Vitis labrusca*. Dr. Thomas Bramwell Welch—the man who started Welch's grape juice—popularized drinking unfermented labrusca juice among teetotalling prohibitionists in the late 1800s. Welch and his compatriots were so successful that they convinced a number of Christian sects to forswear wine in their religious ceremonies in favor of unfermented juice. Labrusca comprises less than 1 percent of the world's grape production.

The *Euvitis* species *Vitis vinifera* from the vicinity of the Caucasus Mountains alone has significant economic value. This species is the wine producer and the producer of most table grapes and raisins. *Vitis vinifera*, like its siblings, is genetically diverse. There are more than 10,000 catalogued varieties, but duplicate names for the same variety abound. The European Union identifies and tracks at least 930 separate nonduplicate varieties. In reality, however, only about fifty have commercial value.

The grape reproduces in three ways: self-pollination, cross-pollination, and vegetatively. The domesticated grape is a self-pollinator whose offspring routinely show mutation. These mutations are usually degenerative and are more susceptible to disease and infection. Cross-pollination results in new varieties. Vegetative reproduction, clonal, is conducted by planting cuttings of existing vines. Clones mutate and may appear different but remain the same variety. As Jancis Robinson states, "Any vine grower will tell you that there can be huge differences between different clones of the same variety. They may be genetically almost identical but some clones will ripen much more easily than others, some will be more vigorous, and the thickness and color of the skins can vary."[4] The two most common examples are the Brunello clone of Sangiovese and the different-colored berries from Pinot cuttings.

PARTS OF THE VINE

Figure 2.1 depicts and labels the above-ground parts of the vine. Central to the vine is the trunk. The trunk of the vine emerges from the soil and extends to the crown or first branch. Below the trunk and embedded in the soil are the roots. Vines produce two types of roots. Surface roots extend for several feet around each vine and are less than three feet deep. Taproots reach deep into the soil. Taproots seek to penetrate to the regolithic layer, directly above the bedrock, where relentless roots find water year round.

Atop the trunk is the crown from which annual growth springs. Spurs, removed or trimmed after harvest, grow anew from the crown during the spring. Each spur will produce leaves, tendrils, flower clusters, and subspurs. Leaves are a rich green when healthy. Leaves have lobes whose depth, smoothness, and crenularity vary by variety. With training one can recognize varieties by examining the leaves.

THE PARTS AND COMPONENTS OF A GRAPE

Each grape connects to the bunch by a woody stalk called the pedicel as shown in Figure 2.2. The pedicel is the fruit's umbilical cord to the vine. Each grape has five interior zones: the skin, the pips (seeds), and three types of interior pulp (exocarp, mesocarp, and endocarp). The pulp consists of thousands of pulp sacks. The endocarp, surrounding the seeds, is the thickest and most gelatinous. The exocarp, under the skin, may have a slimy character. This is most noticeable in New World grapes. The mesocarp comprises the majority of the berry mass. Free-run juice and juice from initial pressing comes from the mesocarp. The juice released by initial and gentle pressing is the highest quality. Higher pressure is required to release the juice

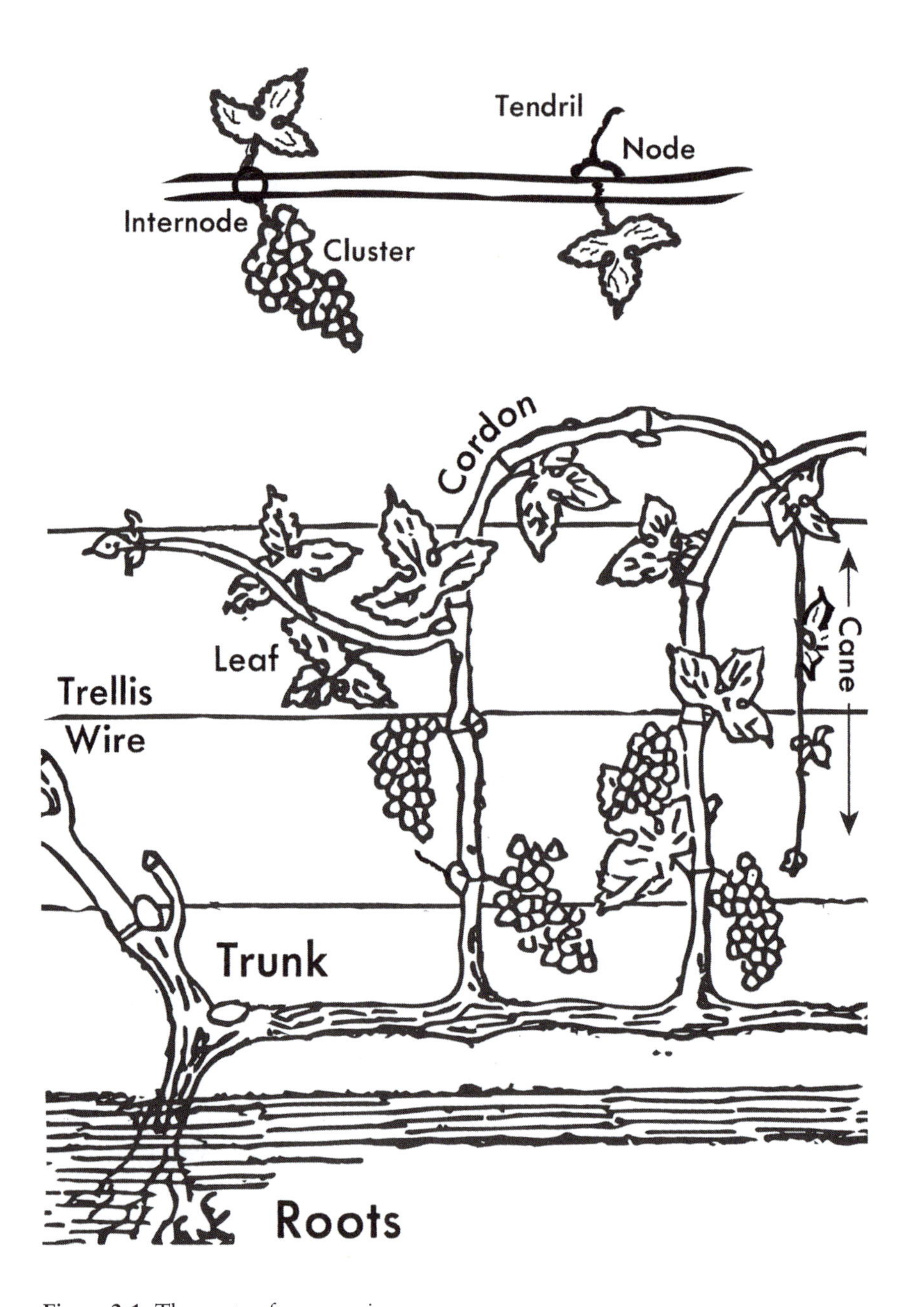

Figure 2.1: The parts of a grapevine.

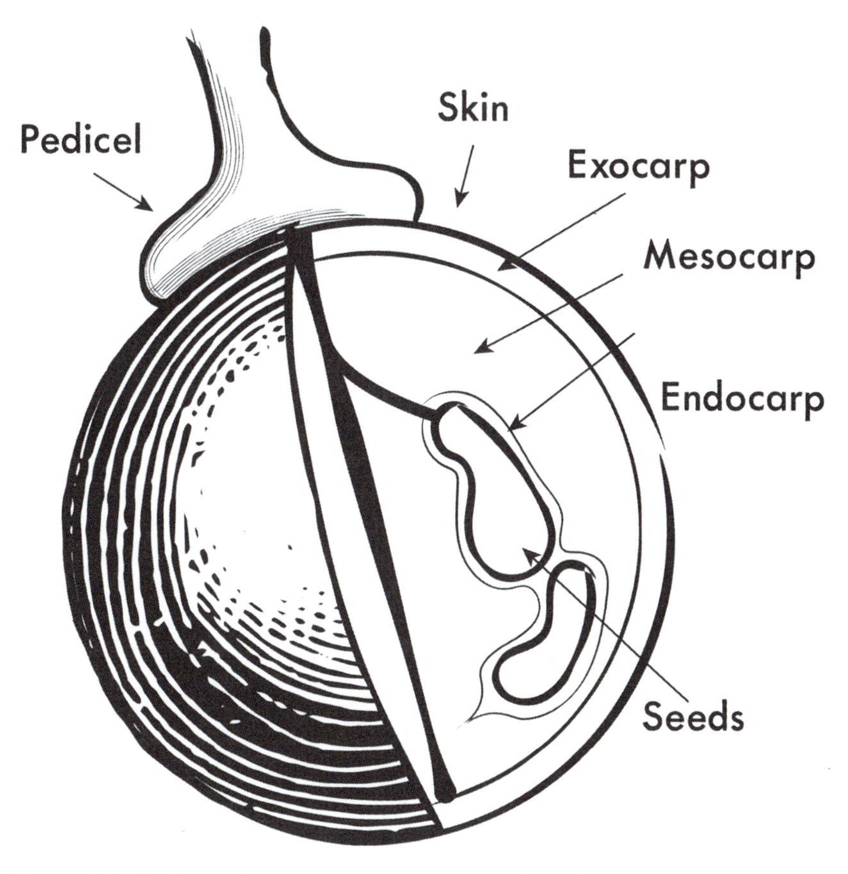

Figure 2.2: The internal components and organization of a grape.

from the endocarp and exocarp. Containing tannins, endocarp and exocarp juices are bitter and should be used to make lesser wines, vinegar, and brandy.

The grape skin protects the interior from drying out and blocks bacterial invasion. The exterior of the grape skin is alive with bacteria (including wild yeasts). These bacteria try to find a cut or bruise through which they can enter the grape. Once inside, the bacteria feed and reproduce inside the grape, destroying it.

The average grape is 76 percent water, 22 percent sugar, and 1 percent acid (mostly tartaric). These three chemicals provide the common background for all wine. The small amounts of mineral salts and volatile organic chemicals that comprise the remainder generate the individual flavors of each wine. Micrograms of mineral salts provide the flavor in the wine, and nanograms of volatile organic chemicals create the aromas and fragrances we perceive from wine. The same variety produces different flavors in different locations largely because of weather differences and the differential uptake of soil nutrients. Differential flavor intensity is primarily a function of water content.

Anthocyanins

Anthocyanins are the water-soluble pigment found in red grape skins. They are responsible for the red to purple color in red wine. Because they are water soluble, extending contact with skins during fermentation increases color extraction. As wine ages, molecules of anthocyanin clump and precipitate out, and the wines assume a brick-red color. Results of this process are visible when comparing ruby port to tawny port.

Flavonoids

Flavonoids and flavonols are common polyphenolic compounds produced by plants. Plants use flavonoid chemicals for flower pigmentation and as a physiological messenger. Flavonoids increase in the grape with exposure to the sun. Widespread plant production of flavonoids means that humans consume large quantities. The antioxidant properties of flavonoids are poorly metabolized.

Tannins

The word "tannin" refers to a group of chemicals found in the skins, seeds, and stems of the vine. Tannins are also present in wood, especially oak. Tannins cannot be seen or smelled, but they can be tasted. Tannins are astringent in the mouth and can improve the mouthfeel of a wine.

WILD VS. DOMESTIC VINE

In their groundbreaking study tracking the genetic history of the grape, This, Lacombe, and Thomas determined that, like the chicken and the egg, there can be no defining resolution of the domestication process, concluding: "The domestication of [the] grape seems linked to the discovery of wine, even if it is unclear which process predated the other."[5] The domesticated vinifera vine differs from wild vines in several key plant characteristics. The domestic grape is bigger and fuller. The thicker peduncle of the domestic grape allows the grapes to hang, waiting for the harvester. Wild grapes fall to the ground early. Domestic stalk internodes are shorter than wild ones. The shorter internodes make the plant more compact. Domestic vines are self-pollinating, rendering more complete fertilization and fuller bunches. Domestic berries are uniform in size.

The genetic characteristics observed in domesticated grape varieties were inherent in the wild vine. Over the past 9,000 years humans selected and reselected *Vitis vinifera* to derive the differences identified in Table 2.1. For centuries vintners relied on spontaneous cross-fertilization and "discovery" of the plants' potential for wine making. Today, grape researchers are steadily crossing varieties to seek new varieties, and genetic engineering projects are underway around the world.

Myles et al. completed a comparative genetic analysis of many vinifera grape varieties.[6] Their work shows the relationship between varieties. The heart of the network is the Traminer grape, the direct parent of at least fourteen other widely consumed varieties. This suggests that the Traminer was

Table 2.1: Wild vs. Domestic Vine Characteristics

Vine Part	Sylvestris (Wild)	Sativa (Domesticated)
Trunk	Often branches, slender, bark separates in long thin strips, branching.	Thick bark separates in wider and more coherent strips. Few branches.
Canes, shoots, laterals	Slender canes, long internodes, flexible hard wood, round in cross-section, tendrils wiry and strongly coiled, dormant buds, small pointed apex, scales tightly sealed.	Large-diameter canes, short internodes, elliptical cross-section; tendrils larger and extended, coiling more at terminal ends; dormant buds large and prominent, apex flattened, scales loosely sealed.
Leaves	Small, usually deep with three lobes; petioles short and slender, dull aspects; petiole sinus wide and open.	Large, many entire or with shallow sinuses; petiole thick, glabrous to downy, many with shiny aspect, petiole sinus partly closed.
Flowering	Sexual reproduction is rule. Male and female plants.	Hermaphroditic, few female plants, self-pollinators, clones.
Fruit Clusters	Small, globular to conical, loose, irregular set, berry maturity in cluster variable, peduncle long and slender.	Large, elongated, compact to well fitted, berries uniform in maturity, peduncle thick and rigid.
Berries	Small, round, or oblate, few elongated black, rarely white; skin highly pigmented, astringent; 4–6 seeds; juice watery, high acid, low sugar.	Large, oval to ellipsoidal, wide range of color, with decrease in pigment; few seeds, numbers vary; moderate acidity and sugar content.
Seeds	Small round body, pointy, highly viable.	Large, pear-shaped body, low viability.

among the first grapes domesticated. The large number of white grapes at the center of their map and the few reds on the edge indicate that our earliest wine-making ancestors had a definite preference for white wine. The red Bordeaux grapes (Merlot, Cabernet Franc, and Cabernet Sauvignon) are at the lower left of their map and the red Burgundy grapes (Pinot Noir, Pinot Meunier, and Gamay) are at the upper right. The peripheral distribution of red wine grapes indicates that they are relatively new. Combined with their geographic distribution, red grapes were most likely parented by wild vinifera grapes growing in Western Europe.

Several keystone varieties are the parents of our current favorite varieties. These keystone varieties are referred to as a *cepage*. The Pinot family of varieties is thus referred to as the Pinot *cepage* and the Muscat family is the Muscat *cepage*. Other parental varieties of significance are the Gouais Blanc, Ungi Blanc, and Cabernet Franc.

VINIFERA VARIETIES

The thousands of vinifera varieties differ in only a few but critical ways. Leaves are differentially shaped, sized, and colored. Some are more cold tolerant than others. Some require longer periods to ripen. Some are vastly more prolific than others. Berry size, shape, and skin pigmentation vary with overwhelming diversity. Grapes also differ internally; skin thickness affects color and tannin transfer. Grape pH ranges between 2.8 and 3.3. Sugar content at ripeness differs between varieties, but local conditions and harvesting date determine a wine's character more. Noble varieties (Cabernet Sauvignon, Sauvignon Blanc, Merlot, Chardonnay, Pinot Noir, and Riesling) grow in multiple countries. Most varieties, however, are limited to a single source region within a European country.

Figure 2.3 shows the preferred temperature regimes for major varieties. This single table represents years of research conducted by climatologist Gregory V. Jones. White grapes cluster in the cooler regimes and red grapes cluster in the warmer regimes. The exceptions are Pinot Noir, a cold-tolerant red, and Viognier, a heat-loving white. Temperature regimes for the Rhine Valley, Burgundy, Bordeaux, and Napa are also indicated. The varieties in the intermediate category produce across a wide range of temperatures, while those on either extreme thrive in a narrower temperature range. Following are individual varietal summaries.

The names associated with grape varieties can be as confusing and varied as the varieties themselves. Many varieties have multiple names. This is a product of the multiplicity of languages spoken in Europe, regional spelling variations, the isolation of most regions before 1850, and the difficulty of distinguishing between varieties without DNA testing. The multitude of regional names is diminishing as winemakers opt for the most marketable

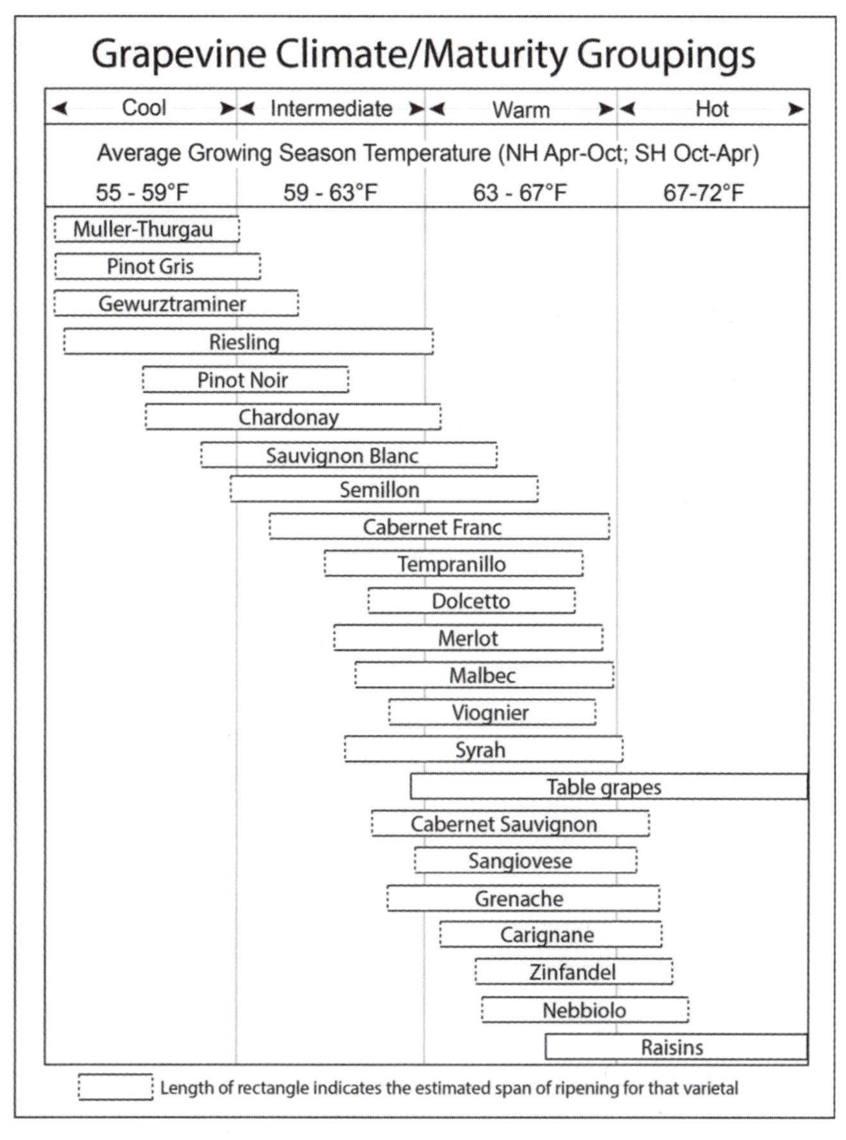

Figure 2.3: Variability in grape variety ripening. Notice that the white grape varieties prefer cooler regimes. (Adapted from Gregory V. Jones)

variety name. The most startling example comes from Chile. For more than 100 years Chileans believed they were growing Merlot when they were actually growing Carmenere. Chardonnay and Pinot Blanc share several alternate names. In Italy the delicate-sounding Trebbiano is the unpalatable-sounding Ungi Blanc in France.

Cabernet Sauvignon

Cabernet Sauvignon is the most highly prized variety. The king of varieties, it is ubiquitous owing to its great flavors and popularity. It is popular with growers because of its hardiness, late bud burst, and high prices. It produces consistently almost everywhere it is grown. Winemakers like it because they can tinker with it to adjust flavor profiles by blending, oaking, and malolactic fermentation. Variations in Cabernet flavors result mostly from grape quality and vinification processes.

Originating in Bordeaux, France, Cabernet Sauvignon resulted from a cross between Sauvignon Blanc and Cabernet Franc during the Middle Ages. Cabernet Sauvignon only became a significant variety in the late 1700s when claret became stylish and legal in England. More resistant to phylloxera than other varieties, it became the darling of the Medoc in the late 1800s.

Flavor bursts out; it tastes of black currants and raspberries. Black currants are an unfamiliar fruit to North Americans; their taste could be likened to the predominant flavor of PEZ candy. If harvested underripe, it presents herbaceous and bell pepper flavors. The Cabernet Sauvignon skins are dense and their tannins permit the wine to age. Aging softens the tannins and renders the mature wine velvety. To bypass aging, most winemakers blend Cabernet Sauvignon with Merlot. A young Cabernet Sauvignon–Merlot blend of 90–10 is as palatable as a three- to five-year-old 100 percent Cabernet Sauvignon.

Pinot Noir

Pinot Noir is the main red wine grape of Burgundy. The grape clusters are said to be shaped like pine cones, hence the word "pinot," which means cone. Its wines usually have a cherry taste. It is delicate, its shoots are thin, and its leaves sparse. It produces small clusters of berries. It prefers cool west coast marine climates in sheltered river valleys. Washington, Oregon, British Columbia, and New Zealand are regions of great Pinot Noir potential. Approximately one-third of the Champagne AOP region of France is planted in Pinot Noir where it and its offspring Chardonnay and Pinot Meunier are blended to make Champagne. The *Blauburgender* is a Pinot Noir clone grown in Germany. Austrian growers call their Pinot Noir clone *Spatburgunder*.

Cultivated and vegetatively reproduced for 2,000 years, numerous spontaneous mutations have occurred. Named separately Pinot Noir, Gris, and Blanc, they are actually the same variety. As Jancis Robinson states, "black, grey and white berries can sometimes be observed on the same vine, or even individual striped berries."[7] Through sexual reproduction, Pinot Noir is the parent of several commercially successful varieties and dozens of others. Pinot Noir and Ungi Blanc have together parented at least twenty varieties including Chardonnay and Aligote.

Chardonnay

Chardonnay is the world's most commonly grown white grape. Chardonnay resulted from a cross between Pinot Noir and Gouais Blanc. It originated in Burgundy. Chardonnay naturally produces a pale green-gold apple-tasting wine. This flavor derives from malic acid, which is present in Chardonnay at higher levels than in any other variety. The malic acid can be transformed to lactic acid by lactobacillus in a process known as malolactic fermentation (see chapter 4 for details of the process). Malolactic fermentation adds a buttery flavor to Chardonnay. Chardonnay is usually oak-aged to add naturally lacking tannins. Oak, through its tannins, makes Chardonnay astringent and adds musty aromas.

Because of the transformations possible, Chardonnay is like clay in the hands of a sculptor. The winemaker can change and manipulate the flavors of the wine dramatically. Generally, Chardonnay from Chablis is citrusy and steely while the heavily oaked Australian Chardonnays possess pineapple and coconut characteristics. The oaked and malolactic fermented wines have an aroma reminiscent of a dirty diaper.

Sauvignon Blanc

The second great white grape of France originated in the Loire River valley where it still dominates. It is the favorite grape of the Central Loire (where it is usually labeled as Sancerre, the principal town of the region). Sauvignon Blanc does not age well and should be consumed young. Sauvignon Blanc assumes the character of the soil well. Some Sauvignon Blanc has grassy, herbal, cat urine, or vegetal aromas, while others produce melon, passion fruit, and citrus flavors. It all depends on where the grapes were grown and how they are vinified. Of all vinifera varieties, Sauvignon Blanc may taste the most like North American grape wine. Sauvignon Blanc from around the world are blended with Semillon, a practice that began in the Bordeaux region.

Sauvignon Blanc is produced in most exporting countries. The Marlborough region of New Zealand is exceptionally well suited for Sauvignon Blanc. Growers there have redefined the variety, producing wines of intense tropical fruit flavors. Amazingly, the flavorful, captivating wines of Marlborough are at their peak when bottled and require no aging.

Riesling

Riesling is the golden white grape of northern latitudes. It is the highest quality white grape of Germany. In land area Riesling represents 20 percent of current German production. Significant production also occurs in neighboring Alsace. The first reliable historical reference to Riesling vines comes from the Rheingau region in 1435 when one monastery sold cuttings to another. Through the years, church and secular authorities promoted Riesling in preference to the Elbling variety. After the Thirty Years' War (1648) Alsace was widely planted with Riesling. The provenance of Riesling was reinforced by physicians A. C. F. Hui and S. M. Wong when their studies of Beethoven's remains and papers revealed, "Beethoven['s] . . . last words referred to the arrival of a case of Riesling, which he realized he was too ill to drink."[8]

Riesling, like Sauvignon Blanc, is sensitive to the chemical composition of the soil. Depending on location, Riesling produces a broad range of aromas and flavors. Highly acidic Riesling wines produce citrus, pineapple, lime, and/or peach aromas. When late harvested (allowed to remain on the vine into October and November), Riesling develops honey, fig, orange, and honeysuckle notes. Noble rot (*Botrytis cinerea*) intensifies Riesling grapes' sweetness.

Riesling occupies a unique position among grape varieties. Unlike any other variety, Riesling has a fan club, the International Riesling Foundation. Although extolled for its qualities and adored by many, Riesling is rarely prominent on wine lists.

Owing to its high acidity, Riesling produces the world's finest ice wines, a dessert wine created from naturally frozen grapes. Harvest usually occurs before dawn on a January morning when temperatures are below 15°F. Vineyards in the Niagara Peninsula of Ontario and Okanagan Valley of British Columbia are major Riesling ice wine producers. Alsace on the French-German border and the Finger Lakes region of New York produce some of the world's finest dry Rieslings.

Syrah (Shiraz)

Syrah wines are deeply colored and rich in tannins. Aromas include those of blackberry, mint, black pepper, and smoke. Syrah is a grape of French birth probably in the twelfth century. Another of the great wine grapes of France, Syrah grows in many of the appellations of France's Southern Rhone valley. The Australians made Syrah their own, renaming it Shiraz. The Shiraz name caught on when Australian wines became globally distributed in the 1980s. As a result, the popularity of this variety has increased dramatically during the last decade. Enjoy Syrah with steak, lamb, or venison.

Muscat

Muscat grapes thrive in their native Mediterranean climates. They prefer long summers and are harvested late, often in October. Members of the Muscat *cepage* (family) are difficult to distinguish. The original Muscat variety is of great antiquity. All Muscat varieties (and many of their descendants) possess a rose-like floral aroma often accompanied by dried fruits or pink grapefruit. The Muscat varieties grow throughout the Mediterranean basin and bear a large diversity of names.

Italians know the Muscat Blanc (Muscat Blanc à Petits Grains) as Muscat Bianco and make Moscato d'Asti and Moscato di Canelli from it. These wines are bottled at about three atmospheres (45 psi). Italian Muscat Bianco sparkling wine, Asti Spumonte, is bottled at six atmospheres (90 psi). The Muscat of Alexandria is of ancient Egyptian origin, as the name suggests. The Orange Muscat with its distinct smoky orange flavor is the grape used to make South African Constantia. Quady Winery of California is a dedicated producer of this variety. The Muscat Ottonel (aka Moskately) is of Eastern European derivation and is often used for dessert wines. The Yellow Muscat (aka Moscato Giallo) grows in northern Italy where it is used in making dessert wines. Several New York Finger Lakes wineries produce dessert wines from the Ottonel variety. Muscat of Hamburg is the red grape of the *cepage*; again Quady is the principal California producer of these delicate sweet wines. The Torrontes and Gewürztraminer are both Muscat descendants.

Pisco, the delicious Peruvian brandy, is distilled from Muscat Blanc grapes. Grappa and Rakia, Italian and Bulgarian brandies, respectively, are often made from Muscat grapes. The Greek brandy Metaxa is distilled from Muscat grapes as well.

Torrontes

First observed in 1867, Torrontes is the white grape of Argentina where it originated. No other country exports its wine. There are three Torrontes varieties, each representing a separate crossing of the Muscat of Alexandria and Mission grapes. In each case the perfume of the Muscat is subdued and subsumed by a citrus crispness derived from the Mission grape.

Nebbiolo

The Nebbiolo takes its name from the fogs common in its growing area in northwest Italy. It was first mentioned in 1268 CE as growing near the village of Rivoli. Nebbiolo wines have the aroma of roses and earthy notes. It is highly acidic and has mouth-drying tannins. Grown in the Piedmont region, it is sold under the names of the towns. The variety name itself is rarely on the label; seek DOCG wines labeled as Barolo, Gattinaria, Barbaresco, and Ghemme instead.

The Nebbiolo buds early and ripens late. It should never be planted in locations with frost potential. The cool fall conditions in the Piedmont caused the fermentation to start and stop over several weeks. The process extracted large quantities of tannin from the skins. The resulting wines were highly tannic and traditionally took decades to mature. Modern temperature-controlled fermentation tanks have largely eliminated the need for protracted maturation.

Malbec

The Malbec was one of the original Bordeaux varieties where it was widely used in making Claret prior to the arrival of phylloxera. It produces a dense red wine with dark fruit flavors. When grown in cooler climates it presents cherry flavors, transforming to raspberry, plum, and blackberry when grown in successively hotter climates.

Phylloxera started a cascade of calamities culminating in the frost of 1956. After 1956 Malbec vines were largely eradicated from Bordeaux. Plantings near Cahors remain. Brought to Argentina by Michel Pouget in 1850, it flourished. The high elevation, dry climate, and sunny days make the Mendoza region of Argentina an ideal ripening environment for Malbec. Malbec ripens early and is highly tannic. April 17 of each year is International Malbec Day.

Sangiovese

Sangiovese is the major component of Tuscan wine production. Sangiovese tastes of strawberries and cherries. The name translates as "the blood of Jupiter" (Zeus). It is the backbone of Chianti, which requires a minimum of 80 percent Sangiovese content. For its fame, surprisingly little grows outside Tuscany. Genetically unstable, several of its clones have distinct identities. The Brunello (meaning little brown one), for example, is smaller and browner than its parent. The Brunello clone is used to make Brunello di Montalcino.

Zinfandel (Primativo, Tribidrag)

Zinfandel! The grape that claimed California! Zinfandel originated in Slovenia, near the Adriatic Sea. How it got to California remains a mystery, but it arrived about a decade before the 1849 gold rush. The Zinfandel grape produces a red full-bodied wine. In cooler climates raspberry flavors dominate, while hotter environments reveal blackberry and pepper notes. It was widely planted and millions of gallons of wine were made from it between 1865 and 1906.

Zinfandel's modern popularity resulted from a mistake. Robert Trinchero accidentally invented White Zinfandel in the mid-1970s. Trinchero wanted to make a deep-red wine. To accomplish this he drained off near clear free-run juice before beginning carbonic maceration of the remaining juice and all the skins. With the excess nearly clear juice Trinchero made a rosé wine.

Sweet, pink, and generally lacking in aromatics, White Zinfandel is nearly impossible to dislike. It represents the latest success in an ancient tradition of rosé wine making. Because of the popularity of White Zinfandel, many older vines were not uprooted. When red Zinfandels became popular again in the 1990s, the 80- to 100-year-old vines produced the most delectable wines.

Merlot

Merlot is the second great red variety of Bordeaux. Quality Merlot vineyards in Bordeaux are concentrated near Pomerol and St. Emillion, both north of the Dordogne River. Lower quality Merlot is produced in the Entre-deux-Mers region. The deeply colored grapes produce a wine with intense color depth. It matures quickly and is usually ready for drinking within two years. Merlot's high pH gives it a smooth and silky mouthfeel. Aromas of berries and plums abound.

A great percentage of Merlot is used for blending. Merlot softens the sharp edges common to Cabernet Sauvignon. Cabs can contain up to 15 percent Merlot without documentation in most countries. The softness made Merlot varietal wines popular in the early 1990s. Outside Europe, Merlot rapidly achieved immense popularity, particularly in California and Australia. In California acreage increased from 2,000 acres in 1985 to 18,000 in 1994. In 2005, Merlot consumption nosedived when it was identified as a "woman's" wine in the movie *Sideways*. Merlot has regained most of its previous popularity.

Semillon

Like Sauvignon Blanc, this white wine variety comes from the Bordeaux region. Semillon thrives in microclimates favorable to morning fogs. It is tricky to grow and uncommon outside France, although growers in Western Australia are having success. In Sauternes, the variety has its home and greatest notoriety as the dessert wine of Chateau d'Yquem. At Chateau d'Yquem, noble rotted Semillon grapes acquire sweet honey flavors that join with the natural aromas of almonds, apricots, butter, and ripe pears. Wines based on the Semillon can age for decades.

Vintners frequently blend Semillon with Sauvignon Blanc to add a buttery character to the otherwise sharp acidity of Sauvignon Blanc. Often, blends of 5 to 8 percent Semillon are adequate to make the subtle change desired.

Cabernet Franc

Cabernet Franc originated in southwest France, perhaps in the Basque region. Cabernet Franc is the parent of a number of other varieties including

Cabernet Sauvignon. Generally used as a blending grape, Cabernet Franc grows well and produces a delicious varietal in eastern North America. It produces a paler crimson wine than Cabernet Sauvignon. It matures quickly. Aromatics include those of violets, raspberries, strawberries, and an earthiness reminiscent of the forest floor. Immature wines might have aromas akin to bell peppers. When aged, Cabernet Franc assumes more cherry-like aromas. Its yield is smaller than its offspring Cabernet Sauvignon.

Gewürztraminer

A clone of the Savagnin, Gewürztraminer literally translates to mean "spicy Traminer." The grape has a peachy pink color. Gewürztraminer presents a floral-rosy aroma coupled with lychee, gingerbread, and citrus. When late harvested and botrytis infected, bold aromas of honey, pineapple, and apricots emerge. It originated near Tyrol, Italy and now grows extensively in the Alsace region and Germany. Its descendant, Traminette (one-quarter *Vitis aestivalis* and three-quarters Gewürztraminer), has great wine-making potential, and plantings in Northeastern North America are expanding rapidly.

Trebbiano Toscano (Ungi Blanc)

There are actually six varieties with Trebbiano in their name. The Toscano is the most important. Grown throughout northern Italy, Trebbiano is used to make Soave, Frascatti, and Orvieto, all refreshing but undistinguished. Transferred to France during the Papal Captivity, it found a home in France's Cognac region and a new name, Ungi Blanc. In Cognac the grape usually produces sugar levels so low that the wine will not keep; distillation offered a way to preserve it. The Dutch seized the opportunity to convert low-alcohol wine into brandy when they built the first commercial stills in Cognac in the 1620s.

Grenache (Garnacha)

Of Spanish origin, the Grenache grape grows well in Spain, Sardinia (Italy), North Africa, and southeastern France. Grenache dominates the Rioja region of northern Spain. Grenache grapes are black and thick skinned. They are high in tannin with aromas of strawberry, raspberry, and cherry. When fermented sealed, Grenache exudes aromas of cherries and blackberries. When fermented in the open air (oxidized), as is sometimes purposely done, it acquires nutty, cocoa, and spice flavors similar to sherry. Vintners often make Grenache into a rosé to limit the tannin content.

Gamay

In the early 1300s the Gamay grape was discovered growing near the village of Gamay in Burgundy. Thanks to the social and economic upheavals resulting from the Black Death, Gamay began replacing Pinot Noir in the mid-1300s. Gamay's economic advantages were enormous. Gamay yields tripled that of the replaced Pinot Noir vines, produced more reliably, and ripened earlier.

The character of the region's wine was changing rapidly, and not for the better according to the powerful. In 1395, Phillipe the Bold, Duke of Burgundy, ordered Gamay to be eradicated in his realm.

Gamay, however, was too economically viable to disappear. Growers simply planted the grape outside his realm to the south where it thrives today. Today, Gamay production centers on the villages of Beaujolais south of Burgundy.

Two wine styles are made from Gamay in Beaujolais and the neighboring regions. There is the typical style common to most red wines and there is Beaujolais Nouveau. Beaujolais Nouveau is ready for consumption only two months after harvest in the fall (ready just in time for the holidays). Five hundred years ago wines shipped in the spring were going bad in the barrels by the fall. The fresh Beaujolais meant clean wines for the holidays. Consumers, tired of the ever more vinegary wines, sought the fresh wine each November. The Beaujolais producers created and still maintain a niche market for this early-release wine.

Airen

Airen vines cover more land than any other variety in Spain, and until 2004 they covered the largest surface area of any grape in the world. Spaniards grubbed up 140,000 acres of Airen between 2004 and 2010. Now it is third in land area. It looks like a small bush and is a small producer, but it produces a crop where little else can grow. Its wine is pale and is mostly distilled into brandy.

OTHER VINIFERA VARIETIES

There are many other varieties used to make wine, and there is a rising appreciation for the uniqueness of these wines. Any number of varieties can be the next consumer darling. When it is "discovered," there will be a windfall for the lucky growers.

The exact relationship between most varieties is obscured in history. With DNA identification of grape varieties well under way, we will shortly be able to trace the roots of the family vine (*vinadae*). This will undoubtedly lead to the discovery of new varieties and end varietal misclassification. It was recently determined, for example, that Chilean Merlot is actually Carmenere and that Argentinean Sauvignon Blanc is actually Sauvignon Vert.

NEW WORLD SPECIES

The grapes of New World species—those from outside of Europe—lack the acidity vital to a quality wine. Their wines lack the sweet–acid balance common to vinifera-based wines. As a result, they taste overly sweet. The most famous of the New World species is *Vitis labrusca*. Its cultivars include Concord, Niagara, Steuben, Catawba, and Delaware. Labrusca wines have a grape taste. In 1867, at a wine competition held in Paris the American Wine Co. of St. Louis submitted its sparking Catawba wine for judging. Christy Campbell relates the events at and results of the competition:

> But the tutored palates of European visitors who tried such beverages could never forgive any product of the labrusca grape's curious animally undertaste. "Musky" they called it—"foxy"—*le gout de renard*. The polite called the taste *framboise*, "raspberry." The more direct called it *pissat de renard*—"fox piss."[9]

The three species of Muscadine grapes, with 40 chromosomes, occupy a separate *Vitis* subgenus. Muscadine grapes are native to the southeastern United States. The grapes of this species usually have a bronze hue. Rather than clusters, the grapes grow in bunches with long individual cherry-like stems. Muscadine grapes were probably used in the production of the first wines in North America. A French colony in Florida made wine (from Muscadine grapes?) in 1565. The hedge grapes described and consumed by Jamestown colonists are even more likely to have been Muscadine. Muscadine species have shown little economic value to date. Demand for Muscadine is on the increase, because their antioxidant capacity exceeds that of all other fruits.[10] Today Muscadine wine, jelly, and so on are sold as a regional novelty in the southeastern United States.

HYBRIDS

The first identified *V. vinifera x V. labrusca* hybrid was discovered growing in Philadelphia near William Penn's abandoned vineyard in about 1740. This hybrid, named the Alexander, became the basis for a short-lived wine boom in eastern Pennsylvania. The Alexander was later called both the Tasker's grape and the Cape grape. The Cape was a major component of the nineteenth-century eastern wine industry and is still grown in Missouri.

The Norton grape, first discovered in the 1830s by Dr. Norton of Richmond, Virginia, is descended from *Vitis aestivalis* and as yet unknown varieties. Norton is rich in anthocyanins, resulting in deeply colored opaque wine. The Norton variety was favored throughout the eastern United States where hundreds of acres were in production before Prohibition. Norton became the signature grape of the Missouri wine industry that flourished before Prohibition. Norton nearly went extinct during Prohibition when the vines were uprooted and replaced with Concord grapes. The Norton variety is making a comeback thanks to the efforts of Dennis Horton of Horton Vineyards of Virginia who reintroduced the variety in 1989.

Other hybrids of note include Chambourcin, Cayuga, Chancellor, Rougeon, Villard Noir, Colobel, Marechal Foch, Rosette, Vidal Blanc, Duchess, and Seyval Blanc. Chambourcin was widely planted in France after the phylloxera invasion. Many of the vines have since been eradicated but significant quantities of Chambourcin table wine are made near Nantes, France. Today, hybrids are most common in the eastern United States where phylloxera prevented the establishment of viniferous grapes.

A few people are responsible for creating most of the hybrids in existence. Albert Seibel was the most prolific of these breeders. Between 1886 and 1936 he developed eighteen hybrid varieties, the most important of which is Chancellor. Joannes Seyve, creator of several varieties with his name, created Chambourcin, probably from a hybrid created by Seibel. Jean Louis Vidal also developed multiple hybrids. Most significant of these is Vidal Blanc, which grows in the Lake Erie region. Vidal Blanc is commonly used to make ice wine.

The work of hybridization continues at land grant universities. Cornell, in Ithaca, New York, is Traminette's source. Brock University in Ontario is home to the Canadian wine research efforts where they specialize in cold-climate varietals. Recently the University of Minnesota released several new cold- and disease-tolerant second- and third-generation hybrids. The Frontenac, La Crescent, and Marquette varieties can produce crops after experiencing temperatures below –30°F. Here is their description of the Marquette variety from the University of Minnesota:

> Marquette is a cousin of Frontenac and grandson of Pinot Noir. It originated from a cross of MN 1094, a complex hybrid of *V. riparia, V. vinifera*, and other *Vitis* species, with Ravat 262. Viticulturally, Marquette is outstanding. Resistance to downy mildew, powdery mildew, and black rot has been very good. Its open, orderly growth habit makes vine canopy management efficient.[11]

THE VINE'S ANNUAL CYCLE

Growing grapes requires intensive year-round management and labor. There are always actions the vintner can take to improve next year's quality and yield. In our generalized Northern Hemisphere annual cycle, the vine's year begins in November when the growing season ends. The general summary offered here has endless variations as each plant of each variety in each microclimate grows and matures at its own individual pace.

In November, when average daily temperatures drop below 52°F, the vines become dormant. Leaves turn red as the trunk extracts nutrients, sugars, and starches, then the leaves drop. Pruning occurs during winter dormancy. The canes from the previous year's growth are removed and the vine is reduced to its trunk and spurs. Most prunings are burned in the vineyards. December is a good time to apply fertilizers. January is a time spent indoors tending the maturing wines. February is spent preparing for the upcoming growing season by grafting rootstock and trunk stock for planting, and refurbishing equipment.

When March arrives, a few days spike above 52°F and the soil begins to warm. In the higher latitudes the lakes unfreeze. It is not until mid-April, however, that the vines show signs of life. Buds appear on the spurs and crown. If the last frost occurs after budding commences, the entire bud may freeze (see the discussion of lake effect and cold air drainage in chapter 3). In this situation, the crippled vine will not produce a proper crop that year. Rocks may be cleared, piled, or stacked depending on local custom. April is a good time for spraying for insects and diseases in preemptive strikes.

In May the buds swell rapidly and burst into fruiting canes. The canes grow rapidly until near the end of June. Canes growing in humid areas with rich soil may achieve lengths of ten or fifteen feet. Workers tie the canes to the trellises to ensure the fruit will be at the proper height. In dry and infertile locations canes may be two or three feet in length and be unsupported.

In early to mid-June, when the mean daily temperature reaches 68°F, flower clusters appear on the canes. After a few days, the flowers self-pollinate. Each will become a grape. After pollination, the vine will produce shoots, lateral branches, from leaf bases along the primary cane. Workers prune the laterals, preventing the vine from producing excess leaves and flower clusters. Suckers, shoots from the trunk, are pinched off.

During late June and throughout July, the berries grow from pinhead to marble sized. Cultivating the soil is again necessary. Weeds and insects require action. The vines themselves need constant attention. Individual leaves are removed to place grapes in direct sunlight. Shoots are removed or tied to the trellis. In late July and early August workers remove excess clusters to improve the quality of the remainder.

In August, little can be done except worry. By this time of the year, growers have done almost everything they can to nurture a successful crop. Weather causes their worry. Will there be a strong thunderstorm? Will hot dry winds rip leaves and grapes from the vine? Will it be too dry? Will it be too wet? Will swarms of insect pests invade the vineyard? Will moist conditions allow the growth of rots, fungi, molds, or rusts? In France, August is the month for wine growers to take vacations, as there is so little they can accomplish.

Veraison (ripening) occurs between late August and early October. As it ripens, the grape softens, becomes less acidic, and herbal flavors disappear. The pH increases from around 2.3 to between 2.8 and 3.3 as acids convert to sugar. Grape flavors intensify and concentrate. As picking day approaches, workers test the grapes to ascertain sugar level and other chemical characteristics.

At harvest, berry acid levels should be between 0.6 percent and 1.0 percent of total weight. The sugar concentration varies by variety. Low-sugar varieties have as little as 10 percent sugar content when mature. High-sugar varieties may have as much as 29 percent sugar content. If the grapes are dried, infected with botrytis, or late harvested, sugar concentrations can double.

September brings a rebirth of human activity in the vineyard. Equipment necessary for picking, transporting, juice extraction, and wine making is cleaned and prepared for use. If it is a damp year, extra leaves may be removed to prevent mold and fungal diseases. Some varieties are ready for harvest in early September, but most need to stay on the vine until after the equinox. Each additional day the grapes stay on the vine the sweeter they become, but the more likely a rainstorm or flock of birds can ruin the crop. This must have been a nerve-wracking waiting game before there were reliable weather forecasts.

Harvest is the busy time. Equipment unused for ten months is cleaned and prepared to receive the grapes. An equipment failure at this time can be disastrous. The grapes are picked, cleaned, pressed, and fermented. Everyone works overtime. By mid-October a killing frost occurs; leaves turn red and begin to drop. Late-harvest grapes are bird netted. The leaves turn yellow and drop. Grapes destined for ice wine are covered in plastic sheeting and harvested in December.

AGRICULTURAL PRACTICES

Most vine growers use modern equipment and rely on fertilizers, pesticides, and herbicides, but not all. Two special techniques for grape production, biodynamics and organic, have devoted followers.

Biodynamics

Biodynamics represents a rethinking of our relationship with the earth. It is a popular socio-political-economic trend among grape growers. Biodynamic wineries strive to achieve sustainability with a spiritual twist. The soil is considered a living entity that requires careful nurturing. Concoctions of natural materials are used instead of synthetic fertilizers and pesticides. The most noteworthy are cow horns stuffed with cow manure and buried over winter.

The horns are unearthed in the spring and sprayed over the vines. Biodynamics also includes other approaches such as growing flowers to attract bees and/or discourage vine-devouring insects and diseases.

Biodynamics was created by Rudolph Steiner, an Austrian, in 1924. The Biodynamics organization, Demeter, certifies wineries as Biodynamic. Many wineries are implementing biodynamic practices as sustainability has become a major issue, even though they may never apply for certification.

Organic Wines

The definition of "organic" varies from country to country. In 1980 Four Chimneys winery in New York became the first certified organic winery in North America. Organic wineries are designated based on their refusal to apply synthetic fertilizers, pesticides, or herbicides. The big difference for the wine consumer is that organic wines are made with limited, and mostly naturally occurring, sulfites. Sulfites are a class of naturally occurring chemicals. To be a sulfite a chemical must contain the sulfite ion, which consists of one sulfur and three oxygen atoms. To complete the molecule, sulfite ions routinely join with sodium, potassium, or calcium. Used as a sterilizing agent and food preservative, sulfites are in many products.

Sulfites are known to produce headaches and facial flushing among susceptible individuals. Organic low-sulfite wines are commonly advertised as not causing these problems. Since 1987 wine sold in the United States with more than 10 sulfite parts per million must include a sulfite warning label. This low level is nearly impossible to reach because sulfites are naturally occurring. Some organic wineries also specify that they use no animal products (for clarification) in making their wines, a consideration for vegetarians. Organic wines are easier to produce in dry regions. Humid regions naturally incur increased insect and fungal damage, making it more difficult for the grapes to achieve maturity.

VINE DISEASES, PREDATORS, AND PESTS

Every species of plant and animal is subject to predators, pests, and diseases. The domesticated grape is more susceptible to attack than most. The vine and its fruit suffer from attacks by mammals, birds, insects, and microorganisms. Few kill vines outright; most merely limit vine vigor, productivity, or ability to take on nutrients. More than sixty vine viruses are known and described. Only a few of the more interesting and historically devastating are highlighted below.

Molds and fungi drift onto the vines, often in excessively moist conditions, and cause the leaves or berries to rot. Many molds and fungi attack the grape including angular leaf scorch, black rot, crown gall, downy mildew, eutypa dieback, grapevine powdery mildew, and phomopsis leaf spot. Those attacking fruit are an annual threat. Those attacking crowns, leaves, roots, and shoots kill.

Odium

Odium tuckerii, or powdery mildew, was discovered by Mr. Edward Tucker who named it for himself. He found it growing on his vines in southeast England in 1846. Appearing as yellow dime-sized spots on leaves, Odium prevents photosynthesis. It reduces vine vigor and greatly reduces yield. It is believed that Odium is of North American origin as it was not previously present in Europe. It might have arrived in England in 1840 when Kew Gardens imported a number of plant specimens for a new exhibit. After its escape, it diffused quickly throughout European vineyards. Odium was spotted in Versailles in 1846. By 1851 it had spread throughout the Mediterranean basin.

The success of Odium during the 1840s can be directly linked to European environmental conditions. During those same years the potato blight (a related fungus) and the subsequent potato famine ravaged northern Europe. With drier conditions in the 1850s both diseases subsided.

A cure for Odium was discovered in 1852; a water-based spray mixed with lime and sulfur killed it. Botanists noted American vines' resistance to odium, leading French nurserymen to import American vines to breed resistance into vinifera grapes. This may have led to the phylloxera disaster, which began in 1865.

Phomopsis viticola

Phomopsis causes small brown spots with yellow halos on leaves and shoots. The spotting spreads when an existing patch is hit by a raindrop, which scatters spores. The wrinkling reduces plant vigor. The disease is most prevalent in years with cool wet springs. When the peduncle is infected, berries wither and drop.

Botrytis cinerea

Also known as noble rot, botrytis attacks many fruits, vegetables, and flowers. It is the asexual, spore-forming variant of *Botryotinia fuckeliana*. The mold spores winter on plant debris. Botrytis is also responsible for the fuzzy white/gray mold found on strawberries and other soft fruits. This fungus draws water from the grapes it attacks. The resulting desiccated grape is proportionally sweeter. Botrytis also modifies sugars in the grape, creating a honeyed flavor.

Botrytis formation requires cool foggy September mornings followed by warm sunny days. These conditions occur frequently in the vicinity of water bodies or at the base of a range of hills. Botrytis made its first documented beneficial appearance about 1500 in the Tokay region of Hungary where the low volcanic hills rise above the Hungarian plain. Austrian vintners near the Neusiedlersee See appear to have picked up the practice shortly thereafter. In 1855 Chateau d'Yquem in Sauternes, France, was declared the finest wine in Bordeaux for its botrysized Semillon wine.

Vine Pests

In terms of number of species and total damage, insects are the worst and most common threat, although this category includes deer, birds, and other small mammals. Only the rose seems to attract more insect attention than the vine. One can often see rose bushes at the end of vine rows. Growers use these roses just as coal miners used canaries. Insects on the rose signal their impending presence in the vineyard.

Some species concentrate on eating leaves, others on shoots, and still others eat the roots. Some of these pests act as vectors, transmitting bacterial and viral diseases to the vine. These pests must be controlled. Most growers use spray insecticides; biodynamic and organic growers encourage predator species. Among the dozens of insect species that enjoy eating grape vines, one stands out beyond all others, phylloxera, and is given special treatment below.

Phylloxera, the Devastator

Daktulosphaira vitifoliae (Fitch 1855), the grape phylloxera (previously *Phylloxera vastatrix*), is a near-microscopic yellow insect that feeds on the entire vine. Vinifera vine roots cannot heal the aphid bites, and the vine bleeds out sap like a hemophiliac. Native to North America east of the Rocky Mountains, phylloxera and American grapes evolved together. American grapes, over the millennia, evolved the capacity to scab over phylloxera bites and thus survived infestation. Phylloxera was one reason for failed attempts at growing vinifera grapes in colonial America. Phylloxera cannot survive in sandy soil and does not like clay-rich soils where its movements are restricted. The pest also does not fare well in cold climates where it must reinvade each summer.

Campbell's careful reconstruction of the initial infestation reveals how one minor and seemingly insignificant transfer helps us all understand the potential impact of invasive species or patient zero in an epidemic. According to Campbell, phylloxera arrived in Europe via steamship with a shipment of vine cuttings. The cuttings were transported in a specially designed botanical specimen preservation case developed by Dr. Nathaniel Ward. The

Wardian case, as it was known, was the forerunner to the terrarium. These vine cuttings were destined for

> [t]he small town of Roquemaure in the department of the Gard on the right bank of the river Rhone, where an obscure wine merchant named M. Borty tended a smallholding of vines in a walled garden behind his modest establishment. In 1861 a "friend from America," a fellow vine-grower called M. Carle, had visited M. Borty. He promised to send back from New York some of his own native vines. . . . One day in the spring of 1862 a case arrived unexpectedly from America.[12]

Borty planted the infected vines. Once in Europe, phylloxera spread contagiously. It quickly devastated one French region after another between 1865 and 1890. The contagious diffusion of phylloxera across France is depicted in Figure 2.4. The map represents how billions of phylloxera spread across Europe while eating and killing millions of vines.

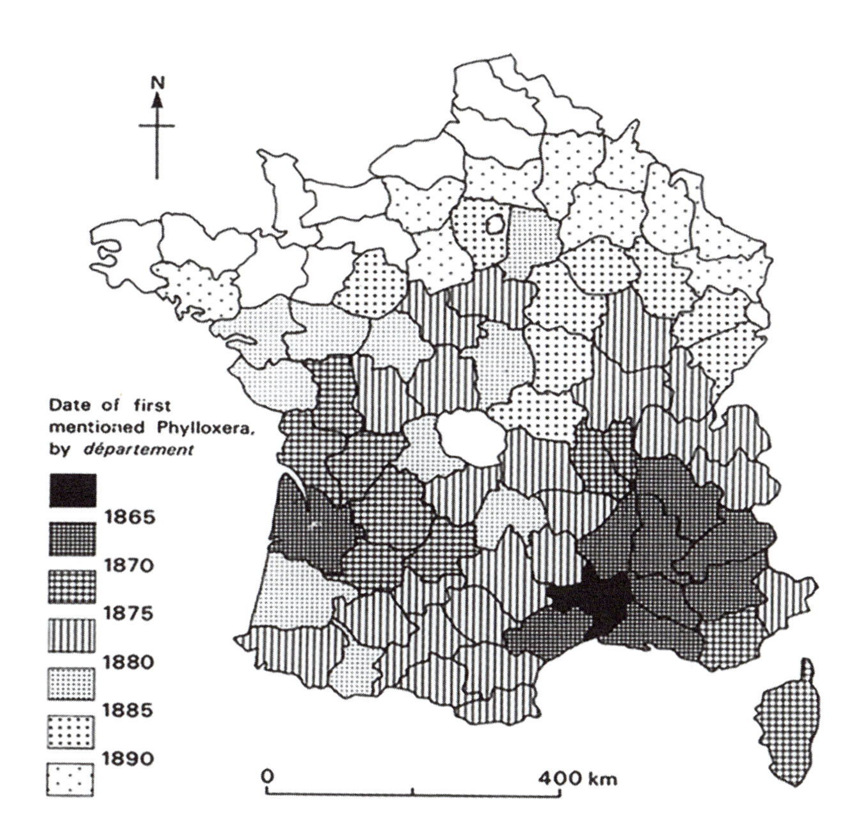

Figure 2.4: The spread of phylloxera across France.

Before 1880, not knowing the cause, most growers believed they would be spared infestation. This refusal to believe greatly slowed reactions and practically prohibited preventive measures. Many vintners simply abandoned their lands and sought new phylloxera-free regions. This migration led to an expansion of grape growing in northern Spain and Algeria, among other places. Unfortunately, wherever they went with their cuttings, phylloxera went with them.

The phylloxera life and annual cycles are complex (see Figure 2.5). There are five forms: winter egg, summer egg, leaf, root, and winged sexual.

The cycle begins with a winter egg deposited in the folds of the vine trunk. When this egg hatches in the spring, the wingless pest wanders to a leaf and begins eating. It is parthenogenetic (born pregnant). Adult leaf and root crawlers lay 400–600 eggs in a gall. The eggs hatch after about fourteen days, depending somewhat on temperature, and become more wingless leaf eaters.[13] After several generations phylloxera with root-penetrating mouth parts are born. They wander down the trunk and commence eating below ground. Chemicals in phylloxera saliva prevent the vine from healing. Again,

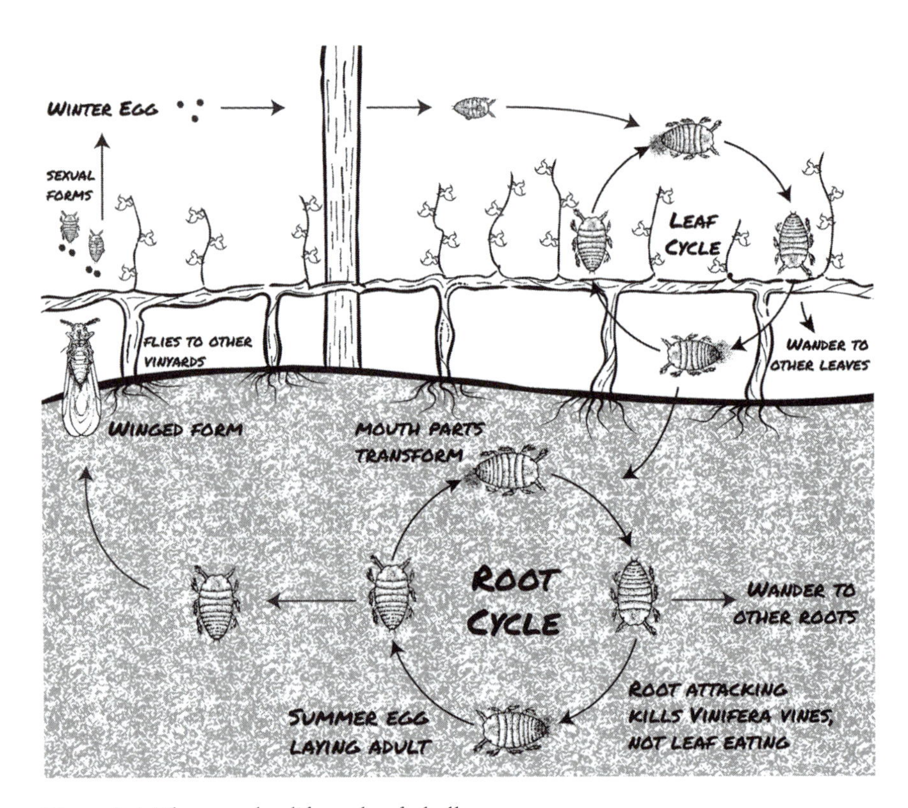

Figure 2.5: The complex life cycle of phylloxera.

this form lays eggs and several more generations hatch before winter. Meanwhile the above-ground type continues to eat leaves and reproduce. As the summer draws to a close, the subsurface form lays eggs that produce winged sexually reproducing forms. The winged forms mate and each female deposits *one* winter egg on vine trunks.

A. E. Bateman in 1884 provided ample evidence of the scale of the problem and the resulting unsatisfactory solutions:

> The effect of the Phylloxera in France had been to diminish the area under vineyards by nearly half a million hectares. . . . the production of wine had fallen off even more. . . . the price of French common wines had consequently risen from 1s. 10d. a gallon in 1876, to 2s.9d. a gallon in 1881, . . . this rise in price had produced three effects, first, the largely increased import of Spanish and Italian wine for mixing with their own diminished supplies; second, the manufacture of wine from imported dried grapes or currants; and third, the manufacture of "*vin de sucre*"—a compound of beetroot sugar, tartaric acid, and hot water poured over the grapes from which wine had already been made.[14]

Bateman continues to discuss resultant production increases in Algeria as thousands of Frenchmen fled phylloxera to the temporarily greener pastures of Africa. Once in Algeria they pushed the indigenous Berber and Arab tribesmen off land suitable for grape production. The tribes attacked the French colonists. The French, in turn, established the famed Foreign Legion to combat the tribesmen. For nearly seventy years, from 1890 to 1960, the French colonies of Tunisia, Morocco, and Algeria exported more wine than France proper. Table 2.2 is a timeline of the spread of phylloxera around the world.

The breakthrough against phylloxera occurred when Jules Planchon and Charles Riley traced the life cycle and origins of the pest. From this information they devised a preventive measure. The "cure" so desperately sought was no cure but a dodge. They demonstrated that European vine tops grafted onto American vine roots can withstand phylloxera feeding on them.

Throughout Europe vines were pulled out and replaced. This, of course, put many vineyards out of production for years. The economic base ruined, many vineyards were never replanted. For the consumer this meant higher prices for real wine and huge amounts of imitation wine were sold.

As the diffusion of phylloxera timeline shows, phylloxera now exists throughout most of the world's vineyards. Exceptional places without phylloxera include Chile and parts of Australia and Argentina.

Disease-Carrying Pests

Pests, while eating the vine, can transmit diseases through their saliva. Two pests are particularly dangerous for this reason, the glassy-winged

Table 2.2: Timeline of Global Phylloxera Diffusion

Year	Event
1835	Ward invents the Wardian case.
1838	Atlantic crossed in 15 days.
1842	Longworth vineyard on banks of Ohio.
1845	Potato blight—Potato-dependent mold mutates and explodes across northern Europe. Turned crops to mush and people to skeletons.
1856	Phylloxera first botanically identified and observed in New York.
1860s	French import American vines for cross-breeding and experimentation.
1862	Wardian case of American vines contaminated with phylloxera delivered to Roquemaure, Gard, France and planted.
1867	Phylloxera found growing on vines in experimental greenhouse gardens in Ireland and Cheshire, England. First reports of new vine disease in Rhone Valley.
1868	The French establish investigative committee. Discovery of phylloxera characteristic defined. It is given the name *Phylloxera vastatrix*, which means the dry-leaf devastator.
1872	Appears in Douro Valley, Portugal, Austria, and Ukraine.
1873	Appears in California.
1874	Appears in Spain and Germany.
1875	Appears in Hungary and Australia.
1879	Appears in Italy.
1880	Appears in Serbia and Slovenia.
1885	Appears in Algeria.
1886	Appears in South Africa.
1890	Appears in Champagne, New Zealand, and Peru.
1900	Phylloxera diffusion through Europe essentially complete.
1955	Discovered in Oregon.
1988	Discovered in Washington.
1990s	New phylloxera outbreak in California.

sharpshooter, a type of insect, and the nematode, a worm. Others such as phylloxera, ladybugs, cane borers, and Japanese beetles simply eat the plant.

Pierce's disease is the vine-killing bacterium *Xylella fastidiosa*. It is transmitted in the saliva of the leaf-eating glassy-winged sharpshooter. Glassy-winged sharpshooters, the Pierce's disease vector, are cold intolerant. In the United States this currently limits Pierce's disease to the Southeast and California. Each summer, seasonal warming sends the sharpshooters racing northward, only to be killed off during the winter.

Once the vine is infected by a sharpshooter, the bacteria multiply. The bacteria concentrate in and block the xylem, the vine's water transport system. The root downstream from the blockage dies from lack of water. With its roots dead, the remainder of the vine dies shortly thereafter. There is no cure and the vines must be eradicated.

There are more than 80,000 described species of nematodes (roundworms). For agricultural purposes there are two kinds of nematodes, predatory and pest. The predatory nematodes kill other garden pests. The pest nematodes attack plants. Of particular interest is *Xiphinema index*, a pest nematode that transmits grapevine fanleaf virus. The disease is characterized by a yellowing of the leaves. There is no cure for grapevine fanleaf virus. Infected plants must be eradicated and the soil treated before replanting.

Other Insect Pests

One or more members of the multitude of insect pests attack the vine at each stage of the growing season. Following are descriptions of only six vine-loving insect species with their timing of attack, type of attack, and resulting economic damage.

Steely Beetle

This small beetle attacks in the spring when buds begin to swell. It enters a bud and feeds on the interior, hollowing it out and killing the bud. The vine may produce a secondary bud from the same location, but it will be several weeks behind and may not produce a crop.

Climbing Cutworm

The worm, really a caterpillar, feeds at night. It climbs the vine until it finds a tender young shoot, no more than three or four inches long, and eats it. As with the steely beetle, the dead bud may be replaced by a secondary, less vigorous shoot in a few weeks.

Grape Cane Borer

The cane borer attacks in the early summer when the shoots are three to six inches long. The cane borer chews two rings of punctures to girdle the tips of fruiting canes. This causes the cane tips to die. Often the canes break at the puncture ring. Between the rings of punctures the borer lays its eggs. Once they hatch, the larvae feed on the inside of the cane.

Grapevine Flea Beetle

The grapevine flea beetle jumps like a flea. There are numerous species that attack grapes and other fruits and vegetables throughout the United

States. All species have similar behaviors. The most common of these is the Japanese beetle. They overwinter below ground. In April and May they emerge. The adult feeds for several days on newly emerging leaves, skeletonizing them, and then lays eggs in the vine bark. Whole vineyards may be stripped if not treated with pesticides. Larvae and pupae develop over the summer and transform into adults in the early fall.

Grape Cane Girdler

Attacking the shoots when they are about a foot long, the girdler acts much like the grape cane borer. This small black weevil makes a series of holes around the shoot, causing it to die. This usually occurs so late in the season that secondary buds may not develop.

Grape Rootworm

This beetle is found in the eastern United States. The adult eats leaves before laying eggs in folds in the bark. The eggs hatch and the larvae crawl down the trunk and follow the roots underground. The larvae perform most of the damage sustained by the vine. They feed on the roots, greatly reducing vine vigor. A heavy infestation results in vine death by destroying the plant's ability to deliver water and nutrients to the shoots.

SUMMARY

The genus *Vitis* contains many species. The *Vitis vinifera* species is economically viable beyond all others. Its global annual production value is in the billions. V. *vinifera* is diverse with more than 10,000 catalogued varieties.

The vine is native to humid continental conditions, yet it thrives in Mediterranean climates as well. Vines like thick soil. The vine sends its tap roots deep into the soil to collect water from above the bedrock.

The vine lives a precarious life. Animals eat the plant and fruit. Birds eat the fruit. Insects and nematodes feed on its various parts and are bacterial and viral disease vectors. Molds and fungi attack it whenever conditions are right.

Beset on all sides, the vine's only friend is the human grower. Humans tend it, nurture it, and protect it from its numerous enemies and hazards. The symbiotic relationship between humans and *Vitis vinifera* is at least 10,000 years old.

Chapter 3

Terroir

No viticultural subject has generated such diverse opinions as the relation of the vine to its environment and to the composition and quality of the resulting musts and wines. . . . The interrelation of the climate, geography, and variety of grape determines the potential for fruit and wine quality.[1]

—Amerine and Wagner

Terroir encompasses the interplay of environmental conditions and human actions to produce a grape crop and wine. Components of environmental conditions consist of local weather, climate, geology, pedology, topography, and geomorphology. Each of these components can assume a multitude of forms. Together, they render each place on earth a near-unique combination of these attributes. Some blends of environmental components, or terroirs, are suitable for grape growing, others to growing onions, citrus, or Christmas trees.

Historically, people grew what long experience had taught them were reliable producers. Adam Smith, author of *The Wealth of Nations*, knew "[t]hat the vineyard, when properly planted and brought to perfection, was the most valuable part of the farm, [and] seems to have been an undoubted maxim in the ancient agriculture as it is in the modern through all the wine countries."[2]

Modern grape growers evaluate a location's terroir before deciding which grape variety to plant. Each variety favors a particular set of environmental conditions, and matching the right variety to the terroir makes the difference between average and spectacular wines. For each piece of land, the grower's goal is selecting the grape variety that will return the highest value and produce the finest wine.

The influences of terroir (in its physical geographic sense) go beyond the grape and shapes resident behaviors in unlikely ways. Thomas Jefferson remarked on these facts while visiting the Burgundy region in 1787, stating:

> At Pommard and Volnay, I observed them eating good wheat bread; at Meursault, rye. I asked the reason of this difference. They told me that the white wines fail in quality much oftener than the red, and remain on hand. The farmer, therefore, cannot afford to feed his laborers so well. At Meursault, only white wines are made because there is too much stone for the red. On such slight circumstances depends the condition of man![3]

At the regional level the terrain component of terroir provides economic advantages to some wine-growing regions and disadvantages others. For example, coastal wine regions experience moderating marine influences and reduced transport costs compared to those upstream. In the case of Bordeaux, the *police des vins* preempted wines from upstream until all Bordeaux was sold.[4] The Chablis region located upriver from Paris was able to deliver wines more cheaply than the remainder of Burgundy, whose wines were sailed around Iberia to reach Paris.

WEATHER

Weather is the most unpredictable component of terroir. Soil, bedrock, and slopes change insignificantly during a lifetime and, for the most part, climate change is indistinguishable from year to year. It is the variability of weather that makes each year's crop different. One year the weather is "good" and the wines are average. Another year the weather is "too wet" or "too cold" and the wines are "poor." In yet another year the weather may be "very dry" and the wines are "excellent." Weather events usher in opportunities for vine pests and diseases. In wet years molds and rots are a threat. Winds disperse weeds and insect pests. In dry years devastation by desperate birds and mammals can increase.

Growers live in fear of "unusual" weather events. In the Northern Hemisphere, growers fear the late frosts of April and May and the early frosts of September and October. Depending on location, they may also fear desiccating spring winds, summer hail, and thunderstorms. Rains immediately before harvest render grapes flabby and encourage rot.

Throughout the world these problems occur in unique mixtures, further serving to make each region's product a unique taste experience. Most vineyards experience several worrisome weather events each season. The growers take what few measures they can. They plant vines on hillsides to avoid frost and flood damage, where they can. They distribute smudge pots to heat frosty air, or they install giant fans to prevent frost from settling. Drip irrigation is widespread, where allowed to compensate for dry conditions.

Increasingly, growers place weather stations in their vineyards. Bill Pregler of Kestrel Corporation, makers of digital weather instruments, claims, "Identifying and tracking weather conditions in the vineyard allows growers to make informed weather-related decisions, saving time and money, and ultimately improving the quality of the fruit."[5]

CLIMATE

On average, one year is much like the last. Daily announcements of average temperatures and conditions are popularly called climate, but climate is actually much more. Climate includes the entire scope of weather events over many years, ranging from single events like the passing of a hurricane to cyclic fluctuations such as the El Niño effect, to long-term droughts and ice ages.

Consider the single weather event that delivered the most precipitation to your location in the last five years. No doubt it caused massive flooding, leading to major sediment transfer and stream channel changes; perhaps people died. Other rare events such as a three-foot snowstorm or a –25°F degree temperature occurring once a decade are similar defining events for plant and animal communities. Dieback resulting from these rare events is common.

Our definition of climate also includes long-term trends and oscillations in precipitation and temperature regimes. Plant communities respond directly to the trends. A decade of "below average" precipitation leads to the territorial expansion of drought-resistant species, while a decade of "above average" precipitation will drown species unable to survive drowned roots. For grapes, this means that shifting climatic conditions require a response in the vineyard, and that varietal selection must be made with care.

Climate has significant impact on the color, acidity, and flavor of grapes. Wines from fruit grown in cool regions are subtle, austere, and elegant. They are characterized by their tartness, light body, crisp acidity, and lower alcohol content. Whites from cool areas produce flavors of apple and pear. Fruit from warm to hot regions produce wines that are lush, full bodied, with smooth acidity, and highly alcoholic. Hot-region white grapes emit mango or pineapple flavors while reds release aromas of fig and prune. In intermediate climates, the fruit ripens fully with medium body, integrated acidity, and moderate alcohol levels.

Climate is the most significant limiting factor for grape growing. The further poleward a vineyard is from 40 degrees latitude, the iffier a successful harvest. Poleward wine regions have shorter growing seasons with longer days; as a result growers plant and tend fast-ripening, highly acidic varieties. Equatorward wine regions are hotter and have longer growing seasons. Their wines are less acidic and more powerful. Tropical regions, those between 23½ degrees north and south latitude, are generally inhospitable to the vine.

In cooler regions, usually found above 50 degrees latitude, vines have not historically received enough summer warmth to permit a crop to succeed every year. With global warming and selective breeding for hardier varieties, the limit crawls poleward each year. Regions populated by northern coniferous forests, steppe, or tundra are not suitable for commercial grape growing. Most are not capable of supporting vines beyond germination.

Tropical regions, those between 23½ degrees north and south latitude, are too hot and often wet for commercial grape growing. The tropical lack of seasonality disturbs the vines' natural rhythm and a crop is not reliably produced, only a gangly fruitless vine. Tropical regions also harbor a wide variety of bacteria and insect life that find various parts of the vine delicious. The vine, unfamiliar with these pests, cannot defend itself and perishes.

Climate influences plant growth at multiple scales. At each scale, different physical geographic components come into play. At the macro-scale (hundreds of square miles), climate defines the limits of grape growing and the growing regimes of large regions. The primary factor at the macro-scale is temperature, but precipitation seasonality is also important. The meso-scale (tens of square miles) climate of a subregion varies with the proximity of that meso-scale region to large bodies of water or mountains compared to other subregions of the macroclimatic region. Micro-scale (less than ten square miles and as small as a few square feet) climatic factors often focus on humidity, slope, aspect, and elevation. Peach and melon flavors abound in white grapes and berry and plum flavors dominate reds.

Whatever climatic scale we examine, the cause is the same: the differential heating of the earth's surface. At the macro-scale the difference between the heat of the summer and the cold of the Siberian winter causes the monsoons of Asia (an annual reversal of wind patterns). Ocean currents redistribute heat. In doing so the currents create permanent climatic zones. At the meso-scale the orographic effect first cools and dries air passing over the mountain range, then it warms the dry air on the leeward side. The lake effect (seasonal reversal of land-water temperatures) and the land-sea breeze (daily reversal of land-water temperatures) represent the same phenomenon, at different scales. Shady areas on mountain slopes are cooler than sunny slopes (differential heating of land surfaces). The shaded areas can create rivers of cold, sometimes frosty air. The cold air can spread across a broad area once it arrives in the flatlands. It is simply a scale difference.

CLIMATE CLASSIFICATION AND VISUALIZATION

Geographers developed several methods to define and classify Earth's climatic zones during the twentieth century. The classification systems are based on a combination of temperature and precipitation seasonality. The first successful, and still highly valued, classification system was created by Vladimir Koppen.[6]

Koppen divided the world into five major climatic zones, Tropical, Subtropical, Continental, Subpolar, and Polar, based on annual temperature statistics. The five major climatic zones are subdivided based on whether the subregion is wet all year, has a dry summer season, or has a dry winter season.

Researchers refined the Koppen climate classification algorithm and associated the climatic zones with biomes.[7] Biomes are large areas occupied by common plant and animal communities. Three biomes are of importance to grape growers. The Mediterranean biome is found throughout the Mediterranean basin, California, southeast Australia, South Africa, and northern Chile. The West Coast Marine biome is found in France, the Pacific Northwest, western Australia, and southern Chile, characterized by cool temperatures and morning fogs with otherwise clear skies. And, finally, the Humid Continental biome is found in Central Europe, the eastern United States, and China. Desert areas with irrigation are of increasing importance in Australia and Chile.

While the Koppen system is adequate for defining macro-climatic zones, grape growers needed a more refined climatic measure to determine what varieties will thrive where. Toward this end, A. J. Winkler developed an algorithm for classifying climate as it pertains to grape production. Winkler's system is based on temperature alone because the vine requires minimal precipitation during the growing season.

The key to Winkler's system is the degree day statistic. The degree day statistic is commonly used to identify appropriate growing season lengths for many crops. Winkler knew that vines (like most plants) are dormant (no growth occurs) below 50 degrees. Only on days when the average temperature is above 50 degrees does growth occur. For every degree increase above a daily average of 50 degrees, one degree day is added to the total. Thus, a day with an average daily temperature of 67 degrees adds (67 – 50 = 17) seventeen degree days to the total. Annual degree days are calculated by summing all the daily totals for the year.

Winkler next identified which grapes performed best in each degree day range. Using these data, he began making varietal and locational vineyard site recommendations based on climate. More importantly, he devised a single table that summarized his recommendations—recommendations that remain the de facto decision-making tool for growers.

Winkler identified five grape-growing regions. Region 1 specifies the coldest realm for wine production and Region 5 the hottest. Table 3.1 summarizes Winkler's defined regions and viticultural recommendations.

Geographers depict climate with a special type of graph called a climatograph. For example, Adelaide, Australia, is in the Southern Hemisphere, so its winter is in July. A climatograph for Adelaide would have three scales:

Table 3.1: Relationship between Regions, Growing Conditions, Grape Varieties, and Planting Recommendations

Winkler Region	Degree Day Range	Locations	White Grapes	Red Grapes	Planting and Growing
1	2,500	Burgundy, Champagne, Oregon, British Columbia, New Zealand	Riesling, Gewürztraminer, Chardonnay, Sauvignon Blanc	Pinot Noir, hybrids	Plant on sun-facing hill-sides, plant near water body
2	2,500 to 3,000	Bordeaux, Sonoma, Asti, Chile	Muscat Bianca, Chardonnay	Cabernet Sauvignon, Merlot, Zinfandel	Hillside planting, fans
3	3,000 to 3,500	Napa, Livermore, Mendoza, Argentina	Chardonnay	Cabernet Sauvignon, Zinfandel, Malbec	Frost unlikely, fans
4	3,500 to 4,000	Capetown, Sydney, Sacramento	Chenin Blanc	Shiraz, Grenache	No frost worries
5	4,000 +	Spain, Portugal, Sicily, Greece	Table grapes, Palomino	Sangiovese, Tempranillo	Irrigation required

Source: A. J. Winkler, J. A. Cook, M. W. Kliewer, and L. A. Lider. *General Viticulture*. Berkeley: University of California Press, 1962. Summary of pp. 58–71.

temperature, precipitation, and time, in months, on the x-axis. The climate of Adelaide is usually described as Mediterranean. In a Mediterranean climate the winter (May, June, July, and August in the Southern Hemisphere) is the wet season.

Macro-Scale Climate

Macro-climatic regions are defined based on potential evapotranspiration. This simply refers to the adequacy of water (precipitation) and the ability of temperature and plant respiration to evaporate water. Macroclimates are expressed on the landscape as major plant communities. Forested lands have a water surplus, grasslands seasonal water, and deserts have a deficit. Hot lands with a water surplus are perennially green, temperate lands contain deciduous trees, cold lands coniferous trees, and really cold lands moss and lichens.

The circular flow of ocean currents from warm equatorial regions poleward creates climatic regions. In the Northern Hemisphere water from the eastern side of a continent flows northward, crosses the ocean, and strikes the west coast of the next continent (the actual latitude varies due to the shape of northern land masses). At high latitudes (50 to 60 degrees north) the current warms the coastal lands. Regions on the west coasts are wet and forested. These climates are called West Coast Marine. Regions include the Pacific Northwest of the United States and Canada, United Kingdom, France, New Zealand, and southern Chile.

After striking the continental west coast, the current turns southward toward the equator. Flowing southward, the current becomes colder than the surrounding land. The current chills the air above it, producing coastal fog. Precipitation occurs in the winter months, and the vegetation is herbaceous. San Francisco, Santiago (Chile), Galicia (Spain), and Cape Town (South Africa) experience these conditions. When the land is much hotter than the water, there are deserts like the Atacama of Chile, California's Mojave, Portugal's Algarve, and the Moroccan Desert. The current completes its loop by crossing the ocean from east to west near the equator.

Meso-Scale Climatic Factors

At the climatic meso-scale, precipitation is usually the defining factor for grape growing. Climatically, within this range grape production and quality are functions of annual precipitation patterns and annual short-term aberrant weather (high winds, thunderstorms, hail, and hurricanes) and seasonal temperature events (late and early killing frost).

Within macro-climatic zones atmospheric phenomena, topography, prevailing airflow, and large bodies of water modify the theme and define mesoclimatic areas. Topography and prevailing winds funnel precipitation into some zones and not others. Mountainsides facing the prevailing winds are wetter than those on the lee.

The Lake Effect

The lake effect commonly comes into play at the meso-scale. Because lakes heat and cool slower than the surrounding land, lakes have a moderating influence on the climate in their vicinity. Larger lakes have a greater area of influence. The area of influence often ends at surrounding ridge tops. Over flat lands the moderating influence rarely extends more than ten miles from the shore line. The lake effect delivers two vital services for fruit growers.

In the spring, land nearest the shore is kept colder than land further away. On early warm spring days fruiting deciduous perennials bud out. If the warm stretch lasts for several days, tender young leaves and shoots might start to emerge. Often these early warm spells are interrupted by late frosts. When this happens the shoots freeze and the plant loses much of its vitality. The cold lake water, by keeping the air cold, keeps the plants from budding out too early and subsequently becoming frosted.

In the fall the reverse happens: the water is warmer than the surrounding land and air. The cap of warm air around and above the lake keeps the fruit from becoming ruined by early frosts and allows some grapes to become overripe without rotting. In select locations the early fall fogs promote the growth of the bacteria *Botrytis cinerea*. Wine made from these specially rotted grapes is highly prized and makes the world's finest dessert wines.

During the summer, large amounts of water evaporate from lake surfaces, producing fogs on summer and early fall mornings. The resulting dew can be an important source of moisture in desert areas. On sunny days, clouds form from the evaporating lake water. The clouds cool the land downwind for many miles. In the winter, lake effect snow showers extend more than 100 miles from the Great Lakes, blanketing vines with a protective layer of snow.

The Orographic and Rain Shadow Effects

As air is forced to climb over mountains, it cools. In cooling, its ability to hold moisture decreases. The measure of the amount of water in the atmosphere relative to the amount the atmosphere can hold at that temperature is called relative humidity. Given constant moisture content, relative humidity increases as temperature decreases. As a result, mountain tops are often wreathed in clouds.

On the windward side of a mountain mists are common and rain falls regularly. On the leeward side the sky is clear and rain unusual. The leeward side is known as the rain shadow. Regions downwind of mountains (lee)

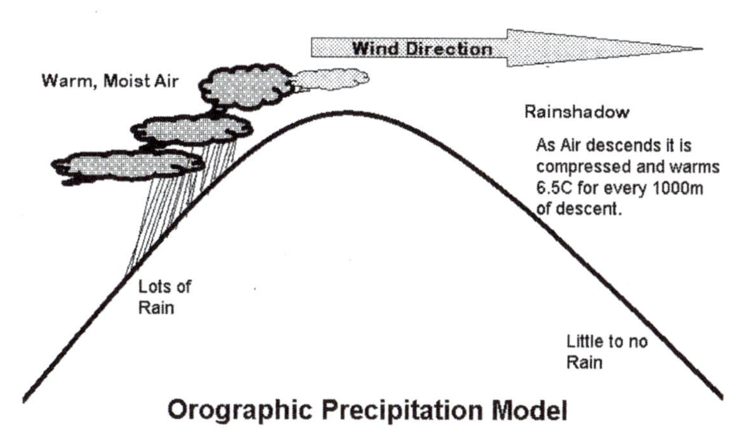

Figure 3.1: The orographic and rain shadow effects are significant factors in mesoclimatology.

receive less rainfall than their upwind (windward) counterparts. Rain shadows are potential wine regions because the dryness limits vigor; regions to the windward of mountains are not. The Willamette Valley of Oregon, Mendoza (Argentina), and the Napa Valley of California are prime examples. See Figure 3.1 for a depiction of the orographic effect. Famous descending winds associated with rain shadows include the chinook of the Pacific Northwest, the foehn of Germany, and the berg of South Africa.

Katabatic and Local Descending Winds

Katabatic is a Greek word that means "traveling downhill" (see Figure 3.2). As with the lake effect, katabatic airflow results from the differential heating and cooling properties of earth surfaces. There are also atabatic winds, those traveling upslope, which are explained by the simple phrase, "warm air rises."

Warm katabatic winds flow downslope on the leeward sides of mountain ranges. As air descends down a mountain slope, it warms. As it warms its capacity to hold moisture increases, but there is no moisture to be added. When the fast-moving dry wind reaches fields and vineyards, it draws the moisture from leaves and can rapidly wilt plants. At the same time, the wind can dislodge feeding insects and desiccate fungal infections. This wind often occurs in the spring when tender leaves and growing shoots are particularly vulnerable. This phenomenon has many names around the world including the sirocco (Atlas Mountains), the zonda (Argentina), and the Santa Ana (California).

Cold katabatic winds are created at high altitudes above mountain ranges. The coldness of the location causes the air to densify and push down the

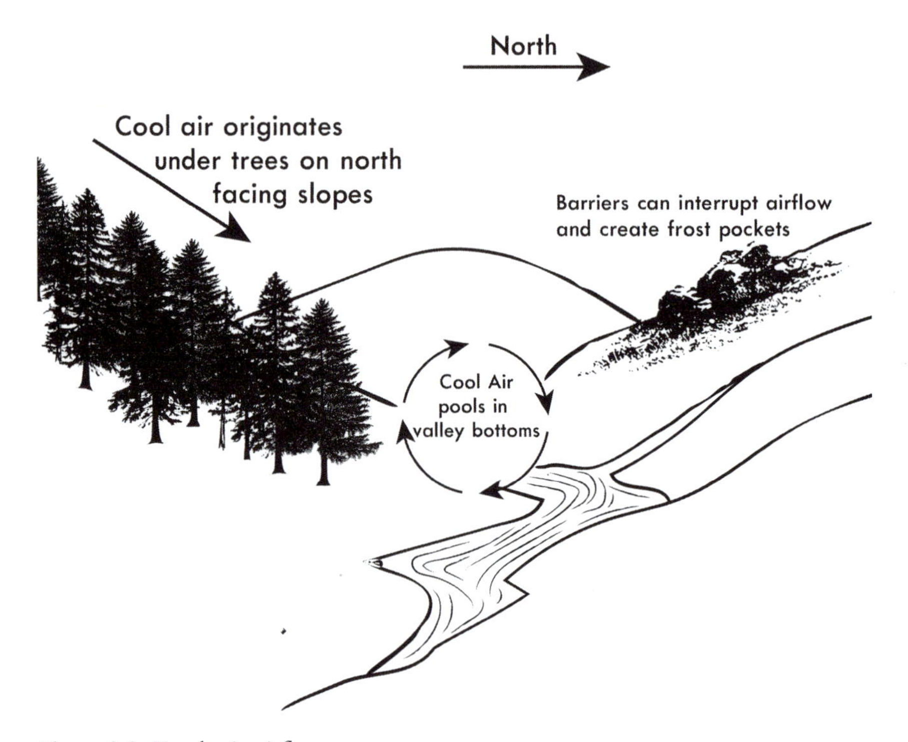

Figure 3.2: Katabatic airflow.

slopes. The flow can be very intense when channeled down a valley. The mistral of France, the bora of Croatia, and the southerly burster of Australia are well-known examples.

Microclimatology

At the microclimate scale temperature again becomes the defining factor. Technically speaking, each vine lives in its own microclimate, a composite of all the atmospheric phenomena that piece of land has experienced. At the micro-scale, proximity to water is crucial, especially in Mediterranean climates where irrigation is often necessary.

There are three phenomena that affect the microclimate in these regions: local airflows, slope angle, and slope orientation. Sunward-facing slopes are warmer and more productive than poleward-facing slopes. Equatorial regions and higher latitude slopes perpendicular to the sun receive the densest photons and are warmest of all. Premature spring budding can occur on sunward-facing slopes, leading to the potential for frost damage, so it is advisable to plant on north-facing slopes to slow bud burst.

Cold Air Drainage

Cold air drainage refers to a localized form of cold katabatic airflow. Bottom lands and narrow valley floors are bad places to plant grapes in Winkler climatic regions 1 and 2. Grapes planted in these localities are subject to a microclimatic phenomenon called cold air drainage. Surface features like walls or hedges paralleling the slope create pools of cold air like the way a dam backs up water. Cold air drainage damages vines in the middle fall when it can result in an "early" frost. As a result, growers in Winkler 1 and 2 regions are careful to avoid planting vines on valley floors. Barriers to air drainage can create additional mid-slope frost pockets.

On a warm sunny day in the early fall or spring, temperatures climb into the 70s. On clear autumn nights, diurnal temperature differentials can exceed 40°F. As evening approaches, the sky remains cloudless. Hillsides facing away from the sun are cool all day and as evening approaches their surface cools rapidly. Air near the surface on shaded hillsides cools as the ground beneath it cools. As the air cools, it becomes denser and begins tumbling down the hillside. These cool spots attract airflow from uphill to fill in the vacuum created. The replacing air cools, and also passes downslope.

When the flowing air reaches the valley bottom, it pools and grows still. The coldest air descends to the lowest elevations. As it pools it quickly reaches its dew point and mist forms. The mist adheres to plants. When the air temperatures drop below 32°F, frost appears in the lowest places first. Frost is the moisture condensed out of the air when the dew point is 32°F or less. Frost freezes plant surfaces to which it adheres. What makes the "early frost" of bottom lands so dangerous for the vine is that the vine is not prepared for the cold. Because of the day's warmth and evening wetness the stoma and chloroplasts are open and the water within the thin plant leaf cells freeze. The frozen water expands and the cell walls explode from within and die.

Frost destroys tender shoots and emerging leaves in the spring and berries in the fall. During the average spring day as the vine grows, the opening leaves are swollen with water and actively synthesizing. With the drop in temperature, the water-laden cells on leaf edges explode as the water in them freezes. The tips of spurs, containing the programming to create a ten- to twenty-foot spur, freezes. The plant may lose up to a month's growth. A few days after the frost, leaves will have brown, burned-looking edges. The vine has lost photosynthetic capability because the starches and nutrients the plant normally pulls back into the trunk during the winter are lost in the frozen tissue. Berry skins, when frosted, also freeze. The skin cells explode, exposing the grape's interior flesh. The skin weakens and provides entry points for fungus and bacteria. Ultimately, the skin splits and valuable juice drains onto the ground.

To combat the problems of cold air drainage, many growers have installed fans to stir the air in frost-prone areas. The fans are usually at least thirty feet tall with ten-foot blades. This creates an artificial wind that pushes the cold air away from the vines. A large vineyard may have a dozen or more fans costing about $30,000 each. In many years, these fans are only used five to ten days per year. Some fans have propane heaters attached to them to help warm the air.

Fog Banks

Some areas are subject to localized daily morning or evening fogs. Fog occurs where cold water meets warm land or warm water meets cool land. In Mediterranean climates these fogs are beneficial to vine vigor by delivering cooler air temperatures and moistening leaves. The resulting lower nighttime temperatures produce more flavorful grapes. In more temperate climates fog provides a beneficial environment for fungal infection. See the discussion in Chapter 2 on the Nebbiolo variety.

The confluence of two streams of different temperatures can, as is the case of Sauternes in the Bordeaux, establish a fog prone region. These consistent and predictable fogs followed by warm sunny days promote the growth of Botrytis molds on the grapes.

GEOLOGY (BEDROCK)

Bedrock is beneath us all. It is the outermost layer of the solid earth. Soil, derived from the erosion of bedrock, covers it to varying depths. The kind of bedrock largely determines the particle size and nutrient content of the soil. The differential weathering properties of the bedrock presents observable geomorphic irregularities.

Bedrock composition is of less concern to the grower than depth to bedrock. The thicker the soil layer above the bedrock, the deeper the vines can root. Grapes prefer thickly overlain bedrock. There are three types of bedrock: volcanic, sedimentary, and metamorphic. Volcanic bedrock is a crystalline assortment of minerals. Volcanic rocks are named for crystal size and mineral composition. Crystal size is based on how slowly the rocks cooled. The more slowly they cooled from the molten state, the larger the crystals. Volcanic glass, obsidian, has microscopic crystals. Geodes, volcanic bombs, often grow inch-long amethyst crystals. Granite and basalt decompose into sand-sized particles.

Sedimentary bedrock is classified as calcareous (containing lots of calcium, usually in the form of limestone) and noncalcareous. Calcareous rocks are soft, cavern rich, and good for grape growing. Noncalcareous bedrocks include sandstone and shale. Sandstone makes for sandy soil and shale for clay-rich soil.

Underground, heat and pressure exerted on sedimentary or volcanic rocks create metamorphic rocks. The heat and pressurization usually results in brittle rocks with a larger crystalline structure than the original rocks. Gneiss, mica, and schist are the most common metamorphic forms.

PEDOLOGY (SOIL)

Soils are formed by the erosion of solid bedrock and the accumulation of surface material via erosion. The process and structure are shown in Figure 3.3. Breaks and cracks in the rocks lead to pulverization. The pulverized particles are then transported by wind or water. They accumulate and plants stabilize them. When plants die, they add organic material to the surface. Minerals in the organic material (the O horizon) are leached from the surface and concentrate in the B horizon. The A horizon is alive, but not so the B and C horizons. The B horizon is occupied by plant roots and burrowing animals that churn and mix inorganic (from the C horizon) and organic (from the A horizon) material. The C horizon consists of inorganic material derived from weathered bedrock. The regolithic rocks are broken from the solid bedrock below.

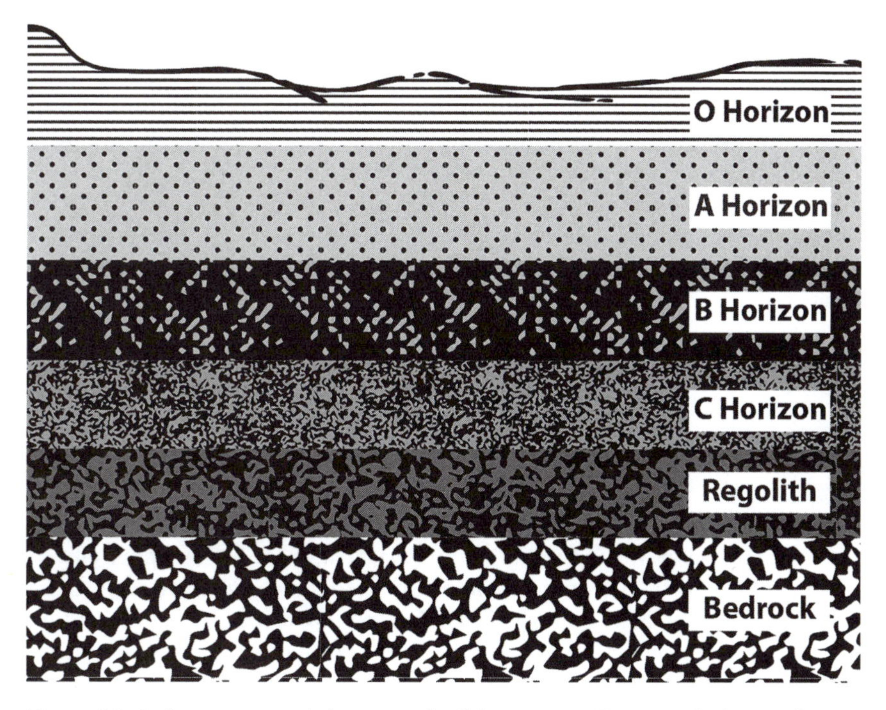

Figure 3.3: Soil structure and elements of soil formation. Horizons O, A, B, and C are identified.

Parent material has a defining character on the resultant wine. Eminent wine soil scientist Scott Burns informs us,

> Calcareous [those derived from limestone] soils produce lemon and citrus flavored wines with a long acid finish. Marl soils and silty calcareous soils tend to generate peppery flavors. Sandstone soils tend to bring a "nervy" character to the wines, and the fine-grained clay soils emphasize tannic characteristics. Schist-rich soils stress the austere nature of the wine, and volcanic soils produce full-bodied wine with smoke-based aromas.[8]

For grape growing, soil pH between 5.5 and 6.5 is desirable. The organic content should be less than 3 percent. Trace elements required by the vine include potassium, phosphorous, magnesium, zinc, and boron. There are three special soils commonly associated with grape growing: limestone, loess, and alluvium.

Calcareous Soils

In Europe, calcareous soils are considered the best grape-growing soils because they drain well and are deeply fractured, providing opportunities for deep rooting. The term *calcareous* refers to soils derived from decomposed limestone soils. Limestone literally melts when it comes into contact with water, and the more acidic the water, the faster it melts. Areas of limestone bedrock have dry surfaces because rainwater enters into channels eaten in the rock and disappears into caverns. Rivers are deeply incised. Caverns and underground rivers are common. Generally, these regions (Jerez in Spain and Chablis in France, for example) are not otherwise agriculturally productive. This is a site characteristic common in many vineyards.

Alluvial Soils

Material from alluvial fans can be hundreds of feet thick and cover areas exceeding a square mile. Alluvial fans result from the deposition of eroded material at the base of a mountain. The eroded material, called alluvium, is a mixture of large and small particles largely sorted positionally on the fan's slope by particle size. Larger, harder to move rocks are at the upper end of the alluvial. The smaller, more easily transported soil particles are further downslope. In the center, stones, often rounded by flowing water, weighing between one and five pounds are common. The smallest particles accumulate at the base slopes because they are most easily and furthest transported by intermittent floods. Microscopic clay particles cluster where the fan merges with the valley floor. In dry regions, the alluvium builds up over hundreds of years with rare rain events moving material downslope. Humus is often layered into the mixture.

Alluvial fans in wetter regions have greater humus content than those of arid regions. Because of the great thickness and rapid drainage of alluvial fans, they are usually excellent places to grow grapes.

What to Do with the Rocks?

The question for the grower is what to do with the rocks! The soil of some regions is nonexistent. In the Chateauneuf du Pape region the "soil" consists of fist-sized river-rounded sandstone rocks. In the Hungarian plain particles larger than a sand grain are unusual. Still, in most vineyards rocks work their way to the surface through winter freeze-thaw action. Depending on local custom, rocks are ignored, piled under vines, or removed. These customs are rooted in local climate and the local populace's perception of the rocks.

In dry areas rocks can serve the dual purpose of reflecting /absorbing solar energy and trap small amounts of moisture on their undersides. The growers weigh these options and make the decisions that seem best. Over thousands of years the opinions changed. The Romans, for example, grew their vines climbing trees and not in the trellised fields we see today.

In Burgundy, the stone walls surrounding the vineyards indicate the one-time local belief that rocks should be removed from the fields. Add a little historical knowledge and we learn that the stone walls were erected by monks to define boundaries for micro-terroir zones. Through generations of steady commitment the monks also provided the labor to lug rocks from the fields, face them, and build them into the substantial walls that remain hundreds of years later. In Germany, rocks are often moved under the vines to retain heat, if dark, or reflect sunlight into the lower leaves, if light in color.

In New World vineyards what to do is not a question. Mechanization overcomes all obstacles. Rocks are moved, removed, and aligned by machine to please the whims and beliefs of individual growers. Growers use the information they gather from the rocks to inform their variety selections.

Aeolian Deposits

Loess is soil deposited by wind (aeolian) action. Loess particles are the extremely small dust and clay particles you see blowing around when the wind kicks up. About 10 percent of the earth's land surface is covered by loess soils.

Like alluvial fans, loess deposits are excellent for grape growing. Loess deposits originate along the southern edge of continental ice sheets. There, small particles of frozen soil (mostly quartz and feldspar) are picked up by the wind and deposited several hundred miles away over thousands of years.

Depressions and valleys filled with this fine-grained material. Loess deposits more than one hundred feet thick are found in China, Ukraine, and Germany. In the United States, loess is common through the Midwest. The Mississippi River bluffs between Vicksburg and Natchez, Mississippi provide excellent opportunities to see it. The eastern half of Washington's Columbia plateau wine region is blanketed in loess.[9]

GEOMORPHOLOGY (TOPOGRAPHY)

At the micro-scale local topography plays a fundamental role in the microclimate. In Winkler regions 1 and 2, proper consideration of slope and aspect make the difference between vineyard success and failure. Slope is the angle tilted from the horizontal. The sloping surface counteracts the effect of angled solar radiation. Slopes drain better than valley floors, making them the preferred micro-environment for grape growing.

Aspect is the direction of a sloped surface. The importance of sunward-facing aspect increases closer to the poles. Slopes facing away from the sun are source areas for cold katabatic winds. On summer days above 45 degrees north there are 66 days longer than 15 hours. Vines planted on a moderately sloping south-facing hillside thus receive solar radiation from near directly overhead.

Recently, planting on poleward-facing slopes has been recommended for cool wine-growing regions subject to cold winters. This scientific recommendation reverses the long-held popular belief that planting on south-facing slopes was the better choice. The premise is that the north-facing slopes are cooler in the spring and are less likely to bud prematurely during an early warm spell. North-facing slopes protect from late frosts in much the same way that a lake moderates temperature. This philosophy has met with great success in the Niagara Peninsula of Ontario where the steep north-facing slopes crumble from the limestone escarpment and tumble toward Lake Ontario.

VITICULTURE—GROWING GRAPES

Establishing a vineyard requires substantial planning. Depending on the region, growers need plans for obtaining the necessary water, drainage, trellising system, and row orientation to maximize the desired effects of terroir. Growers also need to prepare a regimen of insecticides, fungicides, and herbicides. Fertilization is rarely recommended.

Vines are planted in evenly spaced rows with individual vines spaced at regular intervals. Interrow distances are usually determined by machinery more than any other factor. Intervine distances in the same row are often twelve to fifteen feet. Distance is influenced by environmental conditions (such as aridity) or legal limits on yield.

Vines take three to seven years before they produce their first crop. The first crops are often very small. Crops from very young vines may have off flavors as well because the vine is still placing much of its energy into root development. Flower buds are often removed from very young vines to encourage growth.

Naturally, and during antiquity vines grew and dangled from trees. Grape picking was dangerous work because one had to climb high and on thin branches to harvest the fruit. Ancient Roman pickers held one of the first forms of life insurance. Their contracts included the cost of a funeral if they fell and died.

Maintaining a vineyard is hard work. Vines must be tended multiple times during the growing season. Beyond the obvious necessity of harvesting, workers must prune the vines in the winter, graft and replant in the early spring, protect buds from frost, apply chemicals, remove suckers, chase out varmints, and repair trellising systems.

Today, the vast majority of vines are supported and trained on the two-wire trellis system called a cordon. The wires are usually three and five feet above the ground. This system is inexpensive to maintain, provides ample sun, and is easily harvested by hand or machine. Other trellising systems are disappearing around the world as the efficiency of the cordon is realized.

Other trellising systems are generally limited to the region of invention. On Madeira, for example, the regularity of rains has forced the locals to grow their grapes on overhead arbors. An arbor is an overhead framework that grapes grow across, hanging downward through the support structure. This provides plenty of airflow around the grape cluster, reducing the chance of rot. On the Greek island of Santorini, the environment is so dry and windswept that the vines are wound in nests on the ground and many have individual windbreaks built around them. On the steep slopes of German vineyards, each vine is tied to its own pole.

PRECISION AGRICULTURE

Grape growers desiring to maximize quality while maintaining a reasonable yield go to great lengths to ensure the highest quality fruit and best prices. In holdings of less than five hectares growers can be very familiar with their vines because they have time to visually inspect and manage their vines. For the extensive and expansive corporate vineyards of the New World, this is not possible. Instead, growers have turned to geographic information science (GIS) technology.

When applied to farming GIS is called precision agriculture. Precision agriculture is based on knowing what is underfoot throughout the vineyard.

Table 3.2: Typical Weather Conditions for Sonoma County, California, American Viticultural Areas (April–October)

Appellation	Morning	Afternoon	Evening
Sonoma Valley	Cool/fog	Very warm	Cool
Carneros (Sonoma)	Very cool/fog	Windy/warm	Cool
Sonoma Mountain	Cool	Warm	Cool
Bennett Valley	Cool/fog	Warm	Cool
Sonoma Coast	Very cool/fog	Windy/cool	Cool/fog
Russian River	Cool/thick fog	Warm	Cool
Chalk Hill	Cool/light fog	Warm	Cool
Dry Creek Valley	Cool	Hot	Mild
Rockpile	Mild	Very warm	Mild
Alexander Valley	Cool/light fog	Hot	Mild
Knights Valley	Very cool	Hot	Cool late
Green Valley	Cool/thick fog	Warm	Cool/fog

Source: Sonoma County Grape Growers Association. *Exploring the Appellations of Sonoma County* (Sebastapol, CA, 2006).

It starts by systematically measuring the distribution of features in the vineyard including elevation, slope, aspect, soil composition, depth to water table, depth to bedrock, type of bedrock, and seasonality of weather events. Sensors, soil moisture content, temperature, humidity, solar radiation, and wind constantly update the base data. Vine health data are collected and evaluated using aerial imaging and spectral photography. Healthy vines give off a bright red color in infra-red, sick vines and leaves are black, for example. Table 3.2 shows one way to incorporate diurnal temporal and atmospheric moisture into a more regional system.

When herbi-, fungi-, or pesticide treatments are applied, concentrations are adjusted to precalculated doses and delivered with the precision of the global positioning system (GPS). The system records the distribution of the dosages applied and integrates the treatment data as a new component of the database.

SUMMARY

The combined effect of the environmental elements of terroir (weather, climate, soil, geomorphology, and geology) makes each place unique. The uniqueness of each place exhibits itself in the flavors imparted, the color of the wine, and its potential strength.

In regions where terroir is closely studied and its effects on grapes analyzed, fine distinctions are made over short distances. Take, for example, the characterization of the divergent atmospheric terroir characteristics associated with Sonoma County, California, as revealed in Table 3.2. Sonoma County covers more than 1,600 square miles and no two places in Sonoma County are more than 80 miles apart. Sonoma is, however, geologically and topographically complex. As a result, the micro-climate at proximate locations is distinct. The effect of ocean proximity and geomorphology on microclimate are most apparent in Table 3.2. Places near the Pacific (or in the case of Carneros, the San Francisco Bay) are cooler and foggier in the mornings. The foggiest places are coastal river valleys where cold air draining with the river is blocked by the coastal mountains. Interior locations like Alexander, Knights, and Dry Creek valleys become hot in the afternoon. Throughout the county, once the sun goes down temperatures drop quickly because there are usually no clouds to retain the heat of the day.

Chapter 4

How Wine Is Made

Beer is made by men, wine by God![1]

—*Martin Luther*

Wine is sunlight, held together by water.[2]

—*Galileo*

There is no mystique to making good wine that will comply with the requirements of EC and UK laws and be eligible for quality accolades. Crush good quality ripe grapes into juice, treat the resulting natural processes with respect, only make interventions that are within the rules and keep records of all that has happened.[3]

—*Wine Standards Bureau*

Wine making is simple. In fact, wine will, when given the opportunity, make itself. Simply place grapes in a tall food-grade cylinder and in a few days the grapes will burst, yeast will flourish, and wine will result. The rest, as they say, is detail.

Modern wine-making practices place stringent environmental and biological controls on the must to ensure its purity and quality. *Must* is the term for grape juice, with or without skins, that becomes wine. The possible processes required to make red wine are shown in Figure 4.1. Despite its complexity, Figure 4.1 has at its core the idea of putting the grapes in a cylinder and standing back.

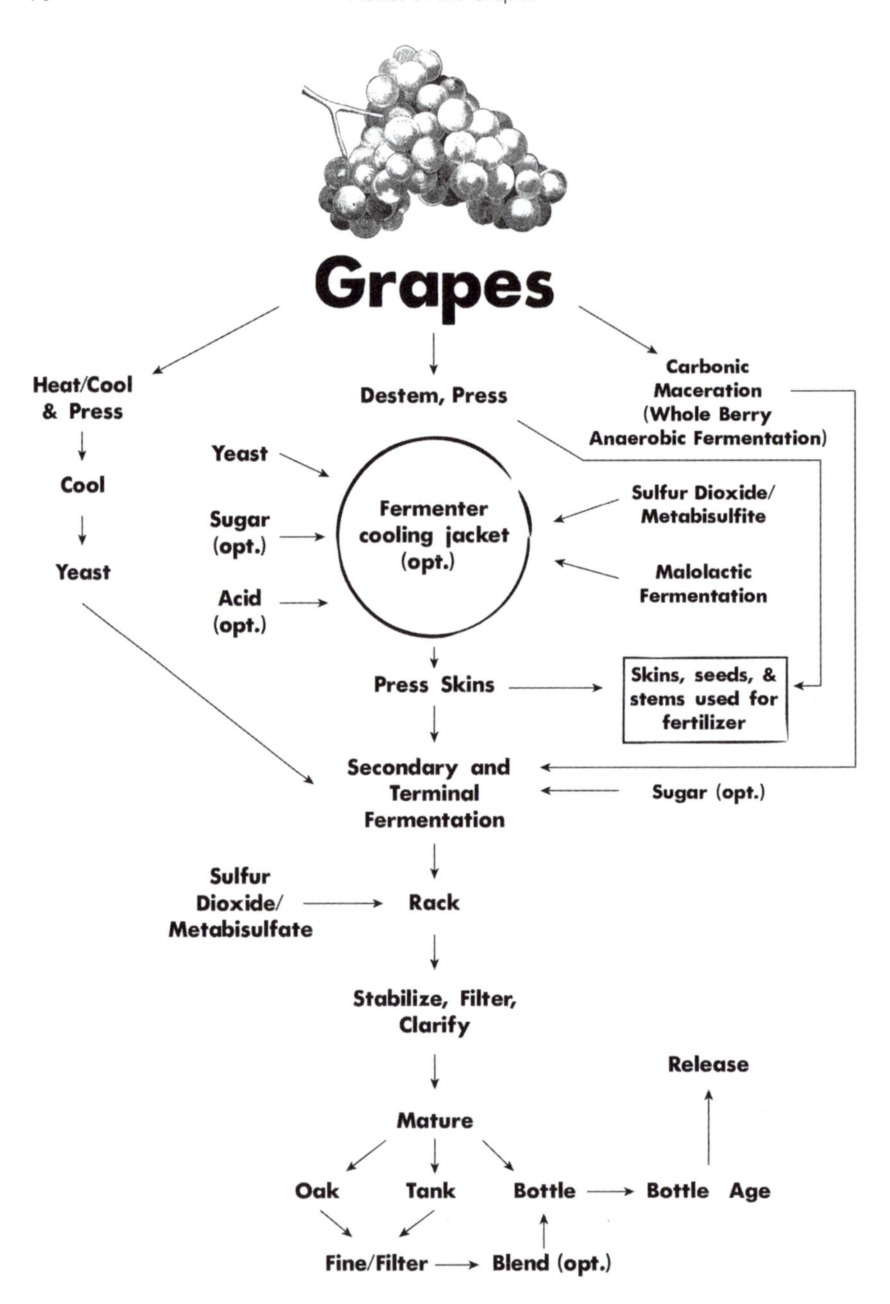

Figure 4.1: Flowchart of winemaker's decision tree for making any wine.

INITIAL PROCESSING

Typically, once harvested grapes are then cleaned, destemmed, and crushed. Historically, after juicing, the must was allowed to ferment in the open air. Wild yeasts fermented the must. The wine was then stored. In the most ancient times clay jars, amphorae, were used. Barrels replaced amphorae during late Roman times. Bottles came into vogue in the 1700s.

Today, must (unfermented grape juice) is usually treated with sodium metabisulfate ($H_2O_6S_2$) to kill wild yeasts and bacteria. Once the sulfur dissipates, the winemaker adds a pure strain of yeast with specific growth characteristics. The must then commences fermentation. Vintners chill the must to retain aromatic chemicals that vaporize during warm fermentation. When initial violent fermentation slows, the wine is racked (separated from the dead yeast and grape particulates) and placed in a secondary fermentation vessel.

Secondary fermentation is a slow process that can take months. After secondary fermentation completes, the wine is given another dose of sulfur compounds to kill remaining yeast. The wine is clarified or filtered and then bottled. At some future date the wine is released for sale.

There are many options and variations to the process as depicted in Figure 4.1. Press the juice right away and white wine results. Ferment whole grapes until they burst from internal pressure (carbonic maceration) and strong vibrant fruit flavors release into the wine. Allow malolactic fermentation to occur and astringent wine takes on a buttery flavor. Sugar, acids, bases, tannin, and pectic enzymes may be added to adjust the final product. Finally, the wine may be blended with wines from other locations, other varieties, and/or multiple vintages. The endless variations make every wine a unique experience for everyone involved. In the following pages we will learn about the major steps involved in wine making.

FERMENTATION

Long before people knew what bacteria were, we employed them to make foods more palatable and increase their longevity. Our knowledge of fermentation dates back thousands of years, probably to the time humans first started living in small villages.

Chemical Process

Wine making is a chemical process. Each sugar molecule converts into two alcohol and two carbon dioxide molecules. The chemical reaction occurs inside the yeast cell. The energy released permits the yeast to thrive. This is specified in the following equation:

Equation 1 $$C_6H_{12}O_6 \Rightarrow 2C_2H_5OH + 2CO_2$$

When wine goes "bad," another chemical process has occurred. In this process, alcohol molecules join with oxygen, where they convert to acetic acid (vinegar) and water. This transformation is given next:

$$\textbf{Equation 2} \qquad C_2H_5OH + O_2 \Rightarrow CH_3COOH + H_2O$$

Other chemical reactions, such as the conversion of malic acid to lactic acid by lactobacillus, may occur.

Biological Process

The primary chemical reaction of converting grape juice to wine is performed by yeast. Yeast is a highly geographically variable life form. The mix of wild yeast varieties is unique to each region. For thousands of years local yeasts consumed the grape sugar and excreted alcohol and carbon dioxide. How it happened we did not know.

For thousands of years no one knew what made grape juice turn into wine. The bubbling action was seen as supernatural and magical. Less than 170 years ago Louis Pasteur discovered that microorganisms make wine. Subsequent researchers identified many strains of yeast and their properties. Through selective breeding they developed strains designed to ferment in a specific manner and instill specific flavors. The same is true for bread and beer yeasts. Thousands of carefully produced strains of yeast are commercially available from companies like White Labs and Scott Labs.

In anaerobic environments yeast grows and multiplies rapidly. Each cell lives for a few hours before budding and dying. Yeasts eat grape sugar and live off the energy released by the conversion of sugar to alcohol and carbon dioxide. The resulting alcohol and carbon dioxide create an environment first friendly and then hostile to yeast. The carbon dioxide expelled from the juice, being heavier than air, forms a cap and perpetuates the anaerobic environment essential to yeast growth. If the carbon dioxide cap is disturbed, aerobic bacteria may enter the juice, contaminating the wine.

YEAST

Yeast (*Saccharomyces ellipsoidues*) performs the chemical reaction. Yeast cells are round or oval in shape. As they grow, yeast cells extend sections called buds, which enlarge and eventually detach to lead a separate existence (Figure 4.2). As long as there is sugar, yeast grows and reproduces until the alcohol concentration reaches between 15 and 20 percent of content. Yeast cells then die from the (alcohol) polluted environment they created. They literally drown in their own waste. Otherwise, the yeast cells die after all the sugar is consumed. By design, the dead yeast fall to the bottom of the cylinder.

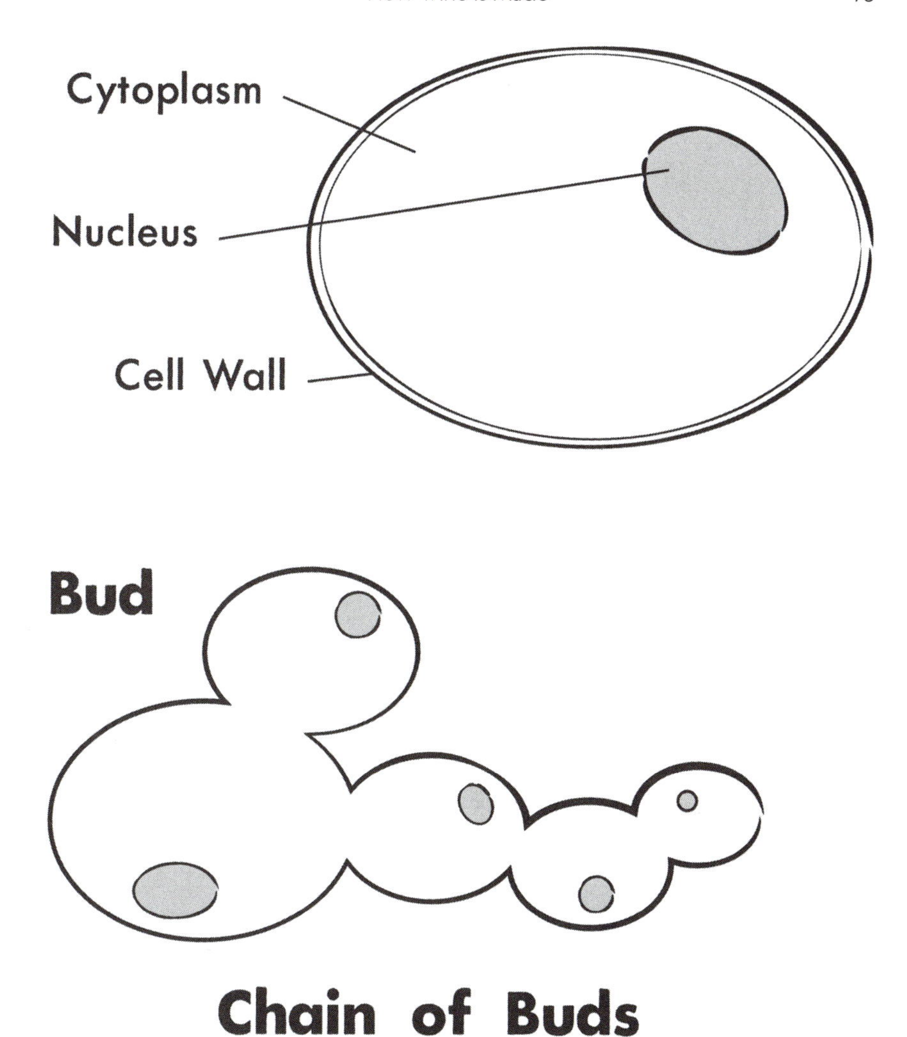

Figure 4.2: Yeast cells grow by budding, unlike one-celled animals that divide by fission. Yet, yeast is not a plant, either.

The growth rate of bacteria in a hospitable environment is truly amazing. Bacterial growth curves typically are *s*-shaped, representing a slow initial start, rapid growth to beyond carrying capacity, and rapid death. Desirable fermentation characteristics of wine yeasts, other than bottom fermenting, include rapid initial fermentation, high efficiency, performance at low temperatures, high alcohol tolerance, and low biomass. Successful modern technology uses genetically stable yeasts that are tolerant of sulfur compounds, will not bind with sulfur, produce small amounts of foam, clump readily, and have low demand for nitrogen.

OTHER BACTERIA

Yeast is not the only bacteria to feed on grape must. Numerous aerobic bacteria thrive there, if given the opportunity. Keeping these microscopic life forms away from the wine must is serious and expensive business. A single bacterium can infect thousands of gallons. Treatments with sulfur compounds kill unwanted bacteria in the must. The carbon dioxide cap created by the fermenting yeast keeps out bacteria during primary fermentation and by a fermentation lock during secondary fermentation.

Acetobactor

Cursed for thousands of years before knowing what it was, *Acetobactor aceti* turns wine into vinegar. *Acetobactor* is an aerobic bacterium discovered by Pasteur in 1864. When exposed to the air, *Acetobactor* spores drift in and contaminate the wine. The bacteria then consume alcohol and oxygen and excrete acetic acid and water as given by the following chemical formula:

Equation 3 $$C_2H_5OH + O_2 \rightarrow CH_3COOH + H_2O.$$

Vinegar has its purposes, but it is neither the grape's proudest nor noblest achievement. To prevent this conversion, keep wine free from oxygen. The seal and cork are the primary defense mechanism. Most wineries prefill bottles with nitrogen, then add the wine. This process fills the ullage (the gap between the bottom of the cork and the top of the wine) with a heavy gas to eliminate contamination of regular air. Open bottles can be saved using spray cans of argon gas (available in consumer packaging for this purpose) to create an oxygen-free cap over the wine.

Lactobacillus

The anaerobic malolactic bacteria (*Leuconostoc oenos*) convert malic acid to lactic acid and carbon dioxide. One gram of malic acid converts to two-thirds of a gram of lactic acid and one-third of a gram of carbon dioxide. The reaction is given in the following equation:

Equation 4 $$C_4H_6O_5 \rightarrow C_3H_6O_3 + CO_2$$

Lactic acid has a higher pH than malic acid. Malolactic conversion results in a softer (less acidic) wine. It reduces and masks fruit flavors. Instead of cherry or blackberry flavors, for example, you taste mixed berry. When applied to Chardonnay, lactobacillus produces pronounced buttery aromas.

For malolactic fermentation to occur, grape pulp must have a pH greater than 3.3 and an alcohol level below 14 percent. For these reasons vintners promote malolactic fermentation shortly after harvest. Malolactic fermentation is appropriate for only a few white wine varieties, most notably Chardonnay. Most red wine varieties are suitable for malolactic fermentation.

Careful attention is required; too much lactic acid can give the wine an unpleasant aroma. Regular mixing is essential while the fermentation proceeds because the grape bodies float and must be pushed down every few hours to help open the skins and let the malolactic fermented juice run free. Malolactic fermentation is generally completed in seven to ten days. If malolactic fermentation takes place in the bottle, it creates a light effervescence. This is a common trait of Portuguese Vinho Verde and other low-alcohol wines.

Leuconostoc and *Pediococcus* are also lactobacilli. They, however, convert sugars to lactic acid. They create undesirable off odors in wine. These bacteria have their place, however; they are the principal agents in fermenting cabbages into sauerkraut.

Candida mycoderma

The aerobic bacteria *Candida mycoderma,* commonly called *flor* in Spain, converts alcohol into acetaldehyde. Highly volatile acetaldehyde is responsible for the nutty flavor present in oxidized wines such as Sherry. Sherry producers encourage *flor* formation by not completely filling barrels. The ever-present bacteria grows in the barrel. *Flor* is a top fermenter and appears as a film on the fermenting wine's surface. The simple molecular rearrangement performed by *Candida*'s metabolism provides it the energy to thrive.

Equation 5 $C2H5O \rightarrow CH3CHO$

Acetaldehyde is produced in a transitory stage during human alcohol metabolization. Our bodies convert alcohol to acetaldehyde. Then acetaldehyde converts to acetate before final processing in the liver. Acetaldehyde is responsible for the facial flushing effect some people, primarily East Asians and Native Americans, experience after consuming alcohol. Flushing occurs because of their body's inability to metabolize acetaldehyde into acetate as quickly as it converts alcohol to acetaldehyde. Thirty times more toxic than alcohol, acetaldehyde may be an agent for addiction.[4]

FINAL PREPARATIONS

Once the wine is "made," a series of optional steps may be taken to "finish" the wine. These actions improve the appearance, aromas, and flavors.

Finishing steps include clarification, acidification, stabilization, and pasteurization. The final process in wine making is maturation. The maturation process includes blending, oaking, and aging.

Clarification

No one wants to drink cloudy wine. Clarification, also known as fining, is the removal of suspended particulates in a wine. These particulates are mostly dead yeast cells and near-microscopic pieces of grape. Fine particulates are suspended in the wine must after fermentation. In dry wines particulates are small and readily precipitate out. Particulates are larger and remain in solution longer when the wine contains residual sugar.

Long ago people discovered that egg whites can accelerate the clarification process. Kosher law requires egg white fining. Egg whites are slightly beaten and spread across the top of the wine must. As they drift downward through the must, the egg whites entrap the particles. After a few hours the whites and particulates are removed via a drain at the bottom of the tank.

Today there are several alternatives to using eggs for fining. The most commonly used commercial fining agent is bentonite. Bentonite is a sodium aluminum silicate that generally exists as clay. It swells when moistened and can absorb its own weight several times over. As it absorbs water, bentonite pores become blocked with particulates. Increasing in weight, the bentonite drifts to the bottom of the tank and is drained off. Simple clear gelatin accomplishes the same task and is a cheap alternative. The most exotic substance, isinglass, is made from dried cod air bladders. The crushed dried isinglass is scattered across the top of the wine and, like the other fining agents, filters down absorbing particles.

Most common of all, however, is simple filtration (aka cold pasteurization). It is by far the cheapest, fastest, and most reliable method for particulate removal. The particles are trapped when pumped through a series of filter pads capable of extracting particles as small as 5 microns. Some complain that filters remove flavoring and aromatic chemicals and reduce color intensity.

Unfiltered wines provide an interesting, denser tannic experience than the same grapes processed with filters. Without filtration wines will produce sediments.

Acidification

Total acidity (TA) is the mass in grams of acid per liter. There are several acids within wine. The two most common are tartaric and malic acids. Not all acids are equally acidic. The pH measures the intensity and activity of hydrogen ions in a water-base fluid. Stronger acids have lower pH because they support more hydrogen ions. Wine generally has a pH between 3 and 3.5.

The best approach to any acid problem is blending with a wine that will ameliorate the problem. There are, however, chemical approaches to adjusting the acid level. If the wine acid level is low, additional acid (preferably tartaric) can be added to bring balance to the wine. Historically, wines that were too acidic were treated with potassium carbonates. Modern winemakers, however, use cold stabilization to remove excess acid.

Stabilization

Chilling the near-finished wine causes tartaric acid to crystallize and precipitate out. The process reduces the overall acidity of the wine. The crystals may then be removed by filtration. Tartaric acid treated with potassium becomes the common kitchen thickening agent, cream of tartar. The chemical formula for tartaric acid, natural grape acid, is

Equation 6 $C^2H^6O^6$

Without cold stabilization tartaric acid crystals can form in the bottle and appear in the glass. The phenomenon occurs most commonly when white wines are served too cold. Occasionally, people mistake these crystals in their wine for broken glass. Tartaric acid crystals do not impair the wine's quality.

Pasteurization and Purification

Sulfur compounds kill bacteria living in the must. After the wine is made, and if the alcohol level is below 14 percent, it may be pasteurized by heating it to 140°F, but this is rarely necessary. Alternatively, the wine may be cold pasteurized by filtration. The potential for irradiating wine has received little attention.

MATURING WINE

All wine, prior to aging, requires a clarifying and settling period, which is facilitated by fining or filtering. Afterward, most wine improves with an additional resting period. This resting period is called aging. Aging is accomplished in vats, barrels, or bottles. Vat aging is the least expensive commercial treatment. The vat, stainless steel or plastic, will also contain a flavoring agent (usually oak chips). The process simulates the preferred oak barrel aging treatment. The longer wine is stored in a vat or barrel, the more it is changed by the experience. Bottle aging is slower and has no outside influences. The wine transforms slowly. Bottle-aged wines take a long time to mature.

The concept of aging wine goes back to the ancients. Aging is satisfaction delayed. One can drink wine now or one can store it for later and greater enjoyment. The longer one delays, the greater the anticipation and the greater the eventual enjoyment. As wine is left to age unmolested in a proper environment, chemical changes take place. The organic chemicals become more complex and nuances of flavor become more pronounced. The sensation of drinking well-aged wine is enhanced by the knowledge that you are consuming a rare commodity, one that can never be recreated, and one that only those sharing the bottle will ever know.

BLENDING

Traditionally blending was done in the field. Multiple varieties grew, were picked, and pressed together. Today, blending is scientifically managed in the lab, but most blends remain traditional; a little Merlot or Cabernet Franc is added to Cabernet Sauvignon, or a little Semillon is added to Sauvignon Blanc. The Australians pioneered the creation of bold exotic blends, which are becoming more common.

Rarely is a commercial wine made from a single variety or the grapes from a single vineyard. The scale of production is too large. Grapes used to make a wine may be trucked from several locations many miles apart and processed together, or processed separately and then blended. Varietal wines showing the name of the variety in big letters do not disclose all. Legally, in most countries, up to 15 percent of any varietal can be juice from another variety without notifying consumers.

THE ROLE OF OAK

The Celts, living in forested northern Europe, had a highly developed wood technology. It was they who invented the barrel and first used wood to store wine. Many species of trees have been used to make barrels, but oak is the least likely to leak and only mildly alters the flavor of the wine. Barrels are toasted internally to reduce the woodiness of extracted flavors and create a seal on the interior. The intensity, flavor, and duration of oak toasting produces different flavors, which might then be imparted to the must. In the lower temperature ranges sweet oaky flavors are enhanced in the barrel staves. Oaky flavors disappear at levels over 280°F and sweetness is lost above 340°F. At temperatures between 340°F and 480°F vanilla flavors are created. Between 380°F and 500°F toasty flavors are also present. Small amounts of almond flavors emerge at temperatures between 460°F and 520°F. Above 500°F acrid flavors overwhelm the senses. Variable toasting levels permit vintners to match toasting to the kind of changes they desire in their wine.

Chapter 5

Tasting Wine

One not only drinks wine, one smells it, observes it, tastes it, sips it, and one talks about it.[1]

—Edward VII of England

The novice focuses on the immediate overall experience. The expert concentrates on specific aspects of the experience.[2]

—Serafin Alvarado

The discovery and popularity of wine caused people to create and build containers and accessories for shipping, storing, dispensing, and drinking wine. In this chapter we investigate the objects used in conjunction with tasting wine and the approved methods and techniques for their use. For information on storing wine refer to chapter 11.

As a novice wine taster, you will have difficulty experiencing many of the wine traits discussed in this chapter. My advice is to concentrate on one trait at a time. Before the first sniff, decide what you will taste for in that wine. Keep it simple. Focus on acidity or sweetness: is it too little, too much, or just right? Alternatively, focus on sweetness or tannin. The rest will come with practice. Do not worry about identifying fruit or other aromas and flavors; that will come with time.

In *Why You Like the Wines You Like*, Master of Wine Tim Hanni explains the "nature and nurture" (or in his terms, "sensitivity" and "genre") factors influencing wine preference. Variability in sensory receptors represent "nature." "Nurture" is more complicated. Familiarity, of course, is nurture's key. Experience with a wine brings memories of taste, aroma, and most importantly the events that transpired. His typology of wine drinkers is insightful in explaining the diversity of the wine market and helpful as a self-learning tool.[3]

Hanni found far too many people drink the wines they think they should drink. "I'm trying to learn to like dry red wines," is heard at many wine tastings. This is by far the biggest mistake one can make. Certainly try every variety and style of wine offered, but enjoy *your* preference. *Your* preference is affected by situational factors, your mood, what you are eating, and your age. Drink what you want; after all, it's all about you.

WINE GLASSES

The original Mesopotamian (modern Iraq) wine-drinking vessel of 4000 BCE was a shallow saucer-like bowl. In bas-reliefs Mesopotamian and Egyptian royalty balance wine saucers on their fingertips. These saucer-like cups were made of clay or precious metals. The low shallow bowl allowed drinkers to separate the wine from the dregs. The ancients flung the dregs away, often at a target.

The first footed cup comes from Egypt and dates to circa 1500 BCE. The Etruscans, circa 600 BCE, drank from oval-shaped two-handled cups. One of the most exciting archaeological finds was the discovery of a gold two-handled drinking vessel in Mycenae (at the southern tip of the Peloponnese in Greece). Named Nestor's cup, it was supposedly the property of Nestor, who fought at Troy alongside Achilles and Odysseus.

Glassmaking dates from about 1400 BCE, but the collapse of Bronze Age civilizations terminated production. Glassmaking revived about 800 BCE; Syrians made the first molded glass cups and decanters about 100 BCE and the first blown glass about 100 CE. The Romans were the first culture to drink wine from glasses. Pliny (23–79 CE) states a growing preference among Romans for the new glass drinking vessels over traditional vessels made from pottery or precious metals.[4] The use of glass, however, remained uncommon because of its great expense and fragility. Before 1675, when George Ravenswood invented leaded glass, most people drank from wood, ceramic, metal, or even leather cups lined with pitch.

Glass stemware became popular in the 1700s after glassmakers developed the necessary skills. The most important characteristic of a wine glass is its inward curvature. This inward curve traps and concentrates the aromas present in the wine. When holding a wine glass, hold it by the stem to prevent heating the wine with your hand. Balloon glasses with large surface areas are better suited to red wines and taller glasses for whites. Tall tulip flutes are the current choice for sparkling wines, as they allow one to watch the bubbles rise long distances from the side. Before 1980, sparkling wine was served in low flat glasses so one could watch the bubbles rise from above.

DECANTING

Decanting is the process of pouring wine from the bottle to a decanter prior to pouring into a glass. Decanting serves three purposes. First, decanting adds a level of aesthetics, mystery, and anticipation to any occasion. Second, if there are sediments at the bottom of a bottle, decanting allows us to pour off the wine, leaving the sediment behind. Third, decanting permits wines to breathe (oxidize). One should decant in one long, slow, continuous pour. Drink older wines immediately after decanting as they can quickly lose their character. Most modern commercial wineries filter before bottling, making decanting no longer necessary. Exceptions are homemade wines, unfiltered wines, and bottle-aged port.

Decanting is the only real method for aerating wine. Simply opening the bottle and letting it sit permits only minimal (below sensory threshold) oxidation. Oxidation immediately prior to consumption releases tannins present in young red wines. After decanting a young wine, allow the wine to sit for a few minutes before drinking.

Sparkling wines should not be decanted.

WINE TASTING CEREMONY

The modern ceremonial restaurant presentation of wine is steeped in tradition. Each element had, at one time, a reason for its existence and inclusion in the experience. Proper presentation allows the wine to be at its best, and consistency in presentation provides a format for comparison. Many of the needs are no longer a concern, but their continued inclusion enhances the overall fine dining experience.

When brought to the table the bottle should be partially wrapped in a towel, clearly showing the label. If the wine has been cellared for some time, the bottle may need to be wiped down. The server holds the bottle out to the customer, who examines the label to ensure that the correct wine and year were brought to the table.

The server should remove the cork at the table. Removal of the cork at the table ensures that the bottle was not previously emptied and refilled with a lesser wine. The cork should not be wiggled, but pulled out straight. The worm (screw portion of the corkscrew) is removed from the cork and the cork presented to the customer. The customer should compare the printed text (or logo) on the cork to the text (or logo) on the bottle. Many countries require printing the winery number on the cork, which must match the winery number on the bottle. A growing number of regions and countries are including a tracking number on the bottle so consumers can trace the wine back to the vineyard of origin.

The customer may examine the wine end (bottom) of the cork. The presence of crystals on the bottom of the cork indicates wine that was not cold stabilized and wine containing excess tartaric acid. Tartaric acid crystals may mean the wine has lost acidity and may be comparatively bland. Crystals may precipitate out if the bottle has been stored upright for an extended period. In such cases the top of the cork may also be dry and flaky. A slimy residue on a white wine cork bottom is a good thing. It indicates that the bottle has been stored on its side. Sniffing the cork can be a precursor to the wine to follow, but is unnecessary.

The next step is called pouring the cork. The server pours a small amount of the wine into a wine glass and serves it to the person who ordered it. At this point, the true process of wine tasting begins. There are eight steps the consumer should experience when tasting wine. These are: see, swirl, sniff, sip, savor, swallow, describe, and score.

When attending a wine tasting event obey the following instructions to maximize the pleasure for yourself and those around you. First and foremost, if the tasting is led, do not drink ahead. You may taste in the wrong order or run out of wine too soon. Do not wear white, especially if you intend to taste reds. Do not wear perfume, lipstick, cologne, or other aromatic substances; they confuse your palate and the palates of those nearby. Assume that the person doing the pouring knows less than you and do not pester them with annoying questions. Save questions for the tasting leader.

See the Wine

To commence your sensory experience, look at the wine in your glass. If the wine looks good, it may be good. When visually examining the wine we look for color, clarity, sediment, bubbles, tartaric acid crystals, and the oft-present piece of cork. Ask yourself the following questions: Is the color of the wine the color you expected? Is the wine cloudy? Is there sediment in the glass consisting of either dark particulates or tartaric acid crystals? Always beware of the floating piece of dry cork broken from the end of an over-penetrated cork.

Swirl the Wine

Swirling wine in a glass releases aromatic chemicals. As the glass has little wine in it when pouring the cork, there is plenty of room for a vigorous swirl without spilling. Pouring the cork wine should cover the bottom, and never more than one ounce, of a six- to eight-ounce glass.

Sniff the Wine

Most of what we call taste comes to us from olfaction (smelling). Sniffing heightens our anticipation of the impending pleasure the wine will deliver. You should sniff the wine in short whiffs from near the top of the glass.

Placing your nose deep in the glass and inhaling deeply mostly results in absorbing the vaporized alcohol in the glass. Sniffing allows us to identify the bouquet of the wine. Bouquet is the mix of aromas released by the wine, like a multispecies bouquet of flowers. An aroma derives from a single source. Considerations for sniffing include evaluating the intensity and aromatic characteristics of the wine. Try to identify the individual aromas of the bouquet. The harmony and relative intensities of the aromas are the wine's bouquet.

If the wine is acceptable, the customer signals the server to pour wine for the remaining guests. While waiting for the next course, you should describe and compare your experience with that of your dining companions. If you are at a tasting, the description might be either written or a group might come to a verbal consensus. At this point one can identify the aromas present using a flavor identification chart such as the one developed by Dr. Ann C. Noble of the University of California, Davis. At a tasting, or judging, you may score the wine using a predefined format for providing a numerical summary for the wine.

Wine Aromas (Fragrance)

Aromas are the individual fragrances within the wine. This is analogous to the smell of an individual species in a bouquet of flowers. When we sniff the wine, we nasally consume the chemicals released by swirling the wine. Each fragrance experienced results from organic chemicals. The complex array of chemical receptors in our noses allows us to identify many fragrances. A principal component of the aroma released is the burning sensation of alcohol. One should train the nose to ignore this aroma to fully experience the more delicate wine aromas.

The fragrances of wine we identify from the aromatic chemicals in wine are the same as those released by other fruits and vegetables. Hence, we say the wine tastes of cherries or lemons or roses. In general we summarize the aromatics as fruity, earthy, spicy, and floral. The wine flavor wheel, developed by Dr. Ann Noble divides the general into specific recognizable flavors associated with known fruits, herbs, flowers, and spices.[5] Aroma may also refer to the intensity of the wine on the olfactory organ.

A recent content analysis study conducted by Rachel Eng analyzed 15,000 wine reviews. Eng reported that the most common flavor mentioned was oak (4.1 percent of cases). Cherry was a distant second at 2 percent of cases. That these flavor terms were present in the reviews does not necessarily make them appealing. Terms such as "rough," "sharp," "tobacco," "grass," and "grapefruit" are commonly thought of as undesirable. "apricot," "chocolate," "coffee," "orange," "smooth," "refreshing," "hazelnut," and "balanced" are desired traits.[6]

Wine Flavors

In addition to specific flavors derived from aromatic chemicals, wines also interact with our taste receptors. Taste receptors can only identify five flavors: bitterness, salty, sweetness, sourness, and umami. Not all flavors are present in all wines. Bitterness is often associated with organic chemicals and is imparted to wine via skins and/or oak. Sweetness is the detection of organic sugars. Saltiness is the detection of ionic solutions. The most common salt is sodium chloride. Umami is the detection of glutamic acid. In its purest form, it is the taste of monosodium glutamate (MSG). Sourness is the detection of hydrogen ions.

Wine acidity is measured in two ways, pH and total acidity. pH measures the concentration of hydrogen ions present in a fluid compared to distilled water. A pH below 7 (distilled water) indicates an acidic solution and those above 7 are bases. Total acidity, measured as the percentage of the total fluid wine volume, rarely reaches above 1 percent. The pH of wine is usually between 3.1 and 3.6. In comparison, lemon juice has a pH of 2, orange juice 3, coffee a pH of 5, blood a pH of 7.4, and milk of magnesia 10. The most prevalent acid in wine by far is tartaric acid. Citric and lactic acids are the next most prevalent acids.

Sweetness is measured in terms of amount of residual sugar present in the wine. The standard unit for measuring sweetness is called "brix." Each brix unit is equal to one-half of 1 percent sugar ($C_6H_{12}O_6$) in the wine compared to the sugar content of distilled water. Five carbon sugars and certain non-sugar chemicals deliver sweetness without calories.

Wine Balance

Balance refers to the blending of the acidity, bitterness, and sweetness flavors present in wine. It is easier to tell if a wine is unbalanced than if a wine is balanced. If the wine is sickly sweet or breathtakingly sharp, it is out of balance. Alternatively, if the wine is delicious and you feel compelled to take another swallow, it is well balanced. Listen to the harmony in your mouth. A wine heavy with sweetness requires an aggressive level of acidity to bring it into balance. Ice wines, which often run over 15 brix of residual sugar, are balanced by a pH of 3.2 or lower.

White Wine Balance

White wines primarily activate two types of receptors in our mouths, sour and sweet. A graph of these two dimensions provides an illustration and a vocabulary for discussing white wine flavors. None of the combinations on the white wine balance table are better or worse than another. Each is appropriate for some occasion. Those hollow in both sweetness and acidity are balanced but without character. They are the products of grapes like the French Colombard, Palomino, Muller-Thurgau, Ungi Blanc, Delaware, and ill-made Rieslings. Balance, balance terms, and character of white wines are revealed in Table 5.1.

Red Wine Balance

Red wines activate three types of oral receptors, sour, sweetness, and bitter. The interrelationship between the three tastes and common flavor terms are portrayed in Figure 5.1. The bitterness comes from the tannic acid present in the grape skins. Darker red wines usually have greater bitterness and require greater sweetness to bring them into balance. If left to age, the astringency of tannins decreases as they evolve into more complex chemicals.

Sip the Wine

After considering the aroma and bouquet, swirl the wine again and sip. The sip should be smaller than a soup spoonful. Slurping is okay. Do not fill your mouth! Leave room for aeration. Open your mouth slightly and drag a breath across the wine to release the aromatic chemicals. Swish the wine around in your mouth and savor the experience. Feel the wine in your mouth. Does it burn with either alcohol or acidity? Is it soft as a cotton ball or rough like sandpaper?

Savor the Wine

Wine is a pleasure and should be savored. Messages from some sensory tissue in the mouth and nose are only evaluated by the brain with the first sip when your brain tries to classify and encompass the experience. On the second sip, these sensory messages are ignored as your brain responds to them with an "Oh, that again." The first taste provides the greatest pleasure and cannot be recaptured. Enjoy each part and location of the stimulation it provides. What is the texture of the wine? What

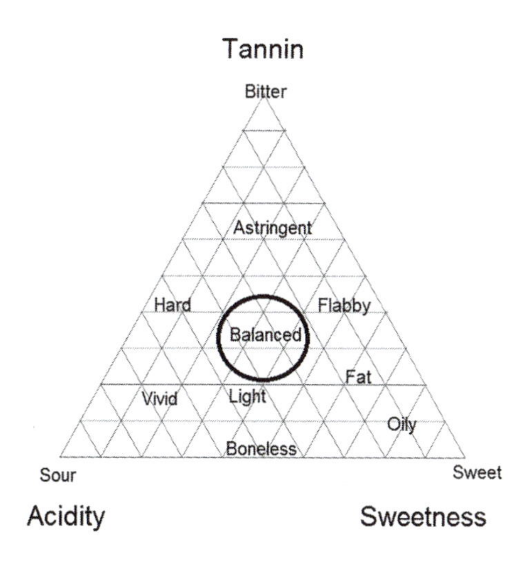

Figure 5.1: A balanced wine harmonizes all three flavors.

Table 5.1: White wine flavor dimensions and terms. Each brix degree represents an increase of 0.5% increase in sugar in the must. Total acidity is the percentage of the fluid comprised of acids (mostly tartaric, malic, and citric).

SWEETNESS	Brix Degrees	Sparkling Wine Term	Still Wine Term					
	+10	Doux	Syrupy					
	6–10	Demi Sec	Heavy	Off				D
	3–7	Sec	Unctuous				E	
	2–4	Extra Dry	Watery			NC		
	1–3	Brut	Little		LA			
	> 1	Extra Brut	Hollow	BA				Off
				Hollow	Thin	Meager	Tart	Aggressive
		pH		4.1	3.8	3.5	3.2	2.9
		TA		0.4%	0.5%	0.6%	0.8%	+1%
				ACIDITY				

tastes do you experience immediately? What tastes develop as you hold the wine? By answering these questions you weigh the body, flavor, and character of the wine. While savoring, evaluate the tastes and textures of the wine. As the wine comes into contact with each part of your mouth, how does it feel? Experience the wine in totality by applying your senses of taste, smell, and feel.

Swallow the Wine

Swallow the wine and feel it travel down your throat. How does it feel going down? Evaluate the intensity and length of the finish (aftertaste). After swallowing, you should continue to feel and taste the wine for ten to twenty seconds. Better wines have long and complex finishes.

Describe the Wine

Try to vocalize your impression of the wine. Avoid the immediate response of "good" and try a basic "yum, yuck, sweet, or sour." Consider each of the wine dimensions: sweetness, acidity, and tannin. Use terms from Table 5.1 and Figure 5.1 to characterize the wine.

Later you can develop the capacity to identify aromas within the wine. Ask yourself the following: What fruits do you taste? Do you taste earthiness? What words do you associate with the experience? Use descriptors from references like Dr. Ann Noble's wine flavor wheel to help ensure you are using standard terminology. Use an aroma kit to develop a base olfactory understanding of common wine aromas, flavors, and tastes.

Score the Wine

We all seem to appreciate a numerical summary, a single number that defines all, and that is what the wine score has become. Shelf flyers, bottle banners, and ribbons adorn high-scoring wines. There are many scoring systems in use. The most common are 100-point systems and are used by the major wine evaluators such as the *Wine Advocate* and *Wine Spectator*. Table 5.2 shows the real range of points awarded by *Wine Spectator* reviewers. The relationship between wine rating points and cost is curvilinear. Prices are set based on cost of production and on what tradition indicates the market will bear. Scores come from blind professional tastings where region and heritage are unknown.

Table 5.2 shows *Wine Spectator* scores categorized by cost for 320 wines. In this study, most wines scored in the 80s. A third of all wines in the sample scored between 84 and 86 points. Three trends are highly visible upon examination of Table 5.2. First, higher price wines do score higher. No wine under $20 scored 90 or above. The median score for low cost ($5–$8) wines was 82 points and the median score for wines over $40 was 88 points. Second, the

Table 5.2: *Wine Spectator* points and cost per liter category for 320 wines. The values indicate the number of wine products in each price range. The modal value in each column is highlighted (N = 320).

WS Points	\$5–\$8	\$8–\$12	\$12–\$15	\$15–\$20	\$20–\$30	\$30–\$40	+\$40	SUM
76	1	1	1					3
77	1		1	1				2
78	1	3		1			1	6
79	4	6			2			12
80		4		4	2		1	11
81	2	7	7	5	1		1	23
82	4	9	4	2	1		1	21
83	**5**	3	10	3	6			27
84	3	10	**17**	6	7	2	2	47
85	1	**12**	12	**14**	8	1	2	50
86		6	9	4	6	**4**	2	31
87		1	5	8	8		7	29
88		1	4	3	6		3	17
89		1	2	3	7	1	4	18
90			2		2		4	8
91					3	3	3	9
92						1	3	4
93							1	1
94							1	1
Total	22	64	74	54	59	12	35	320

Source: Robert Sechrist, "The Expanding International and Varietal Composition of the Wine Spectator 'Top 100,' 1988–2010." Paper presented at the Association of American Geographers Conference, Seattle, Washington, April 2011.

higher the price, the broader the point score range, indicating further that pricing is not directly linked to quality. Two of the $40-plus wines earned scores below 80. Third, scores do not occupy the full range of points; wines scoring below 75 are unlikely participants and rarely worth reporting. In summary, Wine Spectator scores tell us we can usually count on inexpensive wine being lower in quality while expensive wines may be higher in quality.

WINE FAULTS

Wines are subject to several winery contaminates, both chemical and biological. Their actions result in off-flavors or even ruined wine. One common fault, usually resulting from cork failure, is oxidation. Oxidation turns bright red wines into brownish-orange ones.

Faults often arise from misapplication of chemical treatments when processing the wine. The improper addition of sulfur compounds leads to aromas of boiled egg, onion, garlic, burned rubber, and vegetables. Sulfur compounds also appear when the juice lacks nitrogen-rich amino acids.

For the most part, wine faults are caused by microbes ingesting the wine must and cannot be detected until the bottle is opened. *Brettanonyces* bacteria create Band-Aid spicy aromas. Its growth is fostered by low sulfur and high temperatures. Infection usually occurs when used barrels are stored empty. Mousy taint, which is caused by *Lactobacillus brevis*, smells of mouse urine (alternatively, of corn chips or jasmine rice). This flaw is most common in red fortified wines because these wines have a high pH and are barrel aged for extended periods. The presence of vinegar (acetic acid) and ethyl acetate is caused by multiple *Acetobactor* species. These volatile acids are present in all red wines at low concentrations. These chemicals and acetaldehyde accumulate with oxidation. Cork taint is caused by the presence of trichloroanisole, which is usually derived from exposure to *Aspergillus*-infected cork. Tainted cork adds an unpleasant aroma of wet dog, wet newspaper, mushrooms, or compost.

One final source of faulty wine is the consumer. Most wine bottles are colored to protect the wine from ultraviolet radiation. Wine in clear glass bottles, however, can easily develop wet dog notes from exposure. Simply store wine in cool dark places to avoid this catastrophe.

SUMMARY

There is a depth of history and confluence of natural and cultural practices present in each drop of wine. Over the centuries what we find acceptable and pleasant has changed. Sherry, once the favorite drink of the English, was supplanted by port. All the while French wines were unpatriotic. Chardonnay, the darling of the 1990s, has given way to aromatic whites.

Chapter 6

Wine in the Global Economy

The cheapness of wine seems to be a cause, not of drunkenness, but of sobriety.[1]

—Adam Smith

Large-scale capitalist wine-growing cannot fail to run down quality as soon as it is safe to do so.[2]

—Edward Hyams

Wine is unlike most commodities. A bushel of corn is a bushel of corn, but a bottle of wine is not the same as another bottle. Terroir and the vintner's skill make it so. Only the apple comes close to the level of differentiation we find, and treasure, in grapes. Few, however, would be interested in a book on apple geography. Like most commodities the grape and wine industry employs millions in its production and distribution chains. Knowing which bottle is worth a pittance or a fortune employs thousands of Americans in restaurants, wineries, and wine shops. For centuries the Europeans held exclusive control of this prized commodity, but over the past fifty years their control of the market has slipped.

Chile, Australia, South Africa, Argentina, and New Zealand each succeeded in creating modern wine industries imbued with the best science and technology can offer. Through higher quality and lower prices they are establishing themselves in the global marketplace. Eastern European quality enhancement programs are creating similar results and distinctions. Recent international conglomerates acquired holdings around the globe to hedge against crop failure in any one place and to have their foot in the door of the new "hot" wine region.

The wine industry attracts more attention than its scale warrants. Wine grapes cover only 0.2 percent of the world's croplands. Australian wine statistician Kym Anderson provides additional evidence that wine is no ordinary commodity, stating: "Moreover, globally it is not a growth industry in that world wine production and consumption have been declining slightly over the past two decades. But to millions of investors and hundreds of millions of consumers, this industry provides a far more fascinating product than its shares of global expenditure or GDP might suggest."[3]

Consumptive patterns have changed significantly over the past twenty years. Consumption and production are down throughout Europe and Latin America as operating and labor costs skyrocket. Production is up around the Pacific Rim where wine consumption is on the rise. Despite the contraction in European production, exports are increasing from those same places. The proportion of wine exported grew from 25 percent in 1999 to 28 percent in 2005. European domestic consumption declined by 3 percent during the same period. Australia tripled exports between 2000 and 2006. New World wines now account for the majority of United Kingdom imports, despite the proximity of EU wines. Wine quality is also increasing; globally, the per capita consumption of nonpremium wine declined from 2.1 to 1.8 liters between 1999 and 2005. The EU sponsored a major vine eradication program in 2008 to 2010 that reduced overall vineyard area by 161,166 hectares, a reduction of 4.4 percent.[4] During the same period, nonpremium vineyard area declined by 10 percent. Overall, more people are drinking higher quality wine from more diverse locations than ever before.[5,6]

WINE FRAUDS AND COUNTERFEITS

More detrimental to the individual consumer were the lack of control and verification of the source and composition of wine. For centuries adulteration and fraud were a serious problem in the wine industry. The current international system of government control of wine production exists largely to limit counterfeiting and adulteration. Counterfeiting remains a consumer problem at all levels; from the refilled wine bottle ordered at dinner to the millionaire bilked out of thousands for a single fake bottle. Generally speaking, wines selling for less than twenty dollars are not worth counterfeiting. Clear malt beverage manufacturers and their advertising blur the line between these products and wine.

Counterfeiting takes place at every stage of the wine supply chain. At the winery, elderberries and other fruit juices were, and are, mixed with the wine to add color or flavor, which was especially common before 1950. With wars, phylloxera invasions, Odium outbreaks, floods, drought, pestilences, and the occasional geologic event, policing wine merchants was difficult.

In transport, wines from one region frequently became wines from another region. This magical, and illegal, transformation was facilitated in the days of barrel transit, sailing ships, and poor record keeping. After a shopkeeper purchased a barrel, he often bottled the wine on site, giving it whatever popular name he could get away with. Before varietal labeling became the standard in the United States, Americans used regional names to indicate the style of wine they produced to mimic the great wines of Europe.

Efforts to combat this problem led to the creation of careful, and ever increasing, record keeping and tracking by government officials. Current international agreements require countries to label and police their own winemakers and products. Wine-producing nations also established regulations controlling wine production and labeling within their own country. National agricultural quality programs encouraged their wineries to make better wines for the international market. The most significant component of the national regulations is enforcement of the rules and the provision for verifying that wines come from the legally defined subnational region stated on the bottle.

National wine regulators focus on verifying the truthfulness of the wine label, defining wine regions, controlling production levels, and preventing adulteration. This level of policing is necessary to prevent a recurrence of events like the Austrian glycol scandal of 1985 and the murderous results in the Italian methanol poisonings in 1986.

During the 1980s and 1990s the global wine industry rewrote many rules to reduce counterfeiting. Still, frauds and fakes abound. In 2008, makers of Brunello di Montalcino were caught adding nonapproved varieties to their wine. Red Bicyclette Pinot Noir, it turned out, was made from Merlot and Syrah. Eighteen million bottles were sold before the crime was discovered by French officials who realized one company had sold more "Pinot Noir" than was produced in the entire region. Thousands of bottles of fake Jacob's Creek wine were sold in England in 2011, proving even some counterfeiters will make the equivalent of a ten-dollar bill. The crime was spotted when it was realized that Australia was misspelled on the labels. Forgers sold 400,000 bottles of Mont Tauch Fitou to unsuspecting Chinese in 2010. Italian police confiscated 9,200 bottles of fake Moet & Chandon Champagne and labels for 40,000 more bottles.[7]

GEOGRAPHIC INDICATORS

Place of origin has always been associated with rare and exotic products, wine foremost among them. Wine counterfeiting began thousands of years ago when the first amphora was refilled with swill and sold as new. It has been a constant problem ever since. To combat this problem, today's governments label their products and carefully track shipments. Almost 500 years ago Hungary geographically defined the Tokay wine region; 250 years ago Portugal defined the port region and took measures to ensure that only *real* port left Portugal. France began defining its regions about 150 years ago. The United States began geographically defining wine regions, called American Viticultural Areas, in 1980.

The idea of using geographic indicators with commodities is a growing trend. International agreements developed over the past thirty years promote place commodity differentiation by location. The United Nations maintains the list of internationally recognized and trademarked "Geographic Indicators." As a result, Washington is known for its apples, Florida for its oranges, Kobe for its beef, Idaho for its potatoes, Trinidad for its chocolate, Vidalia for its onions, and France for its wines.

Designated wine regions often become focal points for economic development clusters. Economist Michael Porter explains, "Clusters are geographic concentrations of interconnected companies, specialized suppliers, service providers, firms in related industries, and associated institutions."[8] There exist within an economic cluster a pool of skilled workers and suppliers: special service companies (precision agriculture, bottle manufacturers, promotion and tourism, etc.) and skilled workers from a resident labor pool. For a wine region this is critical.

Products and their linked places of origin are protected from infringement by law. With the legal protections in place, consumers buy place-designated products with confidence. Products from these places receive higher prices than the same commodity undifferentiated by place; usually the differences are minimal.

For wine, however, the extent and differences in valuation are extreme with single commercially available bottles of wine ranging from $2 to $20,000 each. Governments carefully track grape harvests, wine production, and shipments to ensure that low-quality or counterfeit wines do not enter the market. Why this is so derives from the unique combination of geography, vine, and culture of each location.

The wine producers of the world agree. Geography matters to them! Among wine grape producers geographic considerations are second in significance only to the grape variety itself. The primary consideration is, Will the terroir permit a variety to thrive?

The Location Matters declaration is widely adopted and supported. The Location Matters program represents a global effort to create a global system of protected wine names and boundaries. The Location Matters declaration shows the importance growers place on vineyard location. This is because grapes acquire flavor nuances from the local environment that make their individual character and taste unique. The place of origin, one aspect of geography, defines the value we place on wine. The declaration confirms centuries of European usage and international agreements for wine regions of renown.

GLOBAL WINE CONSUMPTION PATTERNS

Europeans consume more wine per capita than people from any other world region, according to annual statistics produced by the Wine Institute. In 1998 seven European countries had per capita consumption in excess of forty liters; by 2010 only four countries retained that level of consumption. The five most wine-consuming countries in 2010 were Luxembourg, France, Italy, Portugal, and Croatia. Per capita consumption peaked in Luxembourg in 2000 when it exceeded 64 liters. Consumption by Luxembourgers has since declined to 52.4 liters. French consumption peaked in 1998 at 61.2 liters and hit a low of 45.7 in 2010.

Changes in consumption over this short period vary considerably. Between 1997 and 2010 consumption decreased in only thirty countries. Fourteen of the thirty are in Europe and they include the largest producing countries. The countries of Central Asia experienced the greatest decreases. In Uzbekistan, Azerbaijan, Tajikistan, and Kyrgyzstan consumption dropped by more than 200 percent. Countries experiencing consumption growth of more than 50 percent include Russia (53 percent), New Zealand (54 percent), Ireland (55 percent), Finland (57 percent), China (59 percent), Moldova (59 percent), Czech Republic (68 percent), Mexico (69 percent), and Ukraine (73 percent).[9]

Around the world, radical changes in a country's consumption can result from any number of social and economic reasons. Japan's import reduction in 2000 reflected a significant economic downturn associated with the first Internet bust. China's rapid increase correlates well with the increasing number of Chinese millionaires. In countries like Mexico, the percent increase is great, although actual consumption rose from 0.15 to 0.49 liters per capita and may actually be the result of consumption by tourists.[10]

Thirty-five counties had per capita consumption greater than ten liters per person; only six of these are outside Europe and they are wine-exporting countries. In twenty-one countries, the populace consumed more than twenty liters per capita. Eighteen of these are in Europe, sixteen are wine producers and exporters, and eleven are predominantly Roman Catholic. This cluster correlates spatially with the core of the ancient European wine producing and consuming region.

Global wine consumption patterns continue to change. Bordeaux exports to China in 1995 were practically zero. *China Daily* writer Li Xiang stated in 2013, "China is the third-largest market for French wines, with an estimated value of 800 million euros. The value of the Chinese market accounts for nearly 30 percent of Bordeaux's total wine exports."[11]

The Rise of Varietal Indicators

When Europe was the only source for wine, naming the wines from their place of origin made sense. For the most part, each village's vineyards were a unique varietal combination. Select regions, such as Alsace, implemented varietal labeling long ago.

When vines were planted wholesale in the New World, large tracts were planted with clones of a single parent (or two if grafted). As growers in many places were growing the same few varieties, naming by location became of secondary consideration. Varietal labeling began in the United States in the 1800s with the need to describe grapes of North American or unknown origin such as Longworth's Catawba, Concord, and Zinfandel. Grapes of known origin, Chablis, Burgundy, Rhine, retained their European place of origin identifiers. The emerging need for universal varietal labeling was identified by author Frank Schoonmaker in 1941:

> The fact that we are asking for "Sonoma Sylvaner" instead of "California Rhine Wine" or "Finger Lakes Delaware" instead of "Old Dormitory Chablis" goes far beyond a simple merchandising idea. It is of tremendous sociological significance. It is a symptom of our coming of age. Casting aside our European inferiority complex as regards wines.[12]

Varietal labeling became popular with consumers in the 1970s. Varietal labeling is easier for consumers than recalling the multitude of places with strange names. The French opposed the concept of varietal labeling and took efforts in the 1990s to eliminate the mention of varieties from French labels. In 2010, France reversed policy, and variety names now appear prominently on many French wines.

AMERICAN WINE CONSUMPTION

According to the U.S. Department of Health, in 1996, Americans consumed 500 million gallons of wine. By 2002, this figure grew to 580 million gallons. In the same interval total value increased from $540M to $580M in 2002, with a significant improvement in both quality and varietal diversity. The trend toward higher quality at lower prices continues as more places adopt the new wine-making styles and technology. High-end producers loathe raising prices because new quality wines enter the market every year at ever lower prices. The Charles Shaw winery, for example, produces and markets good California wine for about $2 per bottle, a figure suggesting unprofitability. The increase in value points to an increase in demand for quality wines. Per capita wine consumption in the United States grew from about half a liter in 1935 to more than 2.75 liters in 2015. The rate of increase, as seen in Figure 6.1, is cyclic with an upward trend for the past twenty years.[13]

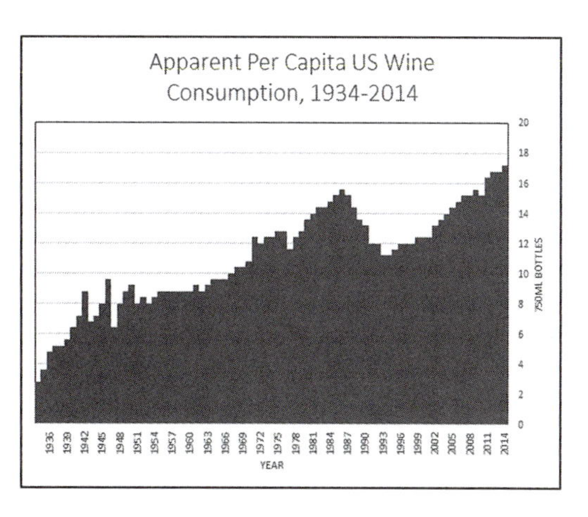

Figure 6.1: U.S. wine consumption.

The District of Columbia maintains the highest national average for wine consumption, 25.7 liters per capita. The numerous politicians, lobbyists, and tourists drive DC consumption. New Hampshire is second with 19.6 liters per capita. People in Vermont and Massachusetts consume more than 15 liters per capita. These three northern states and DC have relatively small populations. Nevada also averaged more than 14 liters per person thanks to the tourist trade.

At the low consumption end we find states where religion, bourbon, or beer renders wine consumption unpopular. Residents of Utah, Iowa, Oklahoma, Kansas, Arkansas, Mississippi, and West Virginia consume less than five liters per capita. The only southern state with substantial wine consumption is Florida, for many of the same reasons that Nevada has high consumption.

States with densely packed populations tend to have higher than average wine alcohol consumption, suggesting an urban factor in predicting the locations of wine drinkers. Northern states, especially those in New England, tend toward higher averages. West Coast states, with their major urban coastal concentrations, also display higher than average consumption. With the exception of Illinois and Wisconsin, the entire central portions of the nation consumes much less wine by comparison with eastern and western states.

For premium sparkling wine, the most festive of beverages, consumption varies widely by state. Considering the top ten sparkling wine–consuming states in 2000, we see a tendency for big populous states to dominate in total and per capita volume. Illinois ranks first in per capita consumption (1.09 L), exceeding both California (0.71 L) and New Jersey (0.71 L) by more than a third of a liter. Texas, Pennsylvania, and Ohio complete the list with less than one-third of a liter consumed per capita each. In the remaining forty states sparkling wine consumption was less than 1 glass per capita annually.

After World War II wine began its transition back to a socially acceptable, enjoyable beverage. Within the past twenty years, the transition has accelerated with the rise of quality American wines; the belief that wine is a sophisticated, healthy fluid; and the dispelling of the "wino" stigma of the past. The rise in wine consumption corresponds to the increase in wine advertising and literature.

Nephew and colleagues have tracked the consumption of alcoholic beverages in the United States for decades. They find that per capita U.S. consumption of alcohol has declined precipitously, at the expense of spirituous liquors, in the past twenty-five years.

Of particular note is the surge of consumption in 1990 and the subsequent abrupt decline in 1991. In 1990, patriotic fervor associated with the first Persian Gulf War brought about an increase in beer and liquor consumption. In 1991, the federal beer tax, established at $9 per barrel in 1951, doubled to $18 per barrel. Rising beer consumption was checked by the tax increase. Revolting beer drinkers switched to ice tea. Liquor, despite not being subject to a tax increase, continued its slow, steady decline in per capita consumption. The event had no apparent effect on wine consumption, which continued its five-year decline.[14]

Second, and more to our point, wine is the least popular of the three classes of alcohol. Wine consumption peaked in the late 1980s and steadily declined during the early 1990s. Since 1995 with the advent of the "wine is healthy" reports, wine consumption has grown slowly. The report by Nephew and colleagues shows a clear tendency for people west of the Mississippi to consume more alcohol of all classes than people from other regions. In 1998, the per capita ethanol wine consumption for Westerners was 0.41 gallons (1.55L).[15] Northeasterners were second with 0.38 gallons (1.44L). Southern and Midwestern values are 0.25 (0.946L) and 0.23 (0.87L) gallons respectively.

Most pertinent to our study, the national per capita wine consumption for 2001 was 7.64 liters and 8.03 for 2002.[16] In terms of ethanol these figures correspond to about one quart (.946 L) per person. Concern over per capita averages, however, will cause us to miss the real dynamics of the wine retail marketplace:

> Of the 268 million people in the United States, there are only eleven million that the Wine Market Council defines as "core" wine drinkers, who drink wine at least two or three times per month. That's the main constituency, and things only get worse; 9 percent of core drinkers consume wine daily; another 49 percent consume "a few times a week"; and 42 percent consume only once a week. But wine consumption is so low in the U.S. that these people account for 88 percent of all wine consumed.[17]

Rephrasing and updating Perdue's statistics for the current 325 million Americans, there might be one million American who drink wine daily and another 20 million Americans who enjoy wine weekly. Perdue's remaining 25 million "core" wine drinkers consume it less than five times a month.

WINE PRODUCTION

Production by Company

Wine is big business. The wine industry is ruled by a handful of major corporations pretending to be small through branding. In 2003, *Wine Business Monthly* began publishing an annual list of the thirty largest wine companies. Table 6.1 lists the thirty largest in 2003, 2008, and 2013. Changes in the list occur routinely. The Wine Curmudgeon, Philip Howard, summarizes corporate wine's strategy this way:

> Consolidation is all. *Wine Business Monthly* included its 2003 top 30 list, and 12 companies on that list are gone, sold or merged into bigger companies. In addition, five companies are on the 2014 list because they bought other companies to get big enough to make the list.[18]

The largest wine companies operating in the United States are presented in Table 6.1. Gallo, Wine Group, and Constellation produce more wine than the remaining twenty-seven companies on the list. The table compares companies in 2013, 2008, and 2003. Membership on the list is highly volatile. Note the loss of eight companies to acquisitions by larger companies in only five years. Three other wine companies, Kendall-Jackson, Wente, and Bonny Doon, dropped off the list. In all, eleven companies not on the list in 2003 made the list in 2008. The most significant change was for Foster's Wine Estates, which came from nowhere to fifth place with 18,000,000 cases.

E. & J. Gallo Winery offers the largest diversity of wine brands, "that total 90 brands and include table and sparkling wines, beverage products, dessert wines and distilled spirits."[19] The list of Gallo wine brands with price points extending from Carlo Rossi to Louis Martini shows the diversity and complexity of products offered (see Table 6.2). Gallo truly tries to offer a wine for every taste at every price point. Most Gallo brands offer Cabernet Sauvignon, Chardonnay, Merlot, and another varietal or two.

Gallo is not the only company with a large stable. Constellation boasts seventy-one wine brands. Trinchero Family Estates maintains thirty brands, including Sutter Home. Treasury Wine Estates offers thirty-four and the Wine Group forty-eight brands. These five companies account for 60 percent of American retail wine sales. According to Wine Curmudgeon Philip Howard, generally these companies "do not clearly indicate the parent company on their label. . . . A company known for producing cheap wine and not quality wine does not necessarily want to be identified with a premium high-end brand."[20]

Table 6.1: Top 30 U.S. Wine Companies by Sales Volume, 2003, 2008, 2013, based on *Wine Business Monthly* reports. Numbers are in '000.[1] Volume values not available for 2003.

	2013		2008		2003
	Company	Annual U.S. Case Sales	Company	Annual U.S. Case Sales	Company
1	E. & J. Gallo	72,000	E. & J. Gallo	67,000	E. & J. Gallo
2	The Wine Group	62,000	The Wine Group	56,000	Constellation Brands
3	Constellation Brands	50,000	Constellation Brands	46,000	The Wine Group
4	Trinchero Family Estates	18,000	Bronco Wine Co.	20,000	Beringer Blass Wine Estates
5	Bronco Wine Co.	17,000	Foster's Wine Estates	18,000	Bronco Wine Co.
6	Treasury Wine Estates	16,000	Trinchero Family Estates	12,000	Robert Mondavi Winery
7	Ste. Michelle Wine Estates	7,500	Ste. Michelle Wine Estates	6,000	Trinchero Family Estates
8	DFV Wines	6,000	Diageo Chateau & Estate Wines	5,700	Brown-Forman
9	Jackson Family Wines	5,500	Jackson Family Wines	5,000	Kendall-Jackson
10	Diageo Chateau & Estate Wines	4,500	Brown-Forman Wines	4,500	Diageo Chateau & Estate Wines
11	Vina Concha Y Toro	3,000	Delicato Family Vineyards	2,500	Ste. Michelle Wine Estates
12	Korbel Wine Estates	2,400	F. Korbel & Bros.	2,500	Allied Domecq
13	CK Mondavi Family Vineyards	1,500	Ascentia Wine Estates	2,000	Delicato Family Vineyards
14	Bogle Vineyards	1,500	Don Sebastiani & Sons	1,750	Golden State Vintners
15	J. Lohr Vineyards	1,400	C. Mondavi & Sons	1,200	Phillips-Hogue
16	Don Sebastiani & Sons	1,300	J. Lohr Vineyards	1,000	C. Mondavi & Sons
17	Francis Ford Coppola Winery	1,250	The Coppola Companies	900	Peak Wines
18	Precept Brands	950	Bogle Vineyards	850	Ironstone Vineyards

19	Foley Wine Group	800	Rodney Strong	800	J. Lohr Vineyards
20	Rodney Strong	800	Hess Collection	650	Chalone Group
21	Boisset Family Estates	750	Precept Brands	600	Don Sebastiani & Sons
22	Wente Vineyards	750	Purple Wine Co.	515	Bogle Vineyards
23	Hess Collection	700	Ironstone Vineyards	500	Rodney Strong
24	Mesa Vineyards	600	Foley Wine Group	500	Barefoot Cellars
25	Domaine Chandon	625	San Antonio Winery	500	San Antonio Winery
26	Vintage Wine Estates	600	Castle Rock Winery	450	Hess Collection
27	Castle Rock Winery	550	Adler Fels Winery	450	Rutherford Wine Co.
28	Purple Wine Co.	400	Domaine Chandon	440	Domaine Chandon
29	Hahn Family Wines	400	Hahn Family Wines	400	Wente Vineyards
30	Michael David Winery	400	Rutherford Wine Co.	380	Bonny Doon Vineyards

Source: *Wine Business Monthly*, "The Top 30 U.S. Wine Companies of 2008." February 15, 2009.

[1] Extracted from *Wine Business Monthly* annual review of the Top 30 U.S. Wine Companies 2003–2013.

Table 6.2: Characteristics of Gallo Wine Brands arranged by price per liter; * indicates once made from grapes.

Brand	Contents	Style	No. of Products	Source Region	Price per 750 ml
Alamos	Varietal	Still	5	Mendoza, Argentina	8.99
Andre	Blend	Sparkling	7	California	4.99
Apothic Red	Red Blend	Still	4	California	11.99
Ballatore	Blend	Sparkling	2	California	9.99
Barefoot Wine	Multiple	Still	20	California	7.99
Barefoot Bubbly	White Blend	Sparkling	10	California	12.99
Bear Flag	Blend	Still	4	California	12.99
Bella Sera	Varietal	Still	4	Italy	6.99
Black Swan	Varietal	Still	4	Australia	8.99
Bodega Elena de Mendoza	Varietal	Still	3	Mendoza	8.99
Bodegas Martin Codax	Varietal	Still	2	Spain	12.79
Boone's Farm*					
Bridlewood Estates	Red Varietal	Still	1	California	14.99
Burlwood Cellars	Varietal	Still	6	California	4.99
Canyon Road	Varietal	Still	7	California	7.99
Carlo Rossi	Blends	Still	11	California	3.00
Clarendon Hills	Varietal	Still	19	South Australia	75.00
Da Vinci	Varietal	Still	4	Italy, Tuscany	11.99
Dancing Bull	Varietal	Still	4	California	8.99
David Stone Vineyards	Varietal	Still	4	California	5.50
Don Miguel Gascon	Varietal	Still	3	Mendoza, Argentina	12.99
Ecco Domani	Varietal	Still/sparkling	3	Italy	8.99
Edna Valley Vineyard	Varietal	Still	2	California	13.99
Fairbanks	Blend	Fortified	2	California	6.49
Five Oaks	Varietal	Still	5	California	2.99

Frei Brothers	Varietal	Still	5	California	20
Gallo Family Vineyards	Varietal	Still	11	California	5.75
Ghost Pines	Varietal	Still	6	California	18.99
Hornsby's	Apple				
La Marca	Varietal	Sparkling	4	Italy	14.99
Laguna	Varietal	Still	1	Russian River, California	28.99
Las Rocas	Varietal	Still	3	Spain	11.99
Liberty Creek	Varietal	Still	4	California	9.99
Livingston Cellars	Varietal	Still	6	California	3.5
Louis M. Martini	Varietal	Still	2	Napa, California	34.99
Madria	Blends	Sangria	2	California	4.5
Maso Canali	Varietal	Still	1	Trentino, Italy	14.79
Mattie's Perch	Varietal	Still	4	Australia	5.00
McMurray Ranch	Varietal	Still	7	Russian River, California	17.99
McWilliams	Varietal	Still/sparkling/ fortified	25	New South Wales, Australia	
Memoir	Blend			California	4.99
Mirassou	Varietal	Still	6	California	12.99
Peter Vella	Varietal	Still	7	California	3
Polka Dot	Varietal	Still	7	Multiple countries	10.26
Rancho Zabaco	Varietal	Still	1	California	14.99
Red Bicyclette	Varietal	Still	3	Languedoc, France	8.00
Red Rock Winery	Varietal	Still	4	California	8.99
Redwood Creek	Varietal	Still	2	California	6
Sebeka	Varietal/blend	Still	41	South Africa	7.49
Starborough	Varietal	Still	1	Marlborough, New Zealand	12.99
The Naked Grape	Varietal	Still	6	California	6
Tisdale Vineyard	Varietal	Still	1	California	6.99

(continued)

Table 6.2: (*Continued*)

Brand	Contents	Style	No. of Products	Source Region	Price per 750 ml
Turning Leaf	Varietal	Still	4	California	7.99
Whitehaven	Varietal	Still	1	Marlborough, New Zealand	19.99
Wild Vines	Beverage	Still	4	California	4.99
William Hill Estate	Varietal	Still	4	California	16
Winking Owl	Varietal	Still	3	California	2.98
Wycliff Sparkling	Varietal	Sparkling	2	California	6.49

Source: www.gallo.com/ portfolio and Pennsylvania Liquor Control Board Price List for Spring 2016.

Production by Country

The amount and value of wine produced by country is the more common approach to discussing the economics of wine. The world's top producing nations and their volume of production are presented in Table 6.3. This table shows that the traditional Western European wine producers make less wine each year, while New World producers make more each year.

Western European countries are facing a serious challenge to their dominance in the global wine trade as seen in Table 6.3. French, Portuguese, and Italian production dropped by more than 20 percent in the past fifteen years. The decline is largely a combination of economic and population dynamics. Europe is aging, there are fewer workers in all fields, and vine dressing is tough physical labor that does not pay well. European rules for grape growing and wine making (such as hand picking and aging regimens) add enormously to production costs—costs not shared by free-wheeling New World producers. Most countries with rapidly growing wine industries are located in the Southern Hemisphere. Eastern European countries show some moderate growth, but big change in the wine industry is coming from China.

As newly discovered prime Chinese vineyard lands begin producing, there is less demand for wine from marginal lands in Western Europe. The Chinese have planted vast areas in their interior. Yantai, the center of Chinese production, had no grapes in 2000. As these vineyards matured over the past decade the volume of wine produced nearly doubled. According to *Herald Sun* reporter Simon Black: "If growth in the Chinese wine industry, driven mainly by red wine, continues at the same rate the country will produce 128 million cases of wine in 2012, an increase of 77 per cent which would see it overtake Australia's forecast of 121 million."[21] Despite the rapid development of their wine industry, the Chinese cannot keep up with domestic demand. In the near future, however, they will likely begin exporting quality products at low prices.

WINE TRANSPORT AND DISTRIBUTION

As we have seen, the wine transportation network is among the world's oldest trade systems. For centuries wine was the world's most commonly transported liquid commodity. In 2006 the global output of wine was 28.4 billion liters. Approximately 9 billion of those liters of wine were shipped from one country to another.

The emphasis over the past few centuries has shifted away from bulk transport to bottling at the winery. The driving factors in this shift are the fight against counterfeit wine and the desire to maximize winery profits by controlling the entire production chain. French *negociants* (grape buyers for wineries without vineyards), for example, face harder times than the growers.

Table 6.3: Wine Production by Country in Thousands of Liters

Rank	Country	2000	2003	2006	2014	% Change 2000–2014	% World Total 2014
1	France	5,754	4,749	5,340	4,670	–23.2	16.54
2	Italy	5,409	4,665	4,712	4,473	–20.9	15.85
3	Spain	4,179	4,037	4,010	3,820	–9.4	13.53
4	USA	2,660	2,350	2,338	3,021	11.9	10.7
5	Argentina	1,254	1,323	1,540	1,519	17.4	5.38
6	Australia	806	1086	1430	1,200	32.8	4.25
7	South Africa	837	956	1013	1,131	26.0	4.01
8	Germany	1,008	819	900	849	–18.7	3.01
9	Russia			655	720		2.55
10	Portugal	784	715	715	624	–25.6	2.21
11	Chile	667	687	845	1,050	36.4	3.72
12	China	534	600	690	1,118	52.2	3.96
13	Romania	545	546	589	511	–6.6	1.81
14	New Zealand	71	76	185	320	450.7	1.13
15	Hungary	430	389	454	294	–46.2	1.04
	Rest of World	3,509	3,823	3,867	3,950	11.2	13.99
	World Total	**28,376**	**26,745**	**28,443**	**28,230**	**–0.5**	**100.0**

Source: Australia Wine and Brandy Corp.

The flow of wine from one region to another is dynamic in both volume and direction. It is a zero-sum game with ever-increasing volumes and value. Conversely, there is an ever-growing volume of unsold wine. Like new restaurants, new vineyard regions explode onto the international market. Old regions lose their luster and exports dwindle. Human trends are responsible for the great shifts in wine exports, while "bad years" limiting crop yield are minor variations. In the early 1950s the French colonies of North Africa were the world's greatest exporters of wine. France, its vineyards once again producing, surpassed the dwindling North African exports in the late 1960s. North African production began declining with independence from France.

In the early 1970s North African, Portuguese, and Bulgarian exports dwindled. German exports expanded with a rising demand for sweet white wine (Figure 6.2). As the French withdrew from North Africa they took the skills and equipment for wine making with them. More importantly, they took the desire to make wine with them. The Portuguese wine Mateus Rosé was the world's most popular wine during the late 1960s and early 1970s. During this period Italy attained its perennial 16 to 20 percent market share.

By the early 1980s France had repaired its devastated vineyards and dominated the world market with more than a 50 percent market share. It was a time of the big name. Small market share countries lost out. In response, Chile and Australia restructured their wine industries, ultimately leading to Australian exports reaching a 4.5 percent global market share in 2000, as Table 6.4 clearly shows.[22]

Table 6.4 gives us a glimpse of the scale of wine exports by country. The data show that Italy produced a third of the imported wine Americans consumed in 2000. Italian wines are even more popular in Pennsylvania where they comprise more than half the imported wine sold.

For many wine-producing countries, wine is a primary export. Table 6.4 shows each nation's market share in the wine trade. As shifting market fads come and go, global attention on a variety or region can flash and fade in a season. The market share losses from these countries become gains for the remainder. Spain was the single largest gainer, capturing 4.7 percent additional market share. Australia followed, capturing an additional 4 percent of the market. Argentina added 1.8 percent and South Africa acquired 1.2 percent.

After the explosive global wine industry growth of the past thirty years, the French/Italian grip on the world wine market is forever gone. These countries will always be the reference point for wine quality and styles, but the industry has now gone global.

After export, wine must be imported to its final destination. Table 6.5 reveals the destination for the wines exported in Table 6.4. In 2000, Germany and the United Kingdom imported 35.4 percent of all wine traded

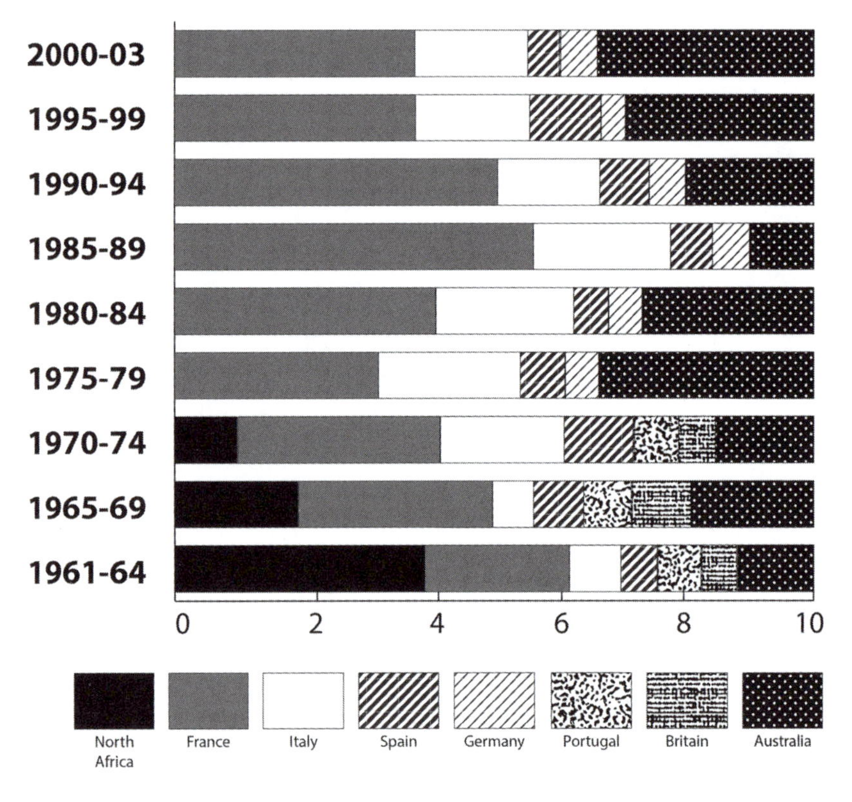

Figure 6.2: Changing structure of international wine trade. Note the reduced importance of Algerian wines.

internationally. By 2006 this figure dropped by 2 percent. Both countries are historically large consumers of wine, but are located outside the grape's optimal growing environment. Many of the other countries listed in Table 6.5 suffer from similar circumstances.

Table 6.5 also shows that French imports fell by 2 percent, and the Netherlands and Belgium each lost 1 percent of the world's share. Japan lost 1 percent owing mostly to its economic problems. Africa and Denmark lost small portions of the market share.

Russia and Italy are the two most significant gainers, each acquiring an additional 1.9 percent of world wine exported from 2000 to 2006. The Other European category expanded by 1.3 percent, followed by the United States with a 1.2 percent increase. Countries whose imports are too small to

Table 6.4: Volume Share of World Wine Exports (%)

Country	2000	2003	2006	2014	Change
Italy	28.2	17.9	20.7	19.6	-7.5
France	24.2	21.1	16.8	29.6	-7.4
Spain	14.5	15.9	19.2	9.8	4.7
Australia	4.5	7.0	8.5	4.8	4.0
Chile	4.0	5.2	5.0	5.4	1.0
USA	4.1	4.4	4.2	4.2	0.1
Germany	3.7	3.6	3.6	3.8	-0.1
South Africa	2.0	4.7	3.2	2.3	1.2
Portugal	2.7	3.4	3.7	2.8	1.0
Argentina	1.5	2.2	3.3	2.4	1.8
Hungary	1.2	0.9	0.6		-0.6
Romania	0.4	0.5	0.2		-0.2

Source: Australia Wine and Brandy Corp.

Table 6.5: Volume share of World Wine Imports (%); total volume is approximately 9,000,000 liters.

Country	2000	2003	2006	Change 2000–2006
Germany	19.2	18.2	16.7	-2.5
United Kingdom	16.2	17.1	15.6	-0.6
USA	7.7	8.3	8.9	+1.1
France	9.3	6.3	7.6	-1.7
Russia	3.8	7.6	5.7	+1.9
Netherlands	5.7	4.6	4.5	-1.2
Belgium	4.6	3.9	3.5	-1.1
Other Europe	3.2	3.7	4.5	+2.3
Canada	3.5	3.6	3.9	+0.4
Denmark	2.8	2.5	2.5	-.3
Japan	3.0	2.4	2.0	-1.0
Switzerland	2.8	2.3	2.0	-0.8
Italy	0.8	3.1	2.7	+1.9
Africa	2.0	2.5	1.9	-0.1
Sweden	1.9	2.2	2.2	+0.3

Source: Australia Wine and Brandy Corp.

appear on this list are also rapidly increasing imports. The California Wine Institute's analysis of sales and production statistics showed that "[m]ajor growth markets outside of Europe and Canada include: China, up 53% by value; Singapore up 68% by value; and Hong Kong, up 19% by value."[23]

WINE AS AN INVESTMENT

Wine is one of the few treasured items that acquire value with longevity, like the Ming vase, ancient map, first-edition book, or rare coin; fragile items, easily broken or lost, that have withstood the ravages of time. They are often treasured for their age outside of all proportion to what they actually are: something to hold water, a book that has been reprinted many times, a small amount of money.

Supply is the key. An aged bottle of wine is something that can never be reproduced. Like a limited run collectible, there will never be any more. Unlike a collectible or painting that thousands may enjoy over many years, a bottle of wine may only be appreciated at one time by those present. The exclusivity of the experience, coupled with the anticipation associated with a rare treat, the chance to share with and impress one's companions, will always draw the rich and powerful.

If you can acquire access to premier wines and cellar them properly for a few decades, you might have something worth much more than it cost. In reality, however, few of us can acquire enough wine to make collecting either reasonable or profitable.

The London International Vintners Exchange (Liv-ex) tracks investment wines. The Liv-ex Fine Wine 100 Index has tracked 100 of the most desired wines, mostly Bordeaux, since 2000. The index is updated monthly. The index is volatile and greatly impacted by conditions in other sectors of the economy. In December 2008 the index stood at $206.81 after peaking in June 2008 at $266.57. The index peaked again in June 2011 at $364.69 but was down to $275.44 in May 2013.[24] As of March 2016 the index value was $243.70.

Collecting and storing wine for personal use, however, is an interesting and enjoyable hobby. The key is to start early and keep a few bottles of each case in the back of your cellar. In five or ten years you can begin sampling the aged wines on special occasions. After fifteen or twenty years you can be the proud owner of an impressive wine cellar. Of almost equal importance, keep a record of the wines you drink and their characteristics.

Acquisition of fine wines as an investment can be a lucrative investment, but near impossible to break into because of the existing high demand. Compared to other precious items, wine value appreciates more quickly than diamonds, gold, or fine art. Wine, however, depreciates when past its peak.

To buy fine wine as an investment, you must purchase it when it becomes available from the vintner. To invest in wine, you do not take possession of the bottles. They are professionally cellared and insured. There are several wine mutual funds that one can buy into to get started. Most collectible wines sell by subscription in cases. It might take several years on the waiting list before an opening on the subscription list becomes available. Once on the list, customers must make annual purchases or be dropped from it. Many vintners also require their customers to purchase several cases of their lower quality wines in addition to the premium wines they want.

SUMMARY

In 2009, the world's vineyards and winemakers produced 27,220,713 tonnes of wine, according to the UNFAO.[25] These vineyards were distributed through 68 countries. Eight countries produced more than 1,000,000 tonnes of wine in 2009. These are (in order): Italy, France, Spain, United States, China, Argentina, Australia, and South Africa. Sixteen countries produce between 1,000,000 tonnes and 100,000 tonnes. The next 23 countries on the list produced between 100,000 and 10,000 tonnes. The final 21 countries on the list produced between 10,000 and 27 tonnes.

Approximately a third of all wine is shipped internationally and the remaining two-thirds are consumed in their country of origin. The network of international shipments is complicated. Several of the largest importing countries also make substantial quantities available for export. France, the number-two exporting country, is also the number-four importing country. Exports are diversifying as more countries outside Europe produce ever larger volumes of wine, while European growers abandon their vineyards. Mostly, however, importer nations are those without an adequate domestic wine industry to satisfy demand, as is the case with Russia, Germany, United Kingdom, Netherlands, Belgium, Canada, Japan, and Sweden.

Chapter 7

Wines of the European Union

Most European wine regions have history and traditions born of twenty to thirty-five centuries of grape growing and wine making. Each European vineyard was repeatedly planted, fought over, fought through, abandoned, reclaimed, tended, and attacked by strange diseases. Through the long years the current distribution of grape varieties and wine making styles emerged. This chapter briefly summarizes the major characteristics of European wine regions. For each wine region described in this and the following three chapters there are books, videos, and articles providing detailed regional history, background, varieties grown, and vinification techniques. The best source for up-to-date information on any wine region is usually the regional association's website.

Europe possesses the world's greatest diversity of grape varieties. In 1999 there were 2,489,453 vineyard hectares, growing to 2,926,878 in 2009. The land area dedicated to each variety ranges greatly. In 1999 there were 95 varieties with more than 1,000 hectares and 101 in 2009. Of the 331 varieties analyzed, the majority existed on less than 75 European Union hectares in 2009, up from a medium of 70 hectares in 1999. Fifteen varieties, with 2 hectares each, not present in 1999 were added in 2009 and 64 varieties present in 1999 were eradicated by 2009. Half the varieties lost between 1999 and 2009 covered less than 10 hectares. The Aranel, at the other extreme, went from 759 to zero hectares. More than 500 hectares of Pineau d'Aunis, Garro, and Grolleau Noir each were removed.[1]

The most widely planted variety in the European Union is Airen, which covered 714,863 dry stony Spanish hectares in 2009. This represents a loss of 60,082 Airen hectares between 1999 and 2009. This is the largest loss by any variety anywhere in the world. The second most widely grown variety,

Tempranillo, covered 151,172 hectares in 1999 and added 284,404 hectares by 2009. This is the largest varietal land area increase during the decade.[2]

WINE RULES OF THE EU

The European Union (EU) is responsible for most of the improvements in European wine quality over the past thirty years. Largely led by French efforts, the EU trademarked wine region names and many production terms. The EU is a major supporter of the Location Matters Declaration because it offers protection from fraud and interlopers.

Officially and within the past thirty years, each EU country adopted a variant of the French geographic and quality control system originally known as *Appellation d'Origine Contrôlée*. Today the system has expanded to the entire EU and much of the world. The official EU designation is Quality Wines Produced in Specified Regions (QWpsr).

Several countries, including France and Italy, have recently changed their system name. The old name included the term *Contrôlée,* emphasizing the controls and restrictions placed on growers and wine makers in varietal selection, how they grew grapes, and how they produced their wines. The new designation uses the term "Protect" instead of "Control." The new French name, *Appellation d'Origine Protégée,* emphasizes protecting the region from outside influences and theft of the region's geographic indicator. The EU continually seeks to further harmonize rules and regulations while respecting local traditions. Table 7.1 gives the official category name for the qualities of wine produced among EU members, the United States, Canada, and (for comparison) ancient Rome.

The EU QWpsr system exists to verify the geographic origin and varietal character, limit quantity, taste test for quality, verify the aging regimen, and define production methods. The EU nations have vigorously enforced stringent production regulations. The system's main function is the elimination of adulterated and counterfeit wine.

The EU is making a concerted effort to reinvent their wine industry; they must. Member countries' market share declines annually. During the 1990s the EU offered modernization grants to poorer member nations, resulting in higher quality and more marketable wines. Before 2008, the EU purchased excess wine and distilled it. With the (low) price guaranteed, growers had no incentive to reduce production. The French were funded to purchase three million hectoliters for distillation, and they wanted more, as did the Italians, Spaniards, and Greeks. Since 2008, the EU has eliminated the practice of "emergency distillation" to use up excess wine, ended restrictions on vineyard expansion, reduced the limits on chapitalization (added sugar), and paid for the grubbing-up of 161,166 hectares of vines. These efforts are transforming the European viticultural landscape.[3]

Table 7.1: European Union Terminology for Still Nonfortified Wine

Country	Table Wine	Table Wine with Geographic Origin Specified	Quality Wine with Geographic Origin Guaranteed	Quality Aged Wine with Geographic Description	Specific Production & Vinification Techniques
Austria	*Tafelwein*	*Landwein*	*Qualitätswein*	*Qualitätswein mit Prädikat, Pradikatswein*	No
France	*Vin de France*	*Indication Geographique Protegee (IGP), Vine de pays*	*Vin Délimités de Qualité Supérieure* VDQS	*Appellation d' Origine Protegee, AOC AOP after 2012*	Yes
Germany	*Tafelwein*	*Landwein*	*Qualitätswein, QbA*	*Qualitätswein mit Prädikat, QmP*	No
Spain	*Vino de Mesa*	*Vino del la Tierra, VdlT*	*Denominacíon de Origen, DO*	*Denominacíon de Origen Calificada, DOC*	No
Italy	*Vino da Tavola*	*Indicazione Geografica Tipica, IGT*	*Denominazione di Origine Controllata, DOC*	*Denominazione di Origine Controllata Garantita, DOCG*	Yes
Ancient Rome	*Posca (slave wine)*	*Mulsum (wine & honey)*		*Falarnian, Opimian*	??
England	*Table Wine*	*Regional Wine*	*Quality Wine*		No
Portugal	*Vinho de Mesa*	*Vinho Regional, Indicacao de Provenienca Reglamentada, IPR*	*Denominacao de Origem DO*	*Denominacao de Origem Controlada, DOC*	Yes

(Continued)

Table 7.1: (*Continued*)

Country	Table Wine	Table Wine with Geographic Origin Specified	Quality Wine with Geographic Origin Guaranteed	Quality Aged Wine with Geographic Description	Specific Production & Vinification Techniques
Czech Republic	*Stolni vino*	*Moravske zemske vino*	*Jakostni vino*	*Jakostni vino s privlastkem*	
Greece	*Epitrapezios Inos Retsina*	*Topikos Oinos, Onomasía Proeléfseos Anotéras Piótitos,*	*Onomasía Proeléfseos Eleghoméni, OPE*	*Appellation of Origion of Superior Quality OPAP*	No
Bulgaria		*Controliran Region*	*Denomination of Controlled Origin*	*Guaranteed and Controlled Origin*	No
Hungary	*Asztali Bor*	*Tajbor*		*Muzealis bor*	Yes
Slovenia	*Namizno vino*	*Dezelno vino PGO*	*Kakovostno ZGP*	*Vrhunsko Vino ZPG*	No
USA	Table wine	Table wine, state specified	AVA specified	AVA specified	No
Slovakia	*Vino bez zemepisneho oznacenia*	*Vino s chranenym oznacenim povodu*	*Akostne vino*	*Akostne vino s privlastkom*	
Canada	Cellared in Canada			Vintners Quality Alliance, VQA	Yes
Brazil		Geographic Indicator			
Israel		Geographic Indicator			

The EU protects terms for sweetness levels for both still and sparkling wines for its member nations. Table 7.2 lists the protected translation of the generic terms *dry*, *medium dry*, *medium sweet*, and *sweet*.

EU sparkling wine sweetness categories, shown in Table 7.3, are not mutually exclusive. All three brut designations (brut natural, extra brut, and brut) may have zero sugar content. The term "dry" is applied to the fourth and fifth driest sparkling wines, creating further confusion for consumers. Historically, demi-sec and doux wines dominated, but they have given way to brut styles in the twentieth century.[4]

Table 7.2: Still Wine Sweetness Categories by Country

Country	Dry < 4 g/l	Medium Dry < 12 g/l	Medium Sweet < 45 g/l	Sweet > 45 g/l
Slovenia	Suho	Polsuho	Polsladko	Sladko
Germany / Austria	Trocken	Halbtrocken	Lieblich, restsuss	Suss, edelsuss
Slovakia	Suche	Polosuche	Polosladke	Sladke
France	Sec	Demi-sec	Moelleux	Doux
Italy	Secco, asciutto	Abboccato	Amabile	Dolce
Portugal	Seco	Meio seco, adamado	Meio doce	Doce
Romania	Sec	Demisec	Demidulce	Dulce

Source: Australian Grape and Wine Authority, http://www.wineaustralia.com/en/Production%20and%20Exporting/Register%20of%20Protected%20GIs%20and%20Other%20Terms.aspx

Table 7.3: European Union Sparkling Wine Sweetness Descriptors

Descriptor	Residual Sugar Sparkling Wine
Brut Natural	Less than 2 gm /l
Extra Brut	0–6 gm / l
Brut	0–12 gm /l
Extra Dry	12–17 gm / l
Dry	17–32 gm / l
Demi-sec	32–50 gm / l
Doux	Over 50 gm /l

Source: Wetherill Company, Sweetness Levels, Champagne411.com/champagne/sweetness.html. Accessed October 10, 2016.

Hundreds of traditional wine-making terms and regional names are protected by international agreement. Protected wine sweetness terms, listed in Table 7.4, identify many of the dessert wines of the European Union.

Dessert wines are created by a variety of techniques. The most straightforward technique is simply to use very sweet grapes, but it is as unreliable as the weather. Chapitalization, adding more sugar than yeast can eat before fermenting, directly ensures the production of adequate alcohol and residual sweetness. Sussreserve is the "reserve of sweetness" attained by adding unfermented, sometimes boiled grape juice to the wine after fermentation.

The opposite approach is the addition of alcohol in a process called fortification. Adding alcohol to fermenting wine kills the yeast and abruptly halts fermentation. Sugar remaining at that point will remain in the wine. Vin du Naturals, Mavrodaphne, Port, and Sherry are made this way as are many others.

A third approach is the partial drying of the grapes before fermentation. Wine made from drying harvested grapes is called raisin or straw wine, referring to the straw mats on which the grapes are placed for drying. Wine made from grapes left to dry on the vine are late harvested and achieve similar results. Finally, *Botrytis cinerea* infection pulls water from the grapes, concentrating the sugars.

Several EU countries and even a few regions have protected terms representing levels of aging. These terms are presented in Table 7.5. Table 7.5 also

Table 7.4: Select European Union Protected Dessert Wine Terminology

Country	Region	Protected Term
Italy	Sicily	Marsala
Greece	Santorini	Vinsanto (one word)
Italy	Tuscany	Vin Santo (two words)
Greece	Patras	Mavrodaphne
Germany, Austria	All	Kabinett
Hungary	Tokay	Puttoyo (1–10) scale
Portugal	Douro Valley	Port
Spain	Jerez	Cream Sherry
Spain	Malaga	Malaga
France	Bordeaux	Sauternes
France	Vins doux naturals	Langedoc-Roussillon
France	Banyuls	Langedoc-Roussillon
Italy	All	Vino Dolce Naturale
Italy	All	Passito

Source: Australian Grape and Wine Authority, http://www.wineaustralia.com/en/Production%20and%20Exporting/Register%20of%20Protected%20GIs%20and%20Other%20Terms.aspx.

Table 7.5: Aging and Alcohol Wine Classification Terms by Country

Country	New	Age Class 1	Age Class 2	Age Class 3	More Alcoholic
Italy		*Riserva*			*Superiore*
Sicilian Marsala		*Fine*	*Superiore*	*Superiore Riserva*	*Vergine*
France	*Nouveau*				*Superior*
Spain	*Vino Joven*	*Crianza*	*Reserva*	*Gran Reserva*	
Portugal (Port only)	*Ruby*	*Late Bottled*	*Tawny*	*Vintage*	
Slovakia	*Mlade Vino*	*Archivne vino*			
Chile		*Especial*	*Reserva*	*Gran Vino*	

Source: Australian Grape and Wine Authority, Register of Protected Names, http://www.wineaustralia.com/en/Production%20and%20Exporting/Register%20of%20Protected%20GIs%20and%20Other%20Terms.aspx.

indicates the special nation-specific term applied to wines that are 1 percent more alcoholic than the standard wine of the indicated region.

The single most important rule across the EU is the protection of geographic origins. Geographic indicators for wine are the most complex of the designations applied to any commodity. The legal and international basis for the protection of place names began in 1958 when the makers of Champagne sued a Spanish sparkling winemaker in English courts over the use of the term Champagne.

Globally, national appellation systems are based on the French national system of 1935. Figure 7.1 shows the generalized, always hierarchical structure of geographic appellation systems. The French system mirrors several other older wine regions including the demarcations of Tokay in Hungary (1730), Douro in Portugal (1756), and Chianti in Italy (1716). Jeffrey Munsie, in his review of international wine regulations, described the origin, character, and extent of the French system this way:

> The hallmark of the Wine Statute was the law of July 30, 1935 that formally created the system of *Appellations d' Origine Controlee* (AOC) that distinguished quality wine from ordinary table wine. This system specifically set out the areas of production, choice of grape varieties, minimum alcohol levels, growing methods, and winemaking techniques required for regions and their wines to carry the honored AOC designation. The *Comitée National des Appellations d' Origine* (CNAO) was also chartered and charged with the job of demarking the vineyards and appellations. Its mission was to study the soil, climate, topography, history, and grapes of any syndicates applying for

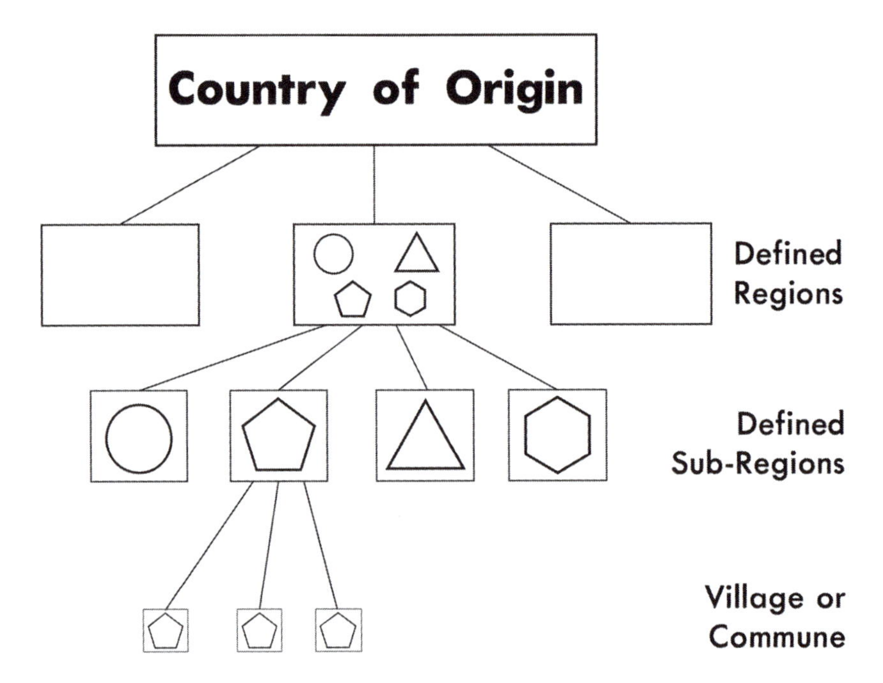

Figure 7.1: Generalized structure of EU hierarchical system for labeling wines.

AOC status and to supervise the practices of those already recognized. The founding members of the CNAO were presidents of syndicates of AOC regions that were at least ten years old and delegates from the agriculture, finance, and justice ministries, all of whom were knowledgeable about viticulture. The classification process was initially slow, such that by 1950 only 10% of French wine was labeled as AOC. A huge expansion later took place in the 1960s that has led to more than 40% of today's French wine being classified under the AOC system.[5]

What makes the geographic appellation system complicated is that each country, while using the same model, has different numbers of levels and unique names for each geographic level and for each wine region. Because of this the geographic indicators for each country are presented in the discussion of each nation. Wines labeled only with a country name mean the grapes are from somewhere in that country.

ORGANIZATIONAL STRUCTURE

Countries bordering the Atlantic Ocean experience the impact of Atlantic storms and the cool southward-flowing ocean currents. Wine regions closest to the Atlantic (Loire, Bordeaux, Douro, Jerez) are sheltered in coastal river valleys. Wine regions further inland (Burgundy, Champagne, Rioja) are sunny, not too warm, and dry during the growing season. Those on the Mediterranean (Rhone valley, Penedes, Malaga, and Jerez) are warm, sunny, and dry during the growing season.

FRANCE

The terroir of each French wine region is distinctive in its geology, topography, and climate. The current French pattern of varieties and production techniques resulted from hundreds of years of practice and experimentation by monks between 500 and 1800 CE. The French wine industry owes its existence to the hard work of generations of these celibate monks. The monks who kept wine culture alive sought to make high-quality wines to glorify God, pay the bills, and obtain political protection from civil authorities. The average vineyard worker was held in high esteem, even if not a monk. In the sixth-century Frankish code, the penalty for killing a wine grower was twice that for killing a farmer.[6]

The four departments comprising the Loire basin show distinctly on the map as does the Rhone valley. Champagne is an isolated northern wine island. The Mediterranean departments of the southwest are the largest producers, but most of their product is destined for distillation. Bordeaux, on the Atlantic coast, produces only 25 percent as much total wine as the southwest, but near equal amounts of quality wine. The remaining wine regions export a great deal of wine to other European countries, but little wine across the Atlantic. The further north one looks, the smaller the volume produced. Departments in Brittany and along the English Channel do not make commercial wines.

There are eleven primary French wine regions, each composed of smaller, higher quality assurance regions. Each of the regions, shown in Figure 7.2 and described in Table 7.6, is discussed below. These regions and their subregions, each representing a village or commune of villages, are defined by the *Institut National des Appellations d'Origine* (INAO), which promulgates and enforces wine regulations in France. The INAO taste tests all wines commercially sold. Those that pass the test are given *Appellation d'Origine Contrôlée* (AOC) status. Wineries with their own vineyards may include the phrase *Mis en bouteille au chateau* (made and bottled at the chateau) on their label.

In 1950, only 10 percent of French wine achieved AOC status. Today, 50 percent achieves AOC status. AOC wines are considered typical of, and

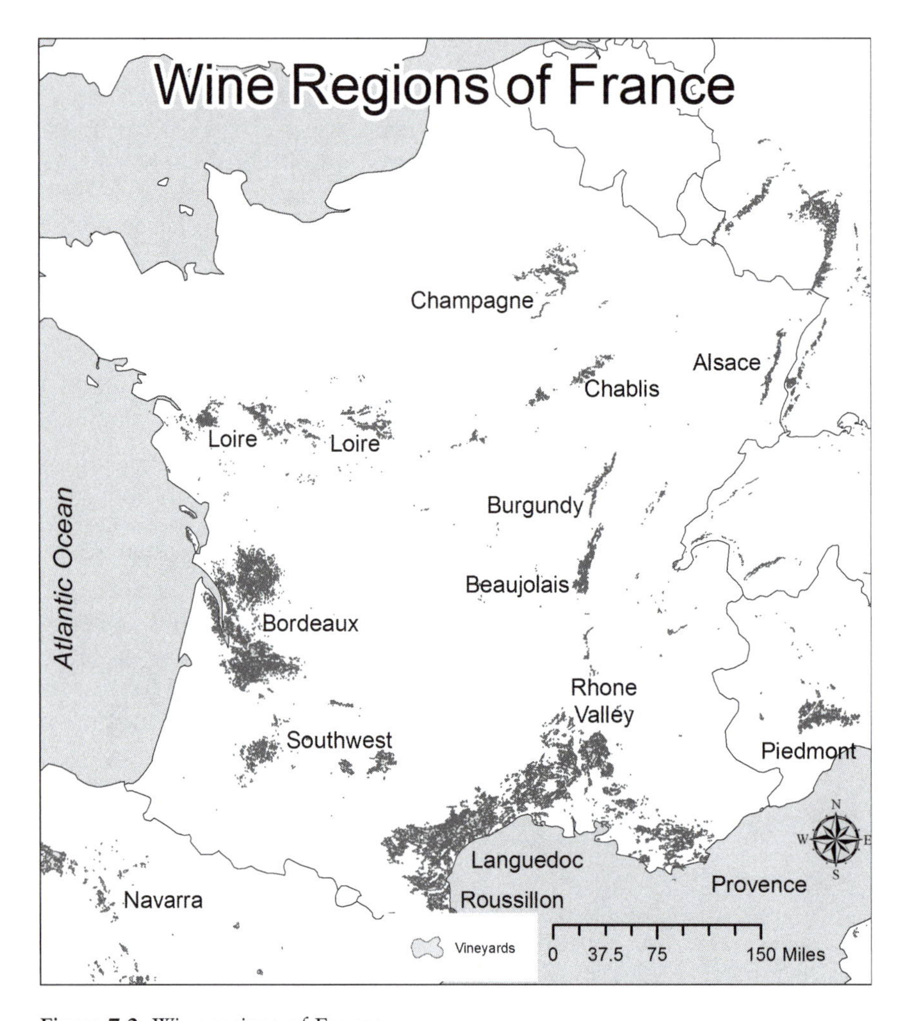

Figure 7.2: Wine regions of France.

generally the best of the designated terroir from which they come. Wines not achieving AOC status receive one of two lower designations: *Vins de Pays* (Country Wine) or *Vins de Table* (Table Wine). The *Vins de Pays* are some of the most interesting and best-priced French wines. About one-third of all French wine is classed as *Vins de Pays*. The remaining 12 to 15 percent of French wines are classed as *Vins de Table*.

In 2011 the French government changed the name of their program to AOP, which stands for *Appellation d'Origine Protégée*. This represents a philosophical change in how the French think about their lands and the values of the wine. No longer controlling production, they now seek to

Table 7.6: French Wine Primary Regional Summary

Region	Vine Area (hectares)	Yield (hecto liters)	Varieties	Soils
Champagne	33,350	2,317,272	Chardonnay, Pinot Noir, Pinot Meunier	River gravels (Marne), chalk and limestone
Bordeaux	117,508	5,745,266	Cabernet Sauvignon, Sauvignon Blanc, Cabernet Franc, Semillon, Merlot	Limestone, River gravels
Burgundy	27,812	1,486,325	Pinot Noir, Chardonnay, Aligote	Limestone, Granite
Beaujolais	18,630	843,032	Gamay	Granite
Chablis			Chardonnay	Limestone
Alsace	15,000	911,951	Gewürztraminer, Riesling	
Languedoc	35,500	1,246,000	Syrah, Grenache	Gravels
Roussillon	19,800	480,000	Grenache	Gravels
Rhone	78,828	3,146,353	Syrah, Mouvedre, Grenache	Gravels
Loire Valley	51,843	2,850,202	Ungi Blanc, Chenin Blanc, Sauvignon Blanc, Gamay, Pinot Noir, Melon de Bourgogne	
Provence	20,400	1,279,000	Syrah	Limestone in west

Source: Inter Rhone, "Key Figures 2011 of Rhone Valley AOCS," http://www.vins-rhone.com/en/5100-WineStatistics.htm, accessed May 30, 2011.

protect their traditions and names from environmental degradation, interlopers, and counterfeiters.

Champagne

Champagne is the northernmost commercially viable wine region in France. North of Champagne the lowlands are boggy and the soil saturated. The climate is heavily influenced by the North Sea, which reduces annual temperature variation. In the Champagne region grapes are difficult to grow because of the short season. The region receives less than 1,800 growing degree days annually. The vines must be planted on hillsides to obtain a minimum degree of ripeness. Sharp, crisp, vibrant wine comes from the region's barely ripened grapes.

Three subareas comprise Champagne. The Vallee de la Marne follows the Marne river bed and is home to Pinot Meunier vines. The Montage de Reims is where Pinot Noir grows. The southernmost section, Cote des Blancs, is where Chardonnay vines concentrate.

Making an acceptable wine from Champagne region grapes became a challenge during the Little Ice Age (1650–1850). Growers responded to the increasingly colder climate by moving vineyards onto the sunniest slopes. Even so, grape clusters did not fully ripen in the coldest years. Winemakers learned to carefully select grapes for inclusion in the wine. Through great diligence Dom Perignon developed a blend of the region's grapes into one of the world's most mystical and valuable fluids. Andre Simon, author of *The History of Champagne*, tells us of the life of this amazing monk:

> Dom Perignon was born at Ste-Menehould in 1639; he renounced the world at the early age of nineteen, and he died in 1715, having served God to the best of his ability for fifty-five years in the Benedictine Abbey of Hautvillers, . . . overlooking Epernay. . . . In 1668, . . . Dom Perignon was appointed Cellarer, an office which he occupied for 47 years, to the day of his death. He had a remarkably sensitive palate and a very keen wine memory, which he retained even in old age, after he had lost his sight. He made better wine than had ever been made before at Hautvillers, rightly blending the wines of different Champagne vineyards in *cuvees*.[7]

Dom Perignon was the first to make champagne-style wine by blending Chardonnay, Pinot Noir, and Pinot Meunier from multiple vineyards and vintages. The Veuve (widow) Clicquot, more than 100 years later, devised the method of purposely putting bubbles into the bottle and removing the bottle sediment created by the action of the yeast during in-the-bottle secondary fermentation. Her method included storing the bottles upside down, disgorgement, and resealing the bottles. In the early years the process was unreliable and dangerous. Poorly formed and overpressured bottles burst routinely. Andre Simon elaborates, "breakage—seldom less than 10% . . . and sometimes as much as 80%—was responsible for the high cost of sparkling Champagne."[8]

Reduction in breakage came about through the improved quality control in bottle manufacture. Better bottles became available when English glassmakers started using coal instead of wood to melt glass. Controlling the sugar content of the grapes was more difficult. It is a simple formula. Sunnier years mean riper fruit. Riper fruit means more sugar. More sugar results in higher alcohol and carbon dioxide. More CO_2 results in greater bottle pressure and increased possibility of explosion. When it was safer to make and had much less breakage, Champagne prices dropped.

Regulating sugar content was overcome when chemist Jean-Antoine Chaptal was appointed Napoleon's Minister of the Interior (think U.S.

Departments of Interior and Agriculture). Chaptal directed "all the wine-making departments [states] to make the *vignerons* [growers and winemakers] add sugar in the *pressoir* [press] at vintage time . . . to raise the alcoholic strength."[9] Ever since, *chaptalization* has been the term applied to the simple act of adding sugar to strengthen any wine.

Market success of Champagne came about in the mid-1800s only after the process was perfected. One of the first international advertising campaigns was launched in the 1830s. This campaign linked the wine to celebrations, ship launchings, seduction, and change of life events.

From the first discovery of "bubbles" in bottles sealed with cork, people in places other than Champagne have made sparkling wines. Wineries around the world used the term "champagne" to describe their wines. "Perlada Spanish Champagne," a product of the Costa Brava Wine Co., was delivered to British shelves in 1956. When publically attacked for using a place name nowhere near the source of the wine, Costa Brava officials dared anyone to do something about it. The Comité Interprofessionnel du Vin de Champagne (CIVC) took up the challenge. Suing Costa Brava in English courts, the CIVC lost in a 1958 jury decision. In 1959, an appeals judge ruled that Champagne had a right to protect their place name. This case established the precedent for establishing geographic indicators for agricultural products.

In 2014 the 300 champagne houses (wineries) produced 307,136,564 bottles, of which 144,870,262 were exported. For 2015 production increased to 312,531,444 bottles. Major recipients of Champagne shipments include the United Kingdom (34,153,662 bottles), United States (20,508,784 bottles), Germany (11,907,887 bottles), Japan (11,799,462 bottles), and Belgium (9,210,659 bottles).[10]

Bordeaux

The Bordeaux wine region extends over the entire *département* (province, state) of the Gironde in southwestern France. The Bordeaux region is centered on 45 degrees north latitude. The vast majority of Bordeaux grapes grow around a large estuary, called the Gironde, and the rivers feeding it. The low coastal hills protect the vineyards from the ravages of Atlantic storms and the estuary moderates temperatures. Frosts and snow outside January are uncommon. Precipitation peaks in December and January at 2.3 inches and is at its minimum in July when they receive about 1.1 inches of rain.

The Gironde and rivers separate the Bordeaux into three subregions. Each of these three consists of multiple AOP-designated sub-sub-areas. These three areas are the left bank, the Entre-Deux-Mers, and the right bank. The Entre-Deux-Mers (literally, between two seas) is the land between the Garonne and Dordogne rivers. On the left bank (south shore of the Garonne River) are the most famous appellations of Bordeaux. Subregions of the left bank include Medoc, Graves, Libournais, and Sauterne. Pomerol and St. Emillion, famed for their Merlots, are on the right bank (north of the Dordogne River).

Bordeaux is the home of Cabernet Sauvignon, Sauvignon Blanc, Cabernet Franc, Semillon, Carmenere, Petit Verdot, Malbec, and Merlot. Each variety grows in specific localized areas. Merlot is the specialty of St. Emillion and Pomerol north of the Dordogne. Lesser quality Merlot comes from the Entre-Deux-Mers. Semillon is the specialty of the Sauternes AOP south of the town of Bordeaux. Cabernet Sauvignon vineyards concentrate on the left bank, between the Garonne and the Atlantic, in a peninsular region called the Medoc. Andre Simon described the Medoc this way:

> The Medoc is a strip of low lying land along the left bank of the River Gironde, some six miles wide and about 50 miles long. The best vineyards are planted on a series of gently swelling elevations of varying heights, which may be likened to some great downs, the soil of which is chiefly composed of siliceous gravel and is sometimes of a calcareous nature.
>
> The principal vineyards of the Medoc, as one leaves Bordeaux and proceeds towards the Bay of Biscay, lie in the districts of Ludon, Macau and Labarde, and on to Cantenac and Margaux, from the stony and gravelly soil of which are produced some of the most delicate and refined of all clarets.[11]

The Medoc section of Bordeaux instituted a classification system for some of its vineyards in 1855. As with all contests, wineries had to enter to win. The classification informed consumers about relative wine quality by dividing the wineries into five categories. The classification system remains in use today when it is beneficial to the producer.

Bordeaux classifications were developed between 1930 and 1960. Graves consists of sixteen chateaux in the Pessac Leognan AOP. Crus Artisans du Medoc also limits membership to the Medoc AOP with the additional requirements that the estates be family-owned, in production for a minimum of 150 years, and less than six hectares in area. The Crus Bourgeois du Medoc classification includes producers in the eight AOPs within the Medoc AOP. Crus Bourgeois is subject to review and modification. The Saint-Emilion classification includes makers of red wine in the Saint-Emilion AOP. Saint-Emilion revises its list every ten years.

The Bordeaux landscape is speckled with chateau. They came about from the 1855 classification. In 1850 Chateau Margaux was the only winery with "Chateau" in its name. In 1855 four wineries were using "Chateau" in their name. By 1900, the transformation was completed when nearly every winery incorporated "Chateau" into its name.

Southwest France

Southwest France is upstream from Bordeaux on both the Garonne and Dordogne rivers. The city of Toulouse marks the region's eastern limit. From the banks of the Garonne to the valley of the Lot, passing through the Pyrenees and the Gascoigne regions, the vineyards of the southwest showcase a wide variety of terroirs and wines. Vineyards cluster in the best locations, which are widely separated.

Growing conditions in southwest France are better for grapes than in Bordeaux. Experiencing greater heat and sunshine in the interior, the grapes ripen earlier and the wines are more alcoholic. The term "Claret" was created to describe the lighter wines of Bordeaux as opposed to the stronger wines from further upstream.

The region remains poorly connected to the international market because of unfair shipping policies established in the town of Bordeaux between 1200 and 1400. Bordeaux harbor was the collecting point for wine prior to shipment abroad. Bordeaux created legal codes requiring that *all* Bordeaux wine had to be sold and shipped before *any* from the interior could be shipped. Many barrels turned to vinegar while waiting for months on the docks.

Southwest France today is home to more than 6,000 vineyards covering 40,000 hectares. They produce about 330 million bottles annually. White wine production exceeds red by 30 million bottles. Only 25 percent of the vineyards have AOP status, the lowest percentage of any French region.

Burgundy

The region known as Burgundy represents a collection of smaller fiefdoms controlled by the house (family) of Burgundy since the 1300s. Today, historic Burgundy cuts across three French departments, Yonne, Côte d'Or, and Saône-et-Loire. Pinot Noir and Chardonnay are the grapes of Burgundy. The decision of which to plant where was determined long ago by Cistercian monks. The monks discovered a complex soil pattern on the slopes leading from the top of the plateau to the river. André Dominé, noted wine author, explains:

> The limy, loamy, and marly soils of the Cote d'Or were created by the erosion of the high plateau of Jura limestone rising above the Cote d'Or, as strata broke off and fell into the valley. This explains why, within a very small area, a variety of terroirs are found with the characteristic properties reflected in the flavor of the wines. In the Cote Chalonnaise and the Maconnnais limestone deposits are more sporadic and mixed with more loamy, sandy soils. Of great importance to all

Burgundy's terroirs is the interaction between good drainage in rocky areas and the water retention of the loamy, clay, or marly substrata.[12]

Within Burgundy there are ninety-nine subappellations, which range greatly in size and character. Major subregions include Cote de Nuits, Cote de Beaune, Cote Chalonnaise, and Maconnais. The Cote de Nuits region is located between Dijon and Nuits-Saint-Georges and is primarily planted in Pinot Noir. This section of Burgundy includes the vineyards of Romanee-Conti and La Tache. Further south, Cote de Beaune has more diverse geology and greater Chardonnay plantings. Cote de Beaune extends from Nuits-Saint-Georges to the River Dheune.

The Cote Chalonnaise region is on the eastern-facing slopes above the Saone and runs southward from the Dheune to the village of St. Gengoux le National. The vineyards in this section of Burgundy are located on three separated patches of limestone. Chalonnaise produces both Pinot Noir and Chardonnay. At the southern extreme of Burgundy and west of the Saone River is the Maconnais region. Chardonnay predominates on the gently rolling limestone hills of the region. Macon is the basic term for any wine made in Maconnais. Chardonnays may be labeled Macon-Villages if produced from a combination of villages. Macon plus an individual village name means the wine came from that village alone. The white wines of Pouilly-Fuissé are the most prized.

Throughout Burgundy *negociants* (merchants) buy grapes from growers and *negociants* produce much of the region's wines. The largest of these, Duboeuf, produces wine from many of the ninety-nine AOPs.

Beaujolais

The wine region south of Burgundy is Beaujolais. The grape of Beaujolais is the Gamay. Vineyards are situated on the eastern-facing slopes west of the Saone River. The Grand Crus are concentrated in the north where granite- and schist-based soils dominate. The Grand Cru appellations are surrounded by the Beaujolais-villages appellation. The Beaujolais-villages AOP includes villages and communes.

The southern portion, underlain by sandstone, produces a fruitier wine suitable for Nouveau. Harvested in early September, the Nouveau wine is ready to drink by mid-November and should be drunk before the end of January of the year following harvest. In other words, harvest it in September, ferment it in October, ship it in November, and drink it in December. Like a white wine, it should be served chilled.

Chablis

Chablis is a small region southwest of Paris where Chardonnay grows on limestone-derived soils. Less than 5,000 acres are planted in vines. Before rail transport, the wines of Chablis were cheaply floated down the Seine River to Paris. The low transport costs meant that Chablis wines were readily available and sold cheaply in the city. Wines of Chablis derive much of their fame from this proximity to Paris. The Chardonnays of Chablis are rarely oaked and are famous for their steely, flinty flavors.

Alsace

Politically, Alsace is the borderland between France and Germany. Wars between these two nations commenced with the crossing of Alsace. As a result, the Alsatian people and their culture are a hybrid, as are the wines. Alsatian vines are planted on the lower slopes of the Vosges Mountains that separate Alsace from the remainder of France.

Environmentally, Alsace has a continental climate with hot summers and cold winters. Snowfall occurs frequently. The soils have highly diverse parentage. The Kastelberg and Rangen Grand Crus grow in volcanic soils. The remaining vineyards reside on sedimentary soils with particle sizes varying by vineyard.

Within Alsace, there are about 42,000 acres (15,000 hectares) under the vine. The vineyards are owned by more than 6,000 wine growers with an average of six to seven acres each. More than 90 percent of all Alsatian wines are white. Riesling, Pinot Blanc, and Gewürztraminer are the dominant grape varieties. Pinot Gris and Sylvaner are also widely grown. Pinot Noir is the primary red variety but is grown in less than 5 percent of the vineyard area. The most important commercial producers are Hugel and Trimbach. Alsatian wines are distributed in tall, slender, shoulderless German-style bottles. Sparkling Alsatian wine is called *Cremant d'Alsace*. Alsatian late harvest wines are labeled *Vendanges Tardives*.

Languedoc-Roussillon

Bordering the Mediterranean, Languedoc-Roussillon is the world's largest wine region. It encompasses more than 40 percent of France's vineyard area and produces a third of all French wines. The wines are often field blends of Grenache, Tempranillo, Syrah, and another twenty varieties. Languedoc is known for its Vins Doux Naturels (sweet, fortified wines). Languedoc-Roussillon produces a large proportion of France's *Vins de Table* and the majority of its *Vins de Pays*. Large proportions of the wine made each year is distilled into industrial alcohol.

Like many places around the Mediterranean with poor stony soils and long dry summers, the Languedoc vineyards have low yields. After phylloxera, the marginal vineyards of this region were abandoned in favor of vineyards in Algeria, Spain, Portugal, and Chile. Remaining vineyards dwindle as fewer twenty-first-century people are willing to live the tough life of a vinedresser. International companies entered the region in the last decade, creating such popular products as Fat Bastard and Red Bicyclette.

Loire Valley

The Loire River drains western central France. The coastal location of the Loire valley made it an ideal place from which to export wine to northern Europe. The wines from the Loire, however, are light in color and weak in character compared to the same grapes grown in Bordeaux. There are eighty-seven distinct appellations approved in the Loire. They are clustered into three major regions: the lower, middle, and upper Loire.

The lower Loire includes the Muscadet region adjacent to Nantes. The Melon de Bourgogne variety has dominated since the mid-1600s. The middle Loire consists of the Anjou-Saumur, Vouvray, and Touraine districts. Viticulturally, this is the most heterogeneous region of France. Varieties grown here include Cabernet Franc, Gamay, Pinot Noir, Chardonnay, Sauvignon Blanc, Semillon, Pinot Gris, and Malbec. Here we find both Bordeaux and Burgundy grapes growing in close proximity. Anjou is famed for its rosé made from Cabernet Franc. Saumur derives its fame from Mousseux, a sparkling wine made from Chenin Blanc. Vouvray makes mostly Chenin Blanc wines, but a wide assortment of grape varieties grow here. Touraine growers focus on Gamay.

The upper Loire is a diverse area. Wine districts are separated by large distances. The most commonly grown grapes are the Sauvignon Blanc and Pinot Noir. The Sancerre district produces red, rosé, and white wines. Pouilly-Fumé only produces whites from Sauvignon Blanc. Flint nodules in the limestone-based soils are believed to instill a smoky flavor in the wine.

The ease of transport made Loire wines cheap and cheap made them popular in England and Holland, although the wines were low in alcohol and quickly turned to vinegar. During the 1630s, when the English were at war against the French and Dutch, Dutch merchants started buying Loire wines for distilling into brandy. Brandy was a new discovery in those days. The Dutch liked it because it was stronger and less bulky to transport than wine. This may be the reason the Loire adopted and adapted itself to its most famous product, Cognac.

Ungi Blanc vines, destined for Cognac, cover 80,000 hectares surrounding the Charente River. The region's chalky soil makes the wines, and resulting Cognac, extra acidic. The wine produced is only about 8 percent alcohol. Cognac is double distilled in a special still. The first distillation brings the alcohol up to about 30 percent. The second distillation takes it to 70 percent alcohol, and it is aged in Limousin oak barrels to mature for a minimum of two years. VSOP Cognac is aged five years. Distilled water is added to reduce the alcohol to 40 percent at bottling.

Rhone Valley

From Vienne (the town's name means wine) to Avignon, the banks of the Rhône River are home to millions of vines and thousands of wine growers. In 2005, the Rhone Valley vineyard lands exceeded 78,000 hectares (197,070 acres or 308 square miles). This land produced more than 3.1 million hectoliters of wine for an average of 40 hectoliters per hectare (428 gallons per acre). The valley is subject to the mistral winds, which may damage the vines, but always reduce mold and insect damage.

The Rhone appellation extends from the merger of the Saone and Rhone rivers near Lyon. The city of Avignon roughly marks the southern boundary of the region. Gravelly granitic and schistous soils cover Cote Rotie and Hermitage. Around Condrieu the soil is decomposed, powdery mica. Coupled with the dry conditions, grapes are the only viable crop throughout much of the region. The wines of the region are big and bold. Predominantly reds, they are deeply colored and highly alcoholic. Syrah and Grenache grow widely. Aromatic whites like Viognier, Roussanne, and Marsonne grow in select villages.

Appellations within the Rhone valley are named for the local hillside (Cotes du Vivarais, Cotes du Ventoux) or town (Vinsobres, Beaumes-de-Venise), but most wines come from one of the several variations of Cotes-du-Rhone. The Cotes-du-Rhone appellation is the most intricate and complex in all France. Nested within the Cotes-du-Rhone is the fragmented Cotes-du-Rhone-Villages, and nested within them are Cotes-du-Rhone Villages with a single village name, made by a village commune.

Provence

The vineyards of Provence are the oldest in France. Archaeological evidence points to numerous coastal communities before the Phoenicians brought eastern Mediterranean civilization to the western Mediterranean. The Phoenicians established Marseilles at the mouth of the Rhone River to trade with the locals. It quickly became a wine-making center. Made famous by rosé wines, Provence offers a large variety of red wines. Provence's white wines carry the personality of Provence's terroirs with aromas of lavender, laurel, and thyme. Vineyards in Provence have the potential to go organic.

Elsewhere in France

Commercial vineyards exist in many other French departments and personal vineyards are found throughout France. For the most part these are classed as table wines. Additionally, northern France produces significant amounts of cider and Calvados (distilled apple cider).

SPAIN

Spanish Wine History

Spain has a rocky and exceptionally dry wine history. The aridity of the Spanish interior supports the vine and little else. In many Spanish vineyards, the vine ekes out a tortured, sunbaked existence. Here vines must be planted widely separated so each may get enough water. Vines look like small bushes. In Spain (as well as southeastern France, Greece, and other places with extremely dry summers) vines do not grow enough to require trellising.

The Basques were probably the first to produce wine in Spain after the Phoenicians brought wine culture to Spain around 1100 BCE. No less than 3,000 years ago Phoenicians established a colony at the present site of Cadiz. Greeks followed 500 years later. Both groups brought domesticated grapes with them. Spain was subsequently conquered by Carthage, Rome, and the Visigoths. All enjoyed Spanish wine. Wine was prodigiously produced in Spain until 711 CE when the Moors conquered the wine-drinking Christian Visigoths.

Wine did not completely disappear from Moorish Spain. Many vineyards were eradicated, but Moors permitted wine making for export to Christian northern Europe. Vines were reintroduced when Spanish troops reconquered their homeland.

In most of Spain, wood is hard to come by. Spaniards therefore stored and transported their wines in skins (sheep, horse, goat, cow, pig) for many centuries. Skin-stored wines did not last long as they were exposed to many forms of bacteria. The bota bag is reminiscent of that era.

In the late 1800s Spanish vineyards expanded after phylloxera invaded France. French vintners, primarily from Bordeaux, moved into Spain looking for lands free of infestation. Many found the Rioja valley well to their liking, and they introduced Bordeaux varieties and techniques, barrel aging, and their market connections.

Spain began modernizing its wine industry after Francisco Franco, president of Spain from 1936 to 1975, died. The growth of the tourist industry in Spain after Franco's death had a lot to do with the transformation of Spain's wine industry. EU membership has also increased the visibility and quality of Spanish wines through grants and technical assistance.

Wine Regions of Spain

Spain has more land under the vine than any other country in the world, more than 2 million acres. Between 2009 and 2012 Spaniards removed 485,000 acres of vines, a 17.8 percent reduction. Despite the reduction Spain remains home to 12.9 percent of the world's vineyards and almost half of the vineyard area in the European Union.[13] Spain, however, is third in overall wine output because the yield of each vine is small. The locations of Spanish wine regions are shown in Figure 7.3.

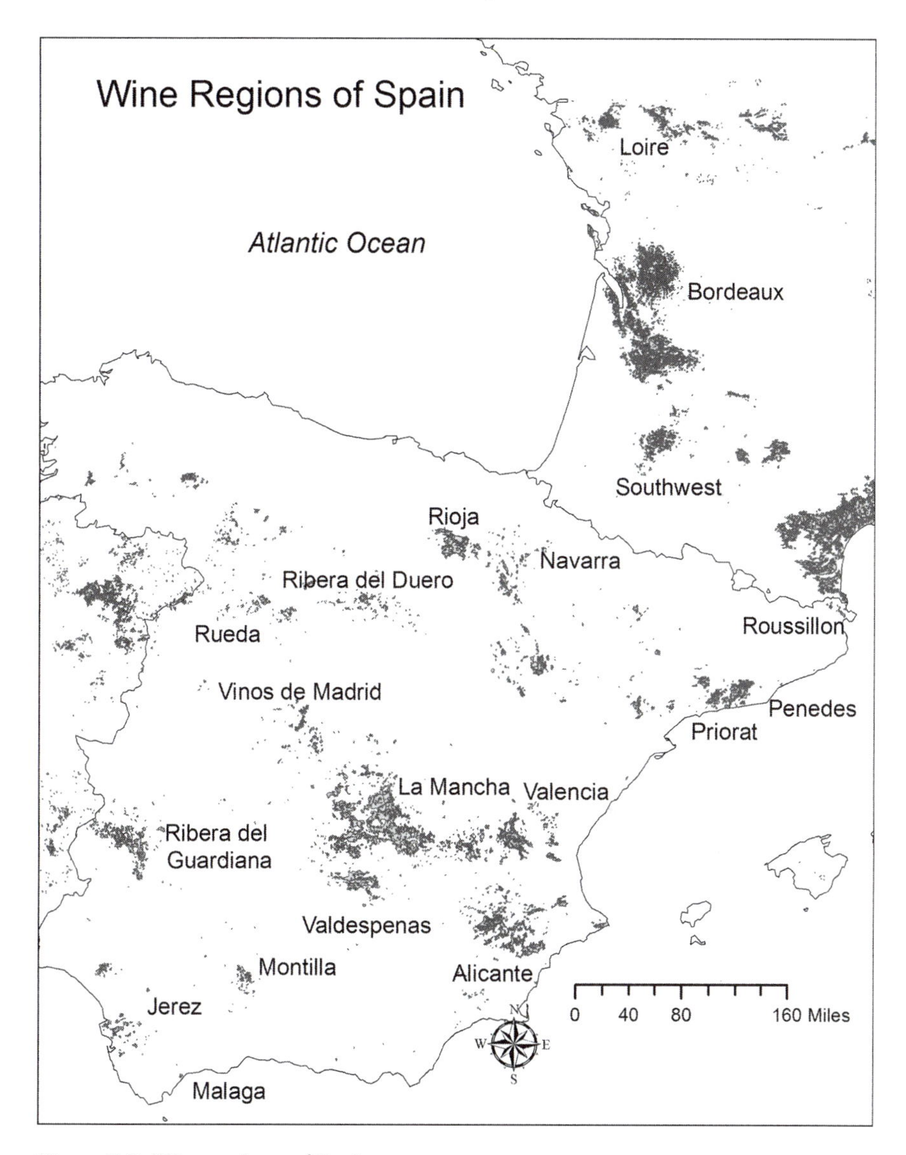

Figure 7.3: Wine regions of Spain.

Rueda

Vines have been in constant production in Rueda since reconquest in the eleventh century. The vineyards of Rueda cover long low alluvial slopes eroded from the nearby mountain. These stony soils permit deep rooting. The region receives less than twenty inches of rain per year, most of it in December. The vines, mostly Verdejo, are bush trained at low density (approximately

1,000 vines per acre). They cover about 28,000 acres of Rueda's 32,000 acres of vineyards. Red grapes cover only 2 percent of the land.

Wines must contain at least 50 percent Verdejo to bear the Rueda Geographic Indicator. Blending grapes are Sauvignon Blanc or Macabeo.

Rueda wines became prized for their geographic significance within Castilian Spain. Much like Chablis, Rueda notoriety derived from proximity to Madrid. "In the fifteenth and sixteenth centuries the town of Medina de Camp, 155 km (96 miles) north of Madrid experienced a time of great prosperity[;] [when] [t]he kings of Castile were in residence here nearly 500 bodegas flourished in and around the town."[14] Production of Rueda wine grew rather steadily from 10 to 100 million kilograms between 1993 and 2013.[15]

Catalonian Penedes

On the Mediterranean coast Penedes vineyards fall into three zones: Penedes Maritim among the coastal hills, Penedes Central in a valley protected by coastal hills, and Penedes Superior in the foothills of the interior mountains. Penedes Maritim produces the region's reds. Elevations for Penedes Superior vineyards range from 1,600 to 2,700 feet. These vineyards experience the large diurnal temperature shifts preferred by white grapes. The soils are generally calcerous.

Penedes is most famous for its sparkling wine called Cava. Sparkling wine production was introduced to Penedes by Josep Raventos I Fatjo of Codorniu winery in the 1870s. Today, Codorniu remains the largest producer. Freixenet is the second largest. The Cavas were traditionally blended from Xarel-lo, Macabeo, and Parellada. Chardonnay is increasingly added to the blend.

Other sections of Catalonia produce still wines, mostly red wines from Grenache, Carignan, and Tempranillo. Most notable is Priorat where Carignan and Grenache vines produce very low yields of incredibly sweet grapes. The high sugar levels result in highly alcoholic wines. Regional certification requires a minimum of 13 percent alcohol and 18 percent alcohol levels are possible.

La Mancha and the Central Plains

Spain's broad central plateau is hot during the long summers, cold in the winter, and dry year round. According to the USDA, "Castile La Mancha has the highest area with 473,268 ha in 2015 or almost 50% of the total Spanish vineyard area . . . with 22.5 million hl of wine production in 2015/16 (–8.7%) and amounting for 60.5 percent of the total Spanish wine production."[16]

Before drip irrigation the Airen grape covered more than 1.2 million acres in La Mancha. With drip irrigation less drought-tolerant and more commercially valuable red varieties are replacing Airen in the vineyards. Airen acreage declined from 760,000 acres in 2004 to 620,000 in 2010.

Rioja

Rioja, located in the north central interior, produces Spain's finest still wines. Part of the region is mountainous and highly inaccessible. Many people of the Rioja are Basque. The broad rolling river valley has numerous south-facing slopes. The slopes are planted in widely spaced Tempranillo vines. The river valley is protected from Atlantic storms by surrounding mountains. The Oja River flows through the center of the Rioja wine region, but it is a small river incapable of floating wine to market. Isolation prevented Riojanos from exporting their wine for centuries, as Hugh Johnson explains:

> Only a trickle of Rioja reached the outside world . . . in the 17th and 18th [centuries] the Basque provinces expanded their trade and population and tempted the Riojanos to grow more wine. What could be done with ox-cart and *borrachas* [drunks] no doubt was done. Miranda de Ebro, nearest to the [Mediterranean] coast, did relative well, but Logrono always had more wine than it could sell. One can imagine the stink of wine-skins getting stronger, until what the locals had not drunk had to be thrown away.[17]

Barrels were slowly introduced into Rioja beginning in the 1780s. Riojan fame outside Spain grew after phylloxera destroyed French vineyards in the last third of the nineteenth century. In 1880 a newly constructed railroad funneled Frenchmen seeking lands free of devastation to Rioja. Millions of gallons of wine shipped out on that same railroad when French vineyards could not meet demand. The economic boom only lasted about twenty years; by then grafted vines were producing and France imposed a protectionist tariff on imported wine. The bust lasted for eighty years. Today, Rioja is a vibrant wine region producing 225 million liters of red wine (85 percent Tempranillo) and 15 million liters of white (88 percent Macerabo) in 2013.

Ribera del Duero

The upper reaches of the Duero River from Gormez to Valladolid (about seventy miles) contain the Ribera del Duero wine region. Situated between Rueda and Rioja, Ribera del Duero produces wines from Tempranillo. There the river crosses the high northern plateau. The river valley slopes and floor are dry with layered alluvial soils. Vineyards run the length and breadth of the valley. Soils are mostly iron-rich limestone eroded from the surrounding mesas. Clay content is highest on the valley floor. The winters are cold and cold air drainage is a problem in both spring and fall.

Tempranillo, Cabernet Sauvignon, Merlot, and Malbec predominate. All red wines must contain a minimum 75 percent Tempranillo. The Albillo is the sole white grape. The area's extreme remoteness kept these wines out of the world's markets until recently. The region's vintners are mostly small plot holders. Hubrecht Duijker in his *Wine Atlas of Spain* reported, "Nearly 90 percent of growers own less than 1 ha, or roughly 2.5 acres. A logical outcome of this is the large number of cooperatives; there are about a score of them. At the beginning of the 1990s only three of them were bottling wine."[18] The other seventeen cooperatives produced and sold their wines by the tanker load.

Vega Sicilia led the development of the region's wine industry. First planted in 1882 to take advantage of the phylloxera-induced wine shortage, the winery fell on hard times between 1900 and 1982. With new ownership Vega Sicilia was reinvigorated and achieved world renown. Success quickly turned the region's bulk wine producers to making quality reds.

Jerez and Sherry

Sherry-producing wineries and storage facilities are called bodegas. There are four major styles of sherry: Fino, Amontillado, Oloroso, and Cream. Making sherry is complex. Lightly pressed grapes yield juice destined for Fino and Manzanilla wines. Intensely pressed grapes become Oloroso. Primary fermentation takes place in stainless steel vats. The wines are then fortified and placed in barrels. Sherry barrels are not completely filled. *Flor*, a strain of top-fermenting yeast, flourishes in the air space left in the barrels. Knowing this, vintners typically place only 500 liters of wine in the 550-liter-capacity barrels. The less fortified Fino (15.5 percent alcohol) grows the thickest *flor*, and the more heavily fortified Oloroso (18–20 percent alcohol) grows the thinnest layer. The region's extreme aridity results in the evaporative loss of about 3 percent of the wine each year. This lost wine is called "the angels' share."

Manzanilla comes from an adjacent community. Palo Cortado results from barrels where the *flor* dies. Fino sherry develops a thick layer of *flor* and Oloroso the thinnest. Fino is pale, dry, and has a nutty and yeasty nose, and a strong citrus finish. Fino is made in stainless steel barrels. Amontillado develops slightly less *flor* and has more color. Amontillados retain slight amounts of residual sugar, usually less than three brix. Oloroso is golden brown in color, largely because it is wood aged. Oloroso has a warming sweetness reminiscent of dried fruit coupled with the nutty flavor present in all sherry. Cream sherry is the sweetest and richest style of sherry. The wine has a pale coloration similar to tawny port and can have the sweetness to match.

To ensure consistency of flavor across decades, sherry makers perform fractional blending in a *solera*. Each *solera* is a large collection of barrels with a defined plan of moving wine between barrels as depicted in Figure 7.4. Here we see a series of four rows of layered barrels. The bottom row, actually called the *solera*, contains the oldest wine. In any one year a sherry house may bottle up to 35 percent of the contents of the oldest (bottom) barrels of the *solera*. In practice, however, most houses never withdraw more than 15 percent. Wine from the second oldest layer, the first Criadera, is used to refill the oldest barrels. Wine from the third oldest layer, the second Criadera, refills the second oldest barrels, and so on. New wine is added to the topmost layer. For this reason there is no vintage sherry. If a date is given, it represents the date the *solera* was laid down. Valdespino laid down its *solera* before 1600. Emilio Lustau, the newest bodega, laid its *solera* more than 100 years ago.

There are two rare forms of aged sherry. The aged sherry terms apply only to sherry aged in the bottle after removal from the *solera*. V.O.S. is very old sherry or *vinum optimum signatum*, a wine that has been bottle

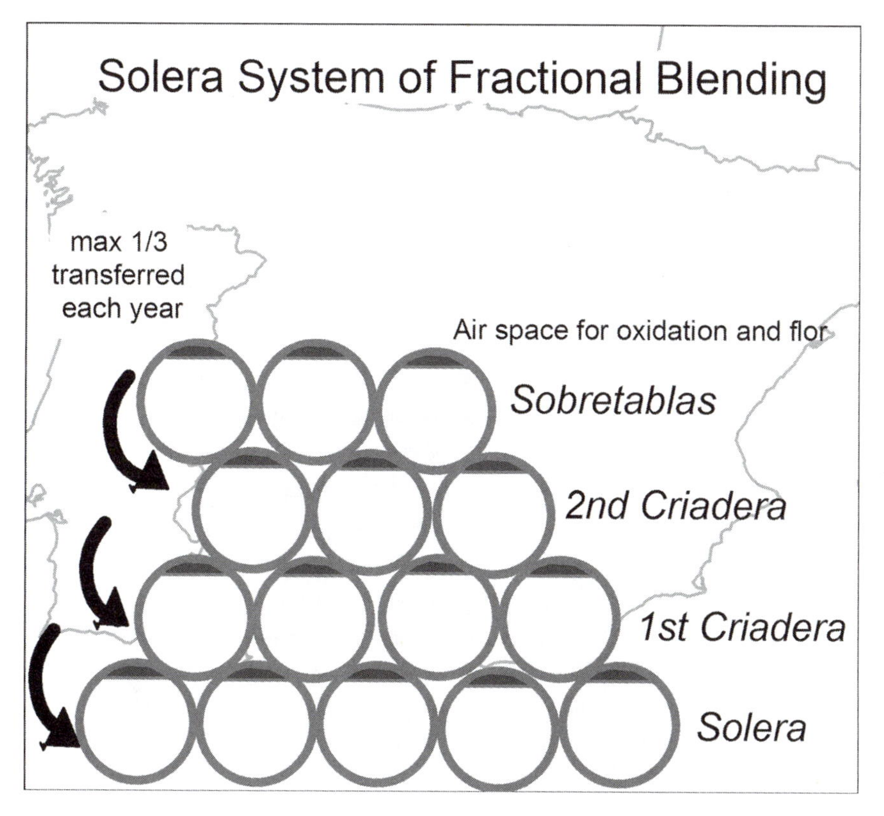

Figure 7.4: Fractional blending creates a long-term consistency of aromas and flavors.

aged for more than twenty years. V.O.R.S. is very old rare sherry, *vinum optimum rare signatum*, a wine that has been bottle aged for more than thirty years.

Jerez has geographic and historical advantages over neighboring cities that also produced fortified wines. The nearby excellent port of Cadiz, the departure point for most of the Spanish conquistadors, made shipping the wine easy. In 1587 Drake "singed the King of Spain's beard" and seized thousands of barrels of sherry. When the British gained control of nearby Gibraltar in 1713, sailors and travelers had ready access to Cadiz and wines from the nearby Jerez de la Frontera. With French wines either banned during war or heavily taxed during peace during the 1700s, sherry was a patriotic alternative.

Rias Baxias

Galicia, located north of Portugal, is Spain's wettest region, receiving more than seventy-five inches of precipitation each year. This coastal area produces white wines, almost exclusively, from the Albarino grape. Soils are granite derived. Use of wild yeast is common.

PORTUGAL

Portuguese Wine History

The Phoenicians "established colonies in Portugal[;] it is difficult to believe that they would not have exploited the vine there had it been present or introduced it had it been absent."[19] Greeks, Carthaginians, and Romans followed, each culture adding to Portuguese wine knowledge, skills, varieties, and techniques. After the collapse of Rome the Visigoths assumed power throughout the Iberian Peninsula. Visigoth Iberia, except for the far northwest, fell to Muslim invaders in 711. Lisbon was captured by Anglo-Norman knights in 1147. The current borders of Portugal were established by 1267.

Portugal, being a small country with a big neighbor (Spain), needed a friend. England became that friend. The first treaty between the two nations promising "perpetual friendship" was signed in 1373. In the 1660s English merchants were expelled from Bordeaux. Seeking wines from elsewhere, the English began purchasing Portuguese wine as an alternative. The wines of Vinho Verde did not suit the British. The British developed a particular fondness, however, for the newly discovered port wine of the upper Douro Valley.

In 1703 the British and Portuguese signed the Treaty of Methuen establishing a trade agreement: English cloth for wine. Over the centuries the British came to the aid of the Portuguese against Spanish and French invasions. In return Portugal delivered two of England's favorite wines, port and Madeira.

Portugal is home to the cork oak. This fantastic tree species is the world's only source of cork. Workers peel the bark off the tree without harm to it. The tree grows a new layer ready for harvesting in about seven years.

Portuguese Wine Regions

There are about one million acres under the vine in Portugal today. The most famous wines from Portugal include port, Madeira, and Vinho Verde. Portugal is home to two wildly popular rosé wines produced by Mateus and Lancers. During the 1970 these wines were the most widely sold in the world and were the first wines consumed by many Americans. The major wine regions of Portugal are highlighted in Figure 7.5 and described below.

Port and the Douro Valley

The centuries-long love affair between the English and port wine is as strong and intense as their love of sherry. The English started their children drinking port early. Charles Dickens wrote, "a glass of Port wine and a biscuit taken regularly at midday, is a capital thing for growing boys and girls."[20]

Port is the great wine of Portugal and among the world's most famous and perilous. Port making is a delicate process. All port comes from the Douro river valley of northern Portugal. At the Douro's mouth is the city of Oporto, from whence all port is shipped (see Figure 7.5). The Douro is home to numerous wineries, each producing wine from a unique varietal blend. Acclaimed red, white, and even rosé wines are made throughout the region from the multitude of Portuguese grape varieties.

The Douro valley consists mostly of steep schist bedrock covered with a thin layer of stony sediment. To create vineyards the Portuguese carved thousands of terraces into the slopes overlooking the river, each holding three or four rows of vines. The small fields and the steep slopes made mechanization impossible before ATVs. Newer terraces, because of bulldozers and other excavating equipment, are broader than older ones.

The abbot of Lamego is celebrated as the first to fortify port by adding brandy in 1678. Fortification kept the wine from turning to vinegar during shipping and in the barrel. That same year the English placed an embargo on French goods, removing the intervening opportunity of Bordeaux.

Counterfeit port, often made from sugared elderberry juice, was a serious problem in the 1700s. Imitators gave port a bad name, resulting in reduced customer confidence and sales. Portuguese growers could not sell their crops. In 1754 the port houses bought no grapes; yet they still made plenty of (counterfeit) port. The first Marquess of Pombal, after overseeing the reconstruction of Lisbon after the 1755 earthquake, was appointed by the

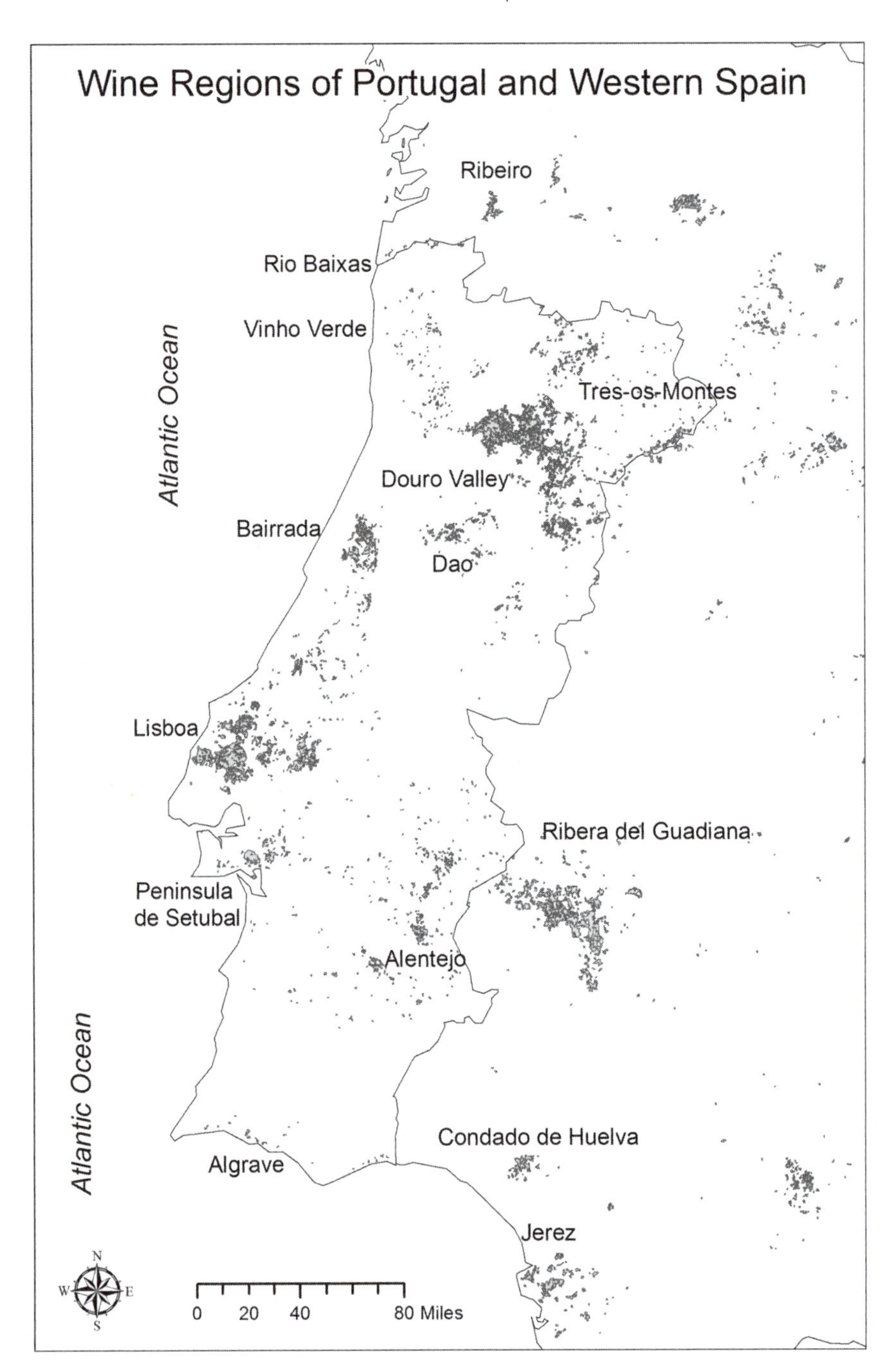

Figure 7.5: Wine regions of Portugal.

king to clean house in the Douro. Godfrey Spence in *The Port Companion* describes the measures taken by Pombal:

> Given almost dictatorial powers by King Jose I . . . [the Marquis of] Pombal took firm control of the port trade. He established the *Companhia Geral da Agricultura das Vinho do Alto Douro* as a monopoly on setting port prices; it was also empowered to rewrite the rules governing port production. In 1756 company officials began mapping the Douro Valley and rating the wines, a task completed in 1761. The best wines went for international export, the middle-ranking wines were destined for Brazil and the lesser wines were reserved for local consumption. Although Tuscans and Hungarians disagree, this is often cited as the earliest demarcation in the wine world, predating the French Appellation Controlee by 180 years.[21]

To combat counterfeiting, Pombal ordered the eradication of all elderberry bushes in northern Portugal. With the origin of the wine guaranteed, the demand and prices increased significantly, and port reemerged as a great wine product and the toast of England.

Pombal's ranking system assigned each vineyard a score identifying its potential for quality. The scoring system is used today to evaluate the potential of new fields. Scores are based on elevation (the lower the better), yield (the lower the better), soil, locality, slope, aspect, variety planted, and vine

Table 7.7: Port Wine Types, Classes, and Descriptive Characteristics

Port Type	Port Class	Wood Aging	Single Year	Blended	Sediment	Description
Wood Aged	Ruby	Short time	No	Yes	Yes	Freshest, fullest, darkest
Wood Aged	Tawny	10 years +	No	Yes	No	Brownish color, nutty flavor, distinct bouquet
Wood Aged	Colheita	20 years +	Yes	No	No	Very old tawny
Bottle Aged	Vintage	2 years max	Yes	No	Yes	Only declared in exceptional years
Bottle Aged	LBV	4 to 6 years	Yes	No	No	Not made in vintage years
Bottle Aged	Crusted	Very short time	??	Yes	Yes	Matures fast
White Port		Short time	No			Clear fluid

age (the older the better). Today, the largest amount of port comes from the lower, western region, Baixo Corgo. The best port comes from the Douro Superior. The Douro Superior is closest to the Spanish border and is the hottest and driest of the three subappellations. Cima Corgo produces wines of intermediate quality.

After harvest grapes destined for port are immediately pressed hard. Traditional pressing required several hours of heavy foot treading to maximize color extraction from the skins. The must ferments in contact with the skins. This vigorous short-term fermentation aids in color extraction. In about three days the sugar and alcohol levels are about 6 percent each and the wine is fortified.

To fortify port the partially fermented must is drawn off the skins and lees. Filtered, the must pours into a vat already containing the proper amount of distilled grape spirits (usually about 77 percent). The alcoholic concentration of the grape spirits immediately kills all yeast. The final product has an alcohol content between 17 percent and 20 percent while retaining the natural grape sugars. The port rests in wooden barrels for several months before going into categorization and final processing.

There are seven categories of port. The three wood-aged-only port categories are ruby, tawny, and of the vintage. Three additional categories, vintage, late bottled vintage (LBV), and crusted, are bottle aged after brief wood aging. White port, the seventh category, is made from white grapes and is usually served as an aperitif (before-dinner drink). The categories and their characteristics are presented in Table 7.7.

English companies operating in Portugal own most of the vineyards and make most of the port wine. Quinta do Cotta claims continuous wine production in the Douro valley since the mid-1300s, long before port was standardized. Churchill, established in 1981, was the first new producer in more than seventy-five years.

In England, after Pombal cleaned house, port became the traditional after-dinner, conversational drink. English walnuts, Stilton (blue) cheese, and dried fruits accompanied the wine. Chocolate and cigars became favored accompaniments much later. Port's consumption became ceremonialized. After dinner, an affair that was designed to take hours, the table was cleared and the table cloth removed to prevent wine stains.

The ceremony began when a servant placed a decanter of port before the host, who poured a glass for himself and for the ladies on either side. The bottle was then passed to the left and guests poured their own. All awaited the completion of the first circuit before toasting and drinking. The first toast was always dedicated to the monarch. Afterward, as the bottle was passed round, conversation, punctuated by song, poetry, joy, and continual toasts, ensued.

The rationale for the rotational passing of the bottle to the left is lost to us. Some say that it is because, on board ship, port is to the left. Others say it is because most people are right-handed. Regardless, the rotational passing prevents any one person from drinking ahead of the rest and conversely it does not permit one to fall behind.

In their genteel manner, the English created several sayings designed to encourage the passing of the bottle. The most direct and common were "Don't let the bottle stagnate!" and "The bottle stands by you!"

Other requests are more subtle and suggestive,

> Such as "Mr. X, your passport is out of order" or "do you know the Bishop of so-and-so, he never *passed* the port," will cause the wine to be circulated again, as no glass should ever [be] left empty. Another well-known cry of despair is "please polish the table," as usually the bottom of the coaster is covered in green baize [felt], and the host will explain that the table is only polished properly when the port is pushed round in its coaster regularly.[22]

Sometimes hosts would take special efforts to ensure that their guests "drank their fill."

> In some houses the decanters [called hoggits] had hemispherical bottoms, so that it was impossible to set them down. . . . this contrivance ensured a rapid passage and quick circulation of the bottle round the table. Another contrivance to effect a similar object, was knocking off the stands of the glasses by giving them a smart stroke with the back of a table-knife, so that it became almost imperative on each person to drink off the contents of his glass as soon as it was filled.[23]

Madeira

Madeira, discovered by the Portuguese in 1419, is a mountainous island off the coast of Africa. The uninhabited island was covered with lush vegetation, which the Portuguese promptly burned to clear fields for crops. Madeira is capable of producing a wide variety of crops. Bananas and sugar cane grow along the coast. The vineyards are located at altitudes on the southern side of the island near Funchal. The rich volcanic soils produce a thin wine.

As the British gained colonies, Madeiran wines grew in popularity. Madeira was popular in colonial America. Madeira wines were present at the signing of the Declaration of Independence and the U.S. Constitution. There were two factors encouraging Madeira's popularity, one geographic and one economic. Geographically, Madeira lies directly on the trade routes between England and the rest of the world. Economically, there was a significant incentive to drink Madeira because it was tax exempt in British colonies. (Recall the friendship between England and Portugal.) For centuries English ships landed at Madeira and loaded fresh foods and wine between England and their final destination.

On board ships Madeira wine, at best an average, transformed to a treasure. The heat and motion of transport made the wine better by simulating decades of cellar aging. The resulting Madeira kept for decades more. Cooking Madeira on ships became a business unto itself. Numerous East India trading ships were loaded with Madeira, then sailed to India before being delivered to London. Madeira thus treated became known as Round Trip wine.

Today, the people of Madeira steam their wine on the island. The highest quality Madeira is heated in large lofts called *estufa,* where the barrels of wine are slow cooked at 110–115 degrees for a minimum of three months before bottling. Lower-quality Madeira is placed in a vat with pipes carrying hot water through them. Both the nautical and modern processes create nutty cooked fruit flavors in the wine. Cooking the wines this way renders them immune to spoilage (conversion to vinegar), making them practically indestructible.

There are four categories of Madeira: Sercial, Verdelho, Malvasia, and Bual. Each is named for the grape variety used to produce it. The Sercial grows at the highest elevations and is harvested last. It is dry and light bodied. It makes a fine aperitif when served chilled. Verdelho is tangy, light, and elegant. It is medium dry. Malvasia is a full-bodied sweet dark wine usually served at dessert. It is the most common Madeira and typifies what most expect Madeira to taste like. Bual is the most full bodied and fruity Madeira. This dessert wine often has a smoky nose.

Vinho Verde

Vinho Verde is produced in northern coastal Portugal. It is the coolest and wettest region of Portugal. The region received geographic designation and established vinification limitations in 1908. Mountainous, the region has exposed layers of schist, granite, and slate. The mountains protect the vine-filled coastal valleys from the worst ravages of Atlantic storms. The vines are traditionally arbored high off the ground to facilitate airflow. The airflow prevents the formation of molds and funguses, which would otherwise be a problem.

Wines from the Vinho Verde region are light in character. Alvarinho is the most commonly used variety. Whether red or white, the wines have an unripe tart flavor and a small effervescence from malolactic bottle fermentation. A number of varieties may be blended in producing a Vinho Verde. Red Vinho Verde is mostly consumed locally. White Vinho Verde is widely exported. The white, like Beaujolais, should be drunk young and cold.

Recently, Vinho Verde producers implemented a tracking system. The tracking system allows consumers to enter a number printed on the bottle to exactly locate the source vineyard.

Dao

The Dao region is located in the highlands of central Portugal. Dao altitudes range from 600 to 3,000 feet. It is surrounded by higher mountains. The best vineyards are below 1,600 feet in elevation. This protects the region from Atlantic storms, but the region is still close enough to experience coastal mildness. Rainfall here is concentrated in the winter (as in all Mediterranean climates). Red wines from Dao must include a minimum of 20 percent of the Touriga Nacional variety. The vast majority of the Dao region is underlain by granite.

Lisbon Area

North of Lisbon are nine DOC regions: Encostas de Aire, Obidos, Alenquer, Arruda, Torres Vedras, Lourinha, Bucelas, Carcavelos, and Colares. Proximity to the capital and coastal locations promoted the success of the region's wineries. Bucelas produces a sparkling wine from Arinto (a Riesling-like variety). Bucelas was popular in England in the 1500s when it was fortified. The Duke of Wellington repopularized the wine in the early 1800s.

THE BRITISH ISLES

According to Hyams, the chalk soils of southern England, like those of Champagne and Jerez, have the potential to produce great wine: "The vine had secured a footing well north of the 50th parallel of latitude by the seventh century at latest, and was firmly established in England for about a thousand years[;] it is a fact that viticulture in Britain . . . is marginal and precarious."[24] Vineyards were plentiful during the monastic period, but church property was confiscated by Henry VIII in 1534.

The confiscation took the vineyards from the monks and gave them to the nobility. The nobility and their serfs did not maintain the vines. The tragic shift in possession was followed by an exceptionally cold period called the Little Ice Age, between 1600 and 1850, which brought English viticulture to an end.

English vineyards are expanding. Many rightly relate the industry's growth to global warming. Peace, tourism, and cold-tolerant varieties are, however, equally significant factors. In 2016 there were 500 vineyards covering 4,500 acres in England and Wales.[25]

ITALY

The Greeks who colonized southern Italy in the sixth century BCE called their new home Oenotria (the land of wine). North of the Greek colonists lived the Etruscans. The Etruscans are a largely forgotten people who lived in what is today Tuscany. We know little of them except that they thought about the afterlife more than the Romans thought acceptable and they liked both horses and wine. They were conquered by the Romans in 264 BCE and blended into the greater pan-Italian Roman civilization.

At about the same time beer-drinking, woodworking Celtic peoples migrated across the Alps and occupied the Po valley of northern Italy. The Romans introduced them to wine after conquering them. The Romans expanded wine culture among Celtic peoples all the way to Hadrian's Wall in England and westward into the Rhine River valley in Germany. After Rome collapsed, the Lombards invaded northern Italy where they also learned to make and enjoy wine.

The three cultural traditions (Greek—south, Roman and Etruscan—center, Celts and Lombards—north) persist in Italian wine making today. The wines of the north are usually light, austere, and fruity. The wines of the south are strong, bold, and a bit rough. Etruscan-Roman-Italian Tuscany is the home of Chianti, the famed wine-producing region of Italy.

Wine Regions of Italy

From north to south and east to west Italy is a wine-producing nation. Practically every part of Italy produces wine. From among the many demarcated regions in Italy only the five largest are introduced here:

1. Piedmont, Lombardy, and Liguria in the northwest
2. Trentino, Friuli, and Veneto in the northeast
3. Tuscany, Umbria, Abruzzo, and Lazio in the center
4. Puglia, Campania, and Calabria in the south
5. Sardinia and Sicily off the toe

Their relative positions shown in Figure 7.6 surprises many who do not expect the high volume and concentration of wine production in the north and the lack of wine production in central Italy.

Northwest

The steep slopes of the Piedmont region of northwestern Italy produce red wines made predominantly from the Nebbiolo and Barbera varieties. The Nebbiolo variety is named for the morning fogs, *Nebbia,* that form throughout the piedmont. Barolo and Barbaresco produce the region's most famous and expensive Nebbiolo. The two communes, located only ten miles apart, produce vastly different wines. Barbaresco is lighter and suppler than Barolo, but both require a minimum of five years' aging before the tannins break down enough to make the wine drinkable. The intensity of tar, berry,

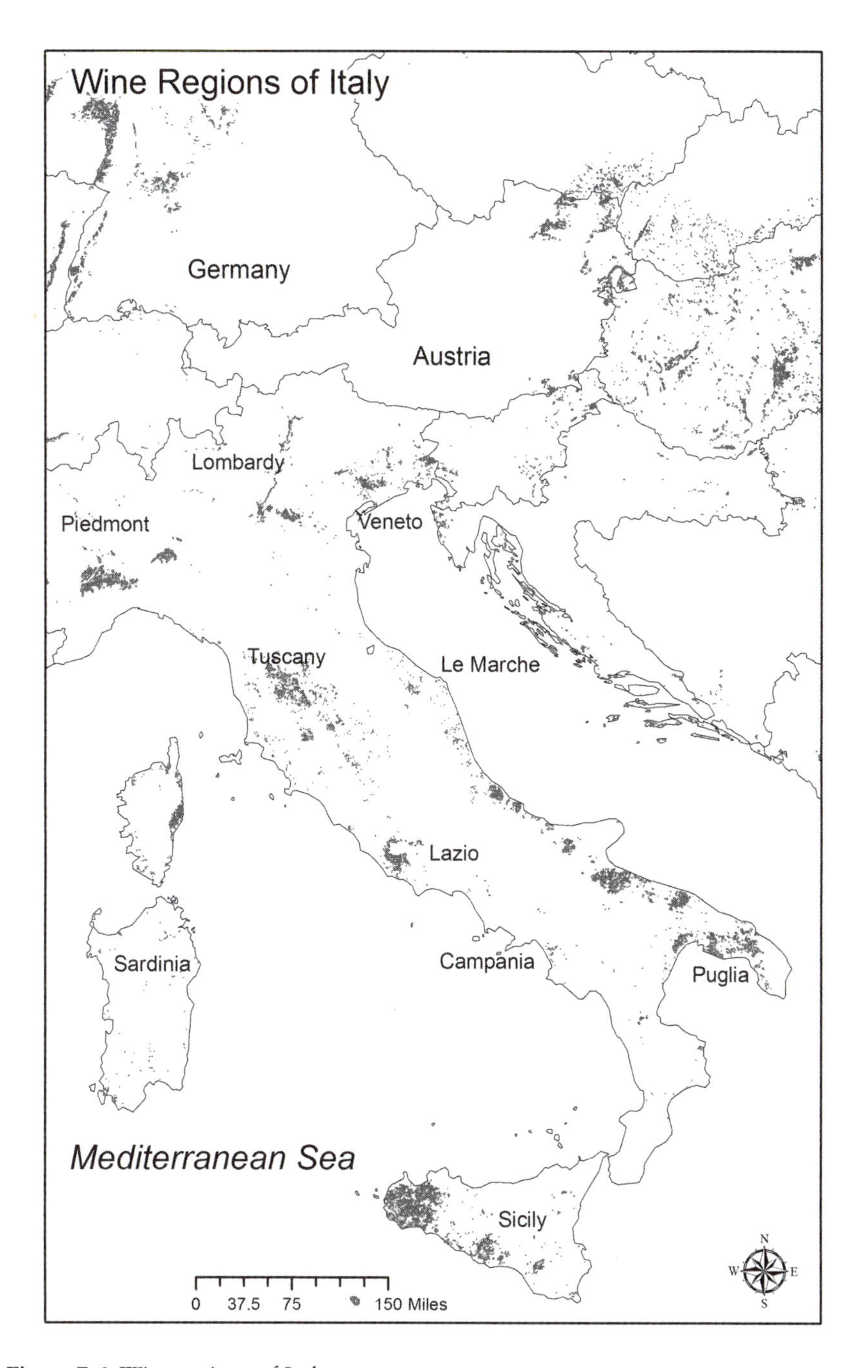

Figure 7.6: Wine regions of Italy.

and licorice flavors are not for the unprepared. Barbera makes a vibrant wine consumed locally with great relish. Thomas Jefferson, during his visit, raved about Nebbiolo, "There is a red wine of Nebbiolo made in the neighborhood, which is very singular. It is about as sweet as the silky Madeira, as astringent on the palate as Bordeaux, and as brisk as Champagne. It is a Pleasing Wine."[26]

The town of Asti is the center for the production of Asti Spumante. Asti Spumante and Moscato d'Asti are sweet sparkling wines made from the Muscat Bianco (aka Moscato) grape. Moscato d'Asti has a much lower alcohol content (under 5.5 percent by law) and is less fizzy (three atmospheres) than Asti Spumante (six atmospheres).

Northeast

Major subregions of northeastern Italy include Veneto, (the area west of Venice), Friuli-Venezia-Giulia (along the Slovenian border), Alto Adige (along the Swiss border), and Trentino (in the center of the region). Northeast appellation zones produce mostly whites. The most common varieties grown here are Riesling, Pinot Grigio, Garganega, Trebbiano (aka Ungi Blanc), and Sylvaner. Soave, a blended wine from Veneto, consists of Garganega and Trebbiano grapes.

The Center

Tuscany's most famous product is Chianti, which is made by blending red Sangiovese juice with white Trebbiano juice. Canaiolo (red) and Malvasia (white) are also blended in small amounts into Sangiovese to make Chianti. There are two quality levels, Chianti and Chianti Riserva. Riserva is heavier in body and oak aged longer, three years minimum, and benefits from bottle aging.

Tuscany contains several famous subregions. The Chianti Classico region is a core surrounded by the more recently planted Chianti region. The Classico region between Florence and Siena was demarcated in 1716. Much of the Chianti region, further, is planted in the less suitable Romagna Sangiovese clone.

Of special interest is Brunello di Montalcino, which is among the richest of Italian red wines. This wine is made from the Brunello clone of Sangiovese. The wine is hard and tannic when young; it takes many years to mature.

Super-Tuscan wines are made by "nontraditional methods" using "nontraditional grapes." "Nontraditional methods" means employing technology and scientific knowledge to produce fresh, fruity, and powerful wines. "Nontraditional varieties" means Cabernet Sauvignon, Merlot, Syrah, and other varieties not originating in Tuscany. Because these wines are nontraditional, they are classed as table wines by the Italian government.

The South

Puglia is the name for the hot dry southern heel region of Italy. Puglia has historically produced 1.2 billion liters of bland, highly alcoholic wine. Much of this wine is distilled to create the base for vermouth. In recent years, however, winemakers in this region have also begun their drive toward producing quality wines for the international market. The Primitivo grape grows extensively in Puglia and is the most important representative of the quality drive.

Sicily

Sicily is the largest island in the Mediterranean. Sicilians make wine from a number of varieties and in a number of styles. Nero d'Avola and Primitivo are hearty reds produced by many Sicilians. Zibibbo is the most popular white grape.

Fortified and Passito wines are common. Englishman John Woodhouse established the first Marsala winery in 1796. Grillo, Catarratto, and Inzolia varieties are used to make Marsala. Marsala is fortified, sweetened, and oak aged for several years before bottling. It's sweet, strong, with a smoky flavor that makes it ideal for cooking. Passito wines, made without fortification, are made viscous and strong by adding dried grapes to the must.

Sardinia, the second largest island in the Mediterranean, grows Grenache and the Vermentino white grapes. Very little is exported. The sole Sardinian DOCG is Vermentino di Gallura. Wines reminiscent of both sherry and port are made on the island.

Italian Wine Regulations

Italy established its appellation control system in 1963 and substantially revised it in 1992 to harmonize it with European Union standards. The rules stress indigenous varieties such as Nebbiolo, Sangiovese, Muscat Bianco, Barbera, Dolcetto, Trebbiano, Canaiolo, and Malvasia.

The rules otherwise are fairly standard. Any statement made on the label must be true. The wine must come from the area indicated. For quality wines maximum production is controlled. Varieties present in the wine must be specified using the 85 percent rule now common in the EU. The Italians also established an inspection system to verify vintners' claims. To receive quality designations the wine must be from a specified region and pass chemical and taste analyses.

Italy has established three quality levels for their wine. In ascending quality order they are *Indicazione Geografica Tipica* (IGT), *Denominazione di Origine Controllata* (DOC), and *Denominazione di Origine Controllata e Garantita* (DOCG). There are 20 IGT regions, 333 designated DOC regions, and 74 DOCG regions. The twenty IGT regions correspond to major political subdivisions of Italy. The number of DOCG regions increased rapidly from 2002 when there were 27 to 2014 when the number rose to 74. The elevation of some regions to DOCG reduced the number of DOC regions. The number of DOCGs will continue to increase as more localities complete the application process.[27] Tuscany contains the largest number of DOCG regions and produces less than 1 percent of the total Italian output, but a much larger proportion of total value. Much of the wine of Apulia is made into brandy or industrial alcohol.

Italian Wine Production

Italian wine production in 2002 was almost 1.3 billion liters. The largest producing regions were Sicily (16 percent of total), Apulia (15 percent), Veneto (14 percent), Emilia Romagna (10 percent), Abruzzo (8 percent), and Piedmont (7 percent). While thousands of varieties are grown in Italy, the list is dominated by the commercially successful varieties shown in Table 7.8. Sangiovese, grown extensively in Tuscany for Chianti, covered more area than any other variety in 2010. The white wine grape Trebbiano, which saw a sharp decline in acreage, grows in northeastern Italy.

Table 7.8: Ten Most Widely Planted Varieties in Italy

Variety	Color	Acres	% Change 2000–2010	Region	Wines
Sangiovese	Red	176,748	+2.5	Tuscany	Chianti
Trebbiano	White	135,649	−42.4	Northeast	Numerous DOC
Montepulciano	Red	86,016	+14.3	Central Italy	Numerous DOC
Catarrato	White	85,942	−46.0	Sicily	Salaparuta DOC
Merlot	Red	69,263	+8.7	Widespread Foreigner	Numerous locations
Barbera	Red	50,901	−39.3	Northwest	Barbera d'Asti, Nizza DOCG
Chardonnay	White	48,682	+40.3	Widespread Foreigner	Numerous locations
Glera	White	48,464	+58.5	Northeast Veneto	Prosecco
Pinot Grigio	White	42,684	+61.4	Northeast Alto Adige, Friuli–Venezia Giulia	Varietal
Nero d'Avola	Red	40,990	+31.2	Avola Sicily	Varietal

Source: Istat, 6th General Census of Agriculture, 2010.

SLOVENIA

There are three major wine regions in Slovenia: Primorje, Podravje, and Posavje. Each has several subregions. Primorje is a coastal region. Podravje and Posavje are both interior regions in the foothills east of the Alps. Podravje is the largest wine region in Slovenia. Central Slovenia is too mountainous and the climate too extreme for commercial vineyards. Slovenia is home to the world's oldest grapevine, which was planted before 1600 CE.

The Slovenians produce wine from more than thirty varieties, most not produced elsewhere. Conversion to noble varieties is occurring rapidly, and Slovenian wines are entering the export market.

GREECE

The Greeks produce wines from varieties found nowhere else. In total the Greeks have about 75,000 hectares of commercial vineyards. More than half of all Greek wines are white and most are sweet. Assyrtiko, Moschofilero, Roditis, and Muscat are the most common white grapes and Agiorgitiko, Xinomavro, Mavrodaphne, and Mandelaria are common red grapes. In the Patras region (see Figure 7.7), Mavrodaphne is fortified and the result is very similar to Port. Attica, the area surrounding Athens, produces for the local market. Crete, previously considered part of Greece, has a concentrated wine district near the center of the island.

Greeks have made and consumed wine for thousands of years. Greek wines were reintroduced into Western Europe during the twelfth century when Crusaders occupied much of Greece. When the knights returned home they took a love of the sweet strong Greek wines with them. A brisk trade was soon established between northern Europe and Greece. The most famous of the wines transported was Malmsey.

After the Christians lost control of Greece to the advancing Ottoman Turks, the Turks ruled Greece for almost 400 years. In the late 1700s the Greeks began to seriously agitate for independence. As the independence movement grew, the Turks became more repressive. One of their principal retaliatory targets was the Greek vineyards. By 1800 all commercial vineyards had been eradicated. When the Greeks won independence in 1832, they replanted.

The windswept Greek island of Santorini was largely destroyed by a volcanic eruption in 1650 BCE. The remnants of the island are covered with lava and ash that provide nutrient-rich soil for grape growing, but the island is dry. Fortunately, over the long cloudless summer steady winds blow and the only source of moisture is the evening dew. The Santorini vines are twisted into wreaths to protect them from drying out and direct the dewdrops toward the vine's roots. Each vine, owing to these conditions, produces a small bunch of grapes.

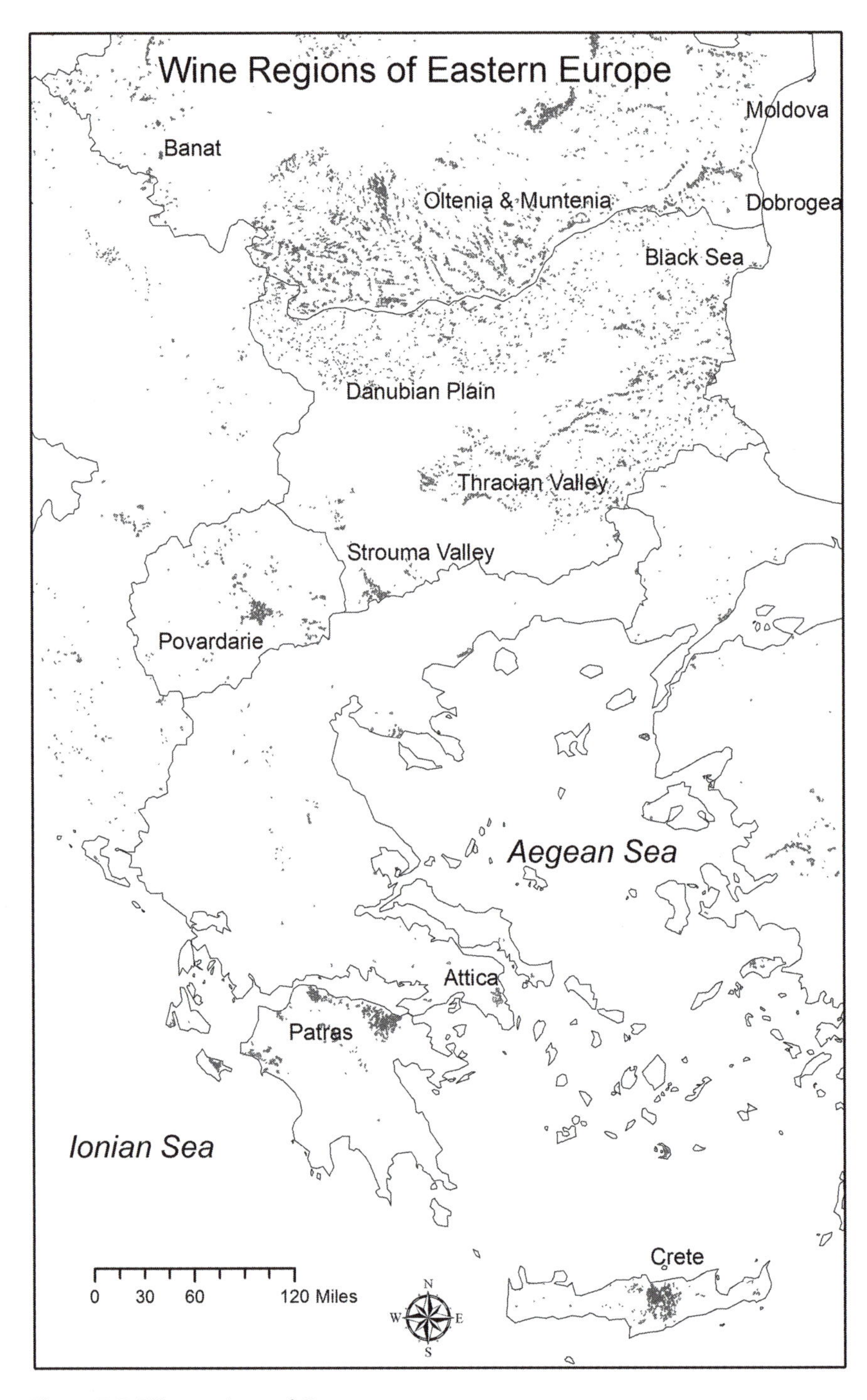

Figure 7.7: Wine regions of Greece.

GERMANY

German wine is based on fortunate geology, long summer days, highly acidic grapes that mature quickly yet hold the vine into the autumn, and political decisions made more than 2,000 years ago. The story of German wine begins with Rome's 400-year-long attempt to keep the beer-drinking Teutonic tribes east of the Rhine River. On the western side of the Rhine and through careful selection the Romans developed Sylvaner, Traminier, and Muscat varieties that ripened in the forested northern edge of the empire. Romans established Trier, which became the northern capital of the empire 2,000 years ago. Dozens of other Roman outposts grew into towns and became regional wine-producing centers.

Since then, German winemakers have discovered varieties and production methods to optimize their wines to suit their palates. In the late 1400s a combination of regional political-religious leaders selected the Riesling over the Elbling. The Riesling wine they prized is amazingly acidic. The Germans developed viticultural methods that produced sweetness to counterbalance the acidity.

Muller-Thurgau, a crossing developed in the 1880s to combat phylloxera, is the source for German bulk wine. Of Germany's 80,000 hectares of vineyards, a quarter are planted with Riesling, a fifth with Muller-Thurgau, and an eighth with Pinot Noir. A half-dozen other cold-tolerant varieties cover the remaining 42 percent of the acreage.

The future for German wine is good. German wine is popular worldwide. Riesling, Germany's signature grape, is more popular than ever. What Muller-Thurgau is not consumed domestically largely goes to Scandinavia as bulk wine. Recent German planting trends reduced Muller-Thurgau acreage. Growers are mostly replanting with high-value Pinot Noir vines with great success.

German rivers twist through soft sedimentary shale and metamorphic schist, and gneiss creating steep-sided, deep, and fairly broad valleys. The hills are rounded because the stone readily crumbles. Were the rocks harder, this would be a region of canyons and spectacular cliffs. Still, hillsides often exceed 50 percent grade and rise 500 feet in many places. On these slopes, vine roots penetrate deeply between book-size chunks of black slate, schist, or gneiss with occasional limestone intrusions.

Low sun angle and the twisting river valleys create an environment with sharp temperature distinctions between sun and shade. The south-facing slopes on the incised meanders create locations that receive high volumes of near-perpendicular summertime solar radiation. For six weeks, from the beginning of June to mid-July, the days are more than sixteen hours long. A short distance away, in the deep shade of a north-facing slope, the day feels short and cool.

German Wine History

The Mosel drainage basin is the historic heart of German wine country, even though it represents only 20 percent of current German grape acreage and wine volume. The Romans brought wine culture to the Mosel and the western banks of the Rhine during the first century BCE. For more than 200 years the Romans transported Italian wine to their Rhineland territories; then, in 212 CE, the emperor permitted viticulture in the provinces. The Romanized populace planted Muscat (and other varieties known to them from the southern flanks of the Alps) on the steep slopes along the Mosel, Nahe, Saar, Ruwar, and Rhine rivers. The grapes presented a more highly acidic, highly refreshing wine than in the land south of the Alps.

Alcoholic beverage preference is one of the few remaining cultural traits distinguishing the old Roman–pagan border. The bounds of today's German wine regions correspond with the extent of ancient Roman control. For the retired Roman soldiers settled on the west banks of the Rhine, a vineyard was a source of revenue and delight. More importantly, a vineyard was an icon of civilization, first Roman and later Christian.

The Romans made the city of Trier, on the Mosel, their northern/summer capital. After Rome's fall the Rhine lands were part of the feudally organized Holy Roman Empire until its demise in 1806 at the hands of Napoleon. For most of the past thousand years the region suffered through innumerable nobles vying to control one another's lands. Despite the turmoil, monks and peasants made wine. As a group, they maintained a remarkable consistency in grape selection, vinification, and wine style. In some cases the wine-making decisions were made by religious and political leaders.

Fabled German wine-making innovations have the moral: Make it sweeter! The first recorded step in this direction was the discovery and acceptance of the Riesling grape. Riesling is a descendant of Gouais Blanc (Weisser Heunisch), Traminier, and an unknown wild variety. The first reliable reference to Riesling vines comes from the Rheingau region in 1435. Through the years, church and secular authorities promoted Riesling production above all other varieties. After the Thirty Years' War (1648) Riesling grapes were widely planted in Alsace. The Little Ice Age, which ended about 1850 and began in the 1600s, intermittently hindered grape production and ripening. In 1787, the Elector of Trier (Mosel region) ordered the eradication of lesser varieties and the planting of Riesling.

In the fabled discovery of Liebfraumilch the devil instructs a soul-seller to plant vines on a rocky knob of a hill near the city of Worms on the Rhine River. The Sylvaner-based wine was so good, the man declared it as good as "milk from the Blessed Virgin." The Virgin Mary was so pleased by his pronouncement that she interceded from heaven and prevented the devil from claiming his prize. In reality the wine derives its name from the Liebfrauenstift monastery. The word for monk was *minch*. It was not long before the wine was renamed and the above fable invented. For the past 300 years the Liebfrauenstift monastery vineyards have been the property of the Valckenberg family. Any wine from the Rheinhessen or Nahe *anbaugebiete* (wine region) can claim to be a Liebfraumilch. Most carry the lower QbA designation (discussed below). Only Valckenberg uses the term MADONNA on its label and is classified at the higher QmP level. These wines are made exclusively for the British and American export markets.

The lower Mosel valley is home to Zeller Schwarze Katz (Zell's Black Cat). In this legend, once upon a time a black cat climbed onto the best barrel of wine and attacked any who tried to broach it. Only wines from the area surrounding the town of Zell are eligible to bear this name. It is usually made from Riesling grapes but may include substantial percentages of Muller-Thurgau. It usually bears the lesser QbA designation.

From the end of this period come three historically documented events: the discovery of how late harvesting affects wine, the incorporation of botrytis-infected grapes in wine, and the discovery of ice wine. Each is portrayed as an accidental discovery and each further secured Riesling's place in German vineyards. The resulting wines from each accident became famous and led to the codification of legends regarding their discovery. In actuality, early season hard freezes, botrytis, and late harvesting occurred repeatedly over the centuries. The existence and character of wines resulting from these conditions must have been known.

The fable for the discovery of Spatlese, and ultimately all the QmP categories, demonstrates the orderly side of German culture. Prior to 800 CE the hillside of Schloss Johannesburg was first planted in grapes. By law, established in 1718, growers around Johannesburg were forbidden to commence harvest before the Prince-Bishop of Feuda decreed it. The decree kept growers from making wine from underripe grapes. More importantly, it kept anyone from getting a jump on the other growers of the region by picking early. In 1775 the Schloss Johannesburg–bound courier with the "Pick the grapes!" decree was waylaid. Couriers to other communities were, apparently, not so delayed. With great concern the people of Schloss Johannesburg awaited the arrival of the courier. As they waited, they watched their ripe grapes succumb to a rot that spread across the vineyards as quickly as the morning fog was burned off. When the poor courier arrived fourteen days later the people quickly picked what grapes remained of their harvest. In desperation, they even pressed the rotted fruit while no doubt cursing the courier. Several months later, their curses supposedly turned to praise when they discovered that the wine from the rotted Riesling grapes was exceptional. Today, there is a statue of the courier in the Schloss Johannesburg courtyard to bemuse the large number of tourists who visit the site annually. It is also why Johannesburg is regularly associated with Riesling.

The first tale of ice wine comes from 1794 or 1835 or 1842. The story is similar to Spatlese. After a cold wet summer some growers near Wurtzbergen,

Franconia (Franken) left their grapes on the vine late into the fall to achieve maximum sweetness. The technique learned at Schloss Johannesburg twenty years earlier had spread across the nation. In early November the first hard freeze of the season struck. To save the vintage they picked and pressed the grapes in the bitter cold. They wisely fermented this must separately, lest it spoil the rest of the vintage.[28]

Undocumented ice wine harvests probably occurred several times previously (Schreiner, 2001). Ice wine production was intermittent and limited quantities were made before the 1960s. The proper environmental conditions to produce it occur naturally only a few times each decade. Germans defined and codified ice wine vinification requirements and production methods in 1961, and modified them in 1973. Under the new rules, the German government encouraged ice wine production. It has been a successful policy; demand regularly exceeds supply.

Wine regions in other countries, with similar environmental conditions, adopted ice wine technology with great success as well. The ice wine concept diffused to Canada in the early 1970s where it has become the staple product of the Niagara Peninsula and Okanagan Valley. New Zealand recently began exporting ice wine to the United States.

When phylloxera arrived in the 1880s, German wine law brought about further ruination. Grafting American roots onto vinifera vines, the solution the rest of the world found so accommodating, was outlawed in Germany. The Germans set their scientists to creating new vinifera and hybrid resistant varieties. Among these newly developed varieties the Muller-Thurgau was the most successful and insidious. Muller-Thurgau was successful because it became widely planted. It was insidious because it destroyed consumer interest in German wines.

The sweetness innovations made German Rieslings prized and expensive wines before a series of events disrupted first demand and then supply. Prohibition and anti-German sentiments swept into Scandinavia, England, and the United States, reducing demand. Internally, three wars between 1870 and 1945 devastated vast portions of the German vineyards. Those not destroyed directly in the fighting suffered from neglect. During the intervening peace periods, German wines were out of favor. When World War II ended the vineyards were restored with Muller-Thurgau and other Riesling-derived varieties. The widespread Muller-Thurgau grape drove many consumers to the wines of other nations.

German Grape Varieties

Germans commercially produce sixteen vinifera varieties. In 2007, there were thirty-one varieties covering more than 250 hectares. The most important are Muller-Thurgau, Riesling, Sylvaner, Gewürztraminer, Spatburgunder (Pinot Noir), Portugieser (Kekfrancos), and Dornfelder. Of these, the Riesling is king with more than 20,000 hectares. Spatburgunder is the heir apparent with more than 10,000 hectares. Muller-Thurgau, the crass cousin everyone wishes would leave, covers more than 13,000 hectares.

There are seven varieties grown in single *anbaugebiete* (German for wine region). Combined they cover more than 9,000 hectares. The smallest of these is Elbling grown on 373 hectares in the Mosel region. The largest is Trollinger on 2,483 hectares in Wurttenberg. Wurttenberg is also the singular home of Schwarzriesling (Pinot Meunier) and Lemberger (Blaufrankish). Baden hosts 1,092 hectares of Gutedel. Franken has 745 hectares of Bacchus and Rheinhessen 1,016 hectares of Scheurebe.

The Riesling grape thrives in the cool German climate. Of the white grapes, Riesling is the most famous and desirable of the German varieties. Harvesting time is especially crucial. Riesling grapes need to remain on the vine as long into the fall as possible to permit the complete creation of sugar in the berries while intensifying flavors by evaporation. The Germans also leave the berries to partially wither on the vine in the dry fall sun, thus increasing the concentration of sugars and flavors. Their wine is usually light-bodied; 8 percent wines are common. Their principle grape, the Riesling, has a crisp acidic flavor reminiscent of Granny Smith apples, pineapple, grapefruit, or lemonade. Riesling, depending on harvesting, can be dry or honey nectar sweet.

Muller-Thurgau was created in 1882 by Hermann Muller who lived in Thurgau, Switzerland. He created it by crossing Riesling and Silvaner. Silvaner is a variety of ancient origin and used to make Liebfraumilch. Muller-Thurgau is a big producer, but inferior to both parents. It is hearty and cold tolerant, which made it popular with growers.

Three red varieties, Spatburgunder, Portugieser, and Dornfelder have significant production levels. Spatburgunder is none other than Pinot Noir. Dornfelder, created in 1955 by August Herold, covers 10 percent of German vineyards. Portugieser is not from Portugal and covers 6 percent of German vineyards, mostly in vineyards outside the Rhine drainage basin. Additional red varieties cover more than a third of Germany's vineyards.

For many years the Germans overcropped (planted too close together and did not drop excess fruit) their Pinot Noir. From the resulting low-quality grapes they made a low-tannin sweet wine. The Ahr valley, the northernmost of the German wine regions, specializes in this product. This style of production began to decline in the mid-1990s as some German vintners sought to enter the quality wine market. These wines have met with consumer approval and demand for them increases. The 2008 Decanter World Wine prize for Pinot Noir was awarded to an Ahr river winery.

In Germany the varietal composition of the vineyards has seen dramatic shifts in the past forty years. In 1970, Riesling, Muller-Thurgau, and Silvaner each represented about a quarter of the total production. Spatburgunder was 4 percent of the total, Portugieser 7 percent, and assorted red varieties constituted the remainder. By 1980 Silvaner had dropped to 10 percent,

Muller-Thurgau was up slightly, and Riesling was down to 20 percent of total production. Portugieser and Spatburgunder were both at 4 percent, and assorted reds only represented 12 percent of the total. Dornfelder made its first commercial appearance in 1979. The year 1990 saw a 2 percent decrease in Muller-Thurgau and Silvaner. There was a 1 percent rise in Riesling, reds, and Spatburgunder. Portugieser and Spatburgunder combined for 10 percent. In 2000, Muller-Thurgau was only 17 percent, Riesling 21 percent, Silvaner 6 percent, Portugieser and Dornfelder 5 percent, and Spatburgunder rose to 10 percent. The big change, however, came in the production of red varieties, which shot up to almost 30 percent of the total. In the last few years the amount of Dornfelder (8 percent), Spatburgunder (12 percent), and other reds (37 percent) shot up dramatically. Riesling and Silvaner declined only 1 percent. Muller-Thurgau declined to 13 percent of the total (German Wine Institute, 2007).

German Geographic Indicators and Wine Regions

The Germans have a legally defined four-tier hierarchically nested system of geographic indicators. At the highest level (covering the broadest area) are the 13 *anbaugebiete* shown in Figure 7.8 (and see Table 7.9). Wineries are not required to include the *anbaugebiete* on the label, but most do. A wine bearing only an *anbaugebiete* designation is usually lacking in character. If the wine carries an *anbaugebiete* designation, 85 percent of the must has to come from that region. Each *anbaugebiete* includes two or more *bereiche* (districts). Jeffrey Munsie interprets the meaning of *bereiche* wines this way: "Wines marketed under the name of a *bereiche* usually consisted of an everyday blend of wine from throughout the district." *Grosslagen*, the next smaller level of geographic distinction, is an area that has similar geologic and climatic conditions. At the finest level of resolution is the *einzellagen,* which is legally recognized and defined. An *einzellagen* must be more than five hectares (12.3 acres). If a wine can pass the requirements, greater accolades fall to wines from a specific *einzellagen* than its surrounding *Grosslagen.*

Eleven of the *anbaugebiete* are in southwestern Germany in the old Roman-occupied territory. The other two, Saale-Unstat and Sachen, are small and in outlying locations to the east. Mittelrhein, Rheingau, and Rheinhessen are on the banks of the central Rhine. The Rheinhessen *anbaugebiete* is by far the largest. From north to south the Ahr, Mosel, and Nahe *anbaugebiete* defined watersheds empty into the middle Rhine. The Pfalz, Hessische Bergstrasse, and Baden *anbaugebiete* are in the broad upper Rhine valley where ox-bow lakes abound. The Wurttemberg *anbaugebiete* is located along a narrow valley formed by the erosive action of the Neckar. Franken is in a similar site on the upper Main. Rheinhessen is the largest *anbaugebiete* with more than 20,000 hectares of vineyards. It is followed in

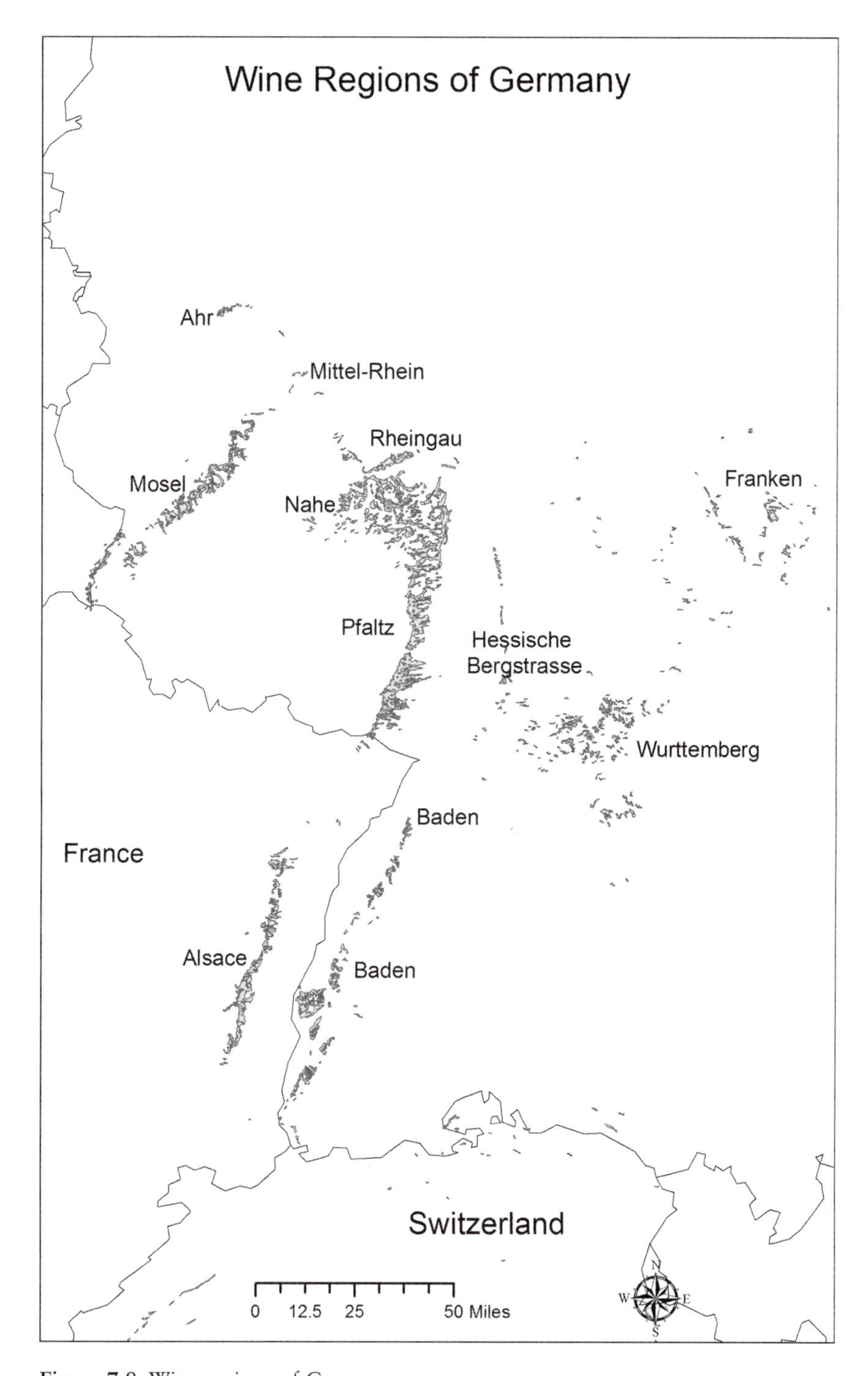

Figure 7.8: Wine regions of Germany.

Table 7.9: German Wine Regions and Select Characteristics

Region	Soil Type	Wine Style	Varieties
Ahr	Volcanic slate	Velvety world class red, racy white	**Spatburgunder,** Portugieser
Mittelrhein	Slate-covered slopes	Racy, fruity acidic, fragrant	**Riesling**, Muller-Thurgau, Kerner
Mosel-Saar-Ruwar	Slate-covered slopes, mineral rich	Piquant, racy, delicate	**Riesling**
Rheingau	Loess, loam, weathered slate	Refined, fruity acidity, racy	**Riesling**
Nahe	Loess, loam, quartzite, porphyry	Subtly racy, fragrant, fruity, crisp	**Riesling**, Muller-Thurgau, Silvaner
Rheinhessen	Loess, limestone, sand	Mild-fruity, round, full-bodied	**Riesling**, Muller-Thurgau, Silvaner
Pfalz	Loam, weathered limestone	Round, full-bodied, aromatic	**Riesling**, Scheurebe, Gewurztraminer
Franken	Loess, sandstone, limestone	Earthy, robust, fresh acidity	**Muller-Thurgau,** Silvaner
Hessische Bergstrasse	Loess	Elegant, fruity, good acidity	**Riesling,** Muller-Thurgau
Württemberg	Shell-limestone, marl, loess	Robust, powerful, high acidity	Trollinger, Lemberger, Spatburgunder, Portugieser
Baden	Loess, loam, volcanic soil	Full-bodied whites, velvety & fiery reds	**Muller –Thurgau,** Gewurztraminer, Riesling, Rulander
Saale-Unstat	Shell-limestone, sandstone	Mild fruity, round	**Muller-Thurgau,** Weissburgunder
Sachsen	Sand, porphyry, loam	Dry, fruity acidity	**Muller-Thurgau,** Weissburgunder, Gewürztraminer

Source: Deutsches Weininstitut, 1997.

size by Pfalz (17,737 hectares), Baden (13,837 hectares), and Wurttemberg (9542 hectares). Five of the regions (Ahr, Hessische Bergstrasse, Mittelrhein, Salle-Unstrut, and Sachsen) are less than 700 hectares (1,750 acres) in size. Three-quarters of Franken and Sachsen wines merit QmP status.

Although in close proximity, each *anbaugebiete* has individual soil characteristics rendering subtle taste variations in the wines. This is especially true for the Riesling variety, which is sensitive to the soil mineral composition. The variations expressed in each region are presented in Table 7.10. The dominant variety is highlighted in the last column.

Table 7.10: German Wine Production by *Anbaugebiete* (Major Region) by Grape Variety, 2006

	Ahr	Baden	Fran ken	Hessische Berg	Mittelrhein	Mosel	Nahe	Pfalz	Rheingau	Rheinhessen	Saale-Unstat	Sachsen	Württemberg	SUM
Bacchus			730											730
Dornfelder							448	3,164		3,535				7,147
Elbling						512								512
Grauburgunder (Pinot Gris)		1,896		43				1,268		1,379				4,586
Gutedel		1,117												1,117
Lemberger													1,666	1,666
Müller-Thurgau		2,559	1,718		26	1,092	532	2,078		4,241	128	80		12,454
Portugieser								1,781		1,439				3,220
Riesling	46	1,119		206	313	5,350	1,170	5,737	2,494	4,267	65	70	2,125	22,962
Schwarz Riesling (Pinot Meunier)													1,539	1,539
Silvaner			1,406							2,371	56			3,833
Spätburgunder (Pinot Noir)	353	5,591	44	7	42		270	1,636	385	1,636			1,300	11,264
Trollinger													2,282	2,282
Weißburgunder (Pinot Blanc)		1,373								102		58		1,533
TOTAL	563	15,822	6,125	450	469	8,776	4,187	23,567	3,166	26,582	765	499	11,373	102,344

Source: Deutsches Weininstitut GmbH. *Deutscher Wein Statistik 2014/2015*. Mainz, Germany: Deutscher Wein Institut. 2015. 7–8. PDF accessed June 9, 2016.

Most wineries follow a bottle color code that dates back hundreds of years. Rhine wines come in brown bottles, wines of the Nahe come in blue bottles, and wines of the Mosel come in green bottles. All of the bottles are tall and slender except the oval *bocksbeutel* bottle of Franken.

German Quality Designations

Germans differentiate their wine using four criteria; locations, variety, quality (interpreted as sweetness), and style. German wines are not differentiated based on vinification techniques. Understanding the terms for each of these categories, and when they are in effect, can seem the task of a lifetime.

There are three defined German quality levels. The lowest quality wines are designated table wines and were a mere 4.1 percent of production in 2006. The intermediary category bears the title *Qualitätswein bestimmer Anbaugebiete* (QbA) and the highest valued quality level is *Qualitätswein mit Prädikat* (QmP). Two-thirds of all German wines attain the QbA designation. QmP wines are further categorized by grape (over)ripeness levels. There are six sublevels of QmP wines based on grape. QbA wines may have sugar added. Sparkling wine, which the Germans produce and consume in copious quantities, is called *Sekt* and categorized separately.

Qualitätswein bestimmer Anbaugebiete (QbA) means "quality wine from a specific place." It is the intermediary quality category. Place is one of the eleven recognized and legally defined wine-producing districts. QbA wine regulations permit the addition of sugar to the must to increase alcohol content. Liebfraumilch (Milk of Our Blessed Lady), Black Tower, Blue Nun, and Zeller Schwarze Katz (Zell's Black Cat) are the most commonly exported presentations of QbA wines. QbA wines may be designated as *trocken* (dry), *halb-trocken* (off dry), or sweet. Sweet is the most common designation for QbA wines. Of the 8.1 million QbA hectoliters produced in 2006, 3.0 million were produced dry, 1.9 million *halb-trocken*, and 3.2 million sweet.

The German phrase *Qualitätswein mit Prädikat* (QmP) means "Quality wine with attributes." Germans attribute (categorize) their wines based on the unfermented (residual) sugar content of the final product. The *Prädikat* (attribute) is natural sweetness. Six levels are legally defined by German law. Natural sweetness is controlled by two factors: first vintners allow grapes to dry on the vine (late harvesting), and second the hope for *Botrytis cinerea* infestation (discussed in chapter 4). The six levels of German wine quality, in order of natural sweetness, are *Kabinett, Spatlese, Auslese, Beerenauslese, Trockenbeerenauslese*, and *Eiswein*. *Beerenauslese* (BA) and *Trockenbeerenauslese* (TBA) are made exclusively from individually selected berries. The other four categories are made in the classic style from whole grape clusters. These wines may also bear the sweetness indicators *trocken, halb-trocken*, or sweet. Table 7.11 relates the technical definitions ascribed to each of these quality levels.

Table 7.11: German Wine Quality Descriptors for *Qualitätswein mit Prädikat*

Name	Translation, Explanation	Style	Usual Sweetness Level	Acidity	Min Alc.	Harvest
Kabinett	From the cabinet, the good stuff	Sharp, light bodied	*Trocken*	High to medium	7.0%	Mid-September
Spatlese	Late harvested	Crisp, light bodied	*Trocken* - *Halb trocken*	High	7.0%	Late September
Auslese	Selected late harvested	Crisp, med. body	*Halb trocken* - Sweet	Medium	7.0%	Late September
Beerenauslese	Berry selected, late harvested with botrytis	Rich, med. to full body	Honey, sweet, viscous	Medium	5.5%	Early October
Trocken Beerenauslese	Selected dry berry, late harvested with botrytis	Richest, med. to full body	Sweetest, honey, viscous	Medium	5.5%	Late October
Eiswein	Frozen selected berries, extremely late harvest	Bright, sharp, acid	Very sweet, no honey Intensely rich	Medium - High	5.5%	December January

Responding to recent pressure from consumers for greater quality in the 1980s, the Germans set maximum yields, in tons, from each vineyard at fairly low levels. As later harvested grapes weigh less, the low yield capacity encourages growers to delay harvesting, resulting in better wine.

At each sweetness level the best of the QmP wines are racy with acidity. The high acid content counters the residual sweetness. The result is an incredibly refreshing drink, both smooth and biting. You find it difficult to pull the glass from your lips. Johannes Selbach of Weingut Selbach-Oster described the German dessert wines: "TBA and BA are like serious classical music, but Eiswein is like jazz, jumping and dancing on your palate."

The AP number or *Amtliche Prüfungsnummer* translates as "official examination number" or "code verifying approval of this wine as a quality wine." An AP number is assigned to each QbA and QmP wine. The AP number must be present and identical on both the label and cork. When ordering German wines, check the AP numbers to verify the wine's origin. AP coding enables the informed to verify a wine's producer and geographic source in one simple numerical code of five blocks:

AP Nr. 2 555 222 8 94

The first of the five blocks identifies the testing center where the wine was evaluated and approved (2). The second block identifies the producer's village (555). The third block specifies the unique number assigned to each producer (222). The fourth block contains the application number for that producer for that year (8). The final block contains the wine's vintage year (94).

German Wine Exports

In 2006, Germany exported 2,700,141 hectoliters of wine; an increase of 250,000 hectoliters over 2005. This represented three-eighths of the amount harvested. Half the wine exported was *Qualitätswein* and two-thirds was white.

Great Britain imports more German wine, 923,655 hl, than any other country. The Netherlands are second with 409,391 hl imported, and the United States is third with 248,826 hl imported. Sweden is fourth, importing 185,216 from Germany. Sales in Great Britain are static, but U.S. and Netherlands sales are increasing by 15–20 percent annually (Germanwineusa.com, 2008).

Value of exports varies greatly by country of destination. Some trading partners buy from the bottom and some from the top. Large volumes of bulk wine from the variety Muller-Thurgau are exported to bottom-end purchasers. The British, Dutch, Danes, Swedes, and Poles, for example, buy from the bottom, each paying an average of less than 150E/hl.

Riesling, because of its high value, is disproportionately exported. More than 90 percent of U.S.-bound exports are Riesling. The Swiss buy quality, paying an average of 640E/hl. High-quality wines are shipped to China (400E/hl), Afghanistan (446E/hl), Austria (314E/hl), United States (333E/hl), and Japan (365E/hl). Substantial proportions of the wines these countries import are at the *Auslese* and higher quality level.

AUSTRIA

Central and western Austria is covered by the Alps. Eastern Austria, on the edge of the Hungarian plain, is wine country. Austria is, on the global scale, the ninth-largest wine producer. The principal city of Austria is Vienna (*Wein* in German), which translates simply to "wine" in English. More than 80 percent of Austrian wines are white and are mostly produced from Riesling, Gruner Veltliner, Gewürztraminer, Sylvaner, and Müller-Thurgau grapes.

Austrians are leaders in sustainable agricultural practices. Between 2005 and 2009 "organic wine cultivation virtually doubled from 1791 to 3,218 hectares. . . . In 2009, organic viticulture accounted for 7.8% of the total cultivated vineyard surface area in Austria."[29]

Austrian wine categories correspond to German categories. Austria has set minimum sugar limits on their *Prädikat Weins* (quality wines with attributes). Table 7.12 shows the precision of Austrian wine categories compared to those of other nations. Table 7.12, when compared with Table 7.11, shows the commonality of terms between German and Austrian wine laws.

Table 7.13 indicates the dominance of Gruner Veltliner, despite its declining acreage and value. Sauvignon Blanc acreage nearly tripled and Muskateller more than tripled. The national varieties of Welschriesling, Muller-Thurgau,

Table 7.12: Austrian Wine Quality Categories and Aging Recommendations

Prädikat Wein Categories	Min. Pre-Farm Brix	Aging Recommendation	Release after Harvest
Kabinett	17	1–2 years	March 1
Spatlese	19	2–3 years	March 1
Auslese	21	3–4 years	May 1
Beerenauslese	25	5–6 years	May 1
Ausbruch	27	8 years plus	May 1
Trockenbeerenauslese	30	8 years plus	May 1
Eiswein	22	10 years plus	May 1

Source: Austrian Wine Marketing Board, Vienna, Austria, http://www.austrianwine.com/our-wine/wine-law/overview-of-quality-levels/, accessed October 10, 2016.

Table 7.13: Austrian Wine Grape Area and Value for Varieties with More Than 500 Hectares

Austrian Variety	Hectares 1999	Hectares 2009	% Change	Total Value % 1999	Total Value % 2009
Gruner Veltliner	17479	13518	–22.7	36.0	29.4
Welschriesling	4323	3597	–16.8	8.9	7.8
Muller-Thurgau	3289	2102	–36.1	6.8	4.6
Chardonnay	2935	3426	9.9	6.5	7.4
Riesling	1643	1863	13.4	3.4	4.1
Sauvignon Blanc	314	933	196.7	0.6	2.0
Gemischter Satz	1371	807	–41.1	2.8	1.8
Neuburger	1094	652	–40.4	2.3	1.4
Fruhroter Veltliner	626	424	–32.2	1.3	0.9
Scheurebe	529	398	–24.8	1.1	0.9
Muskateller	143	527	267.7	0.3	1.1

Source: Statistics Austria, Vineyard Land Survey, 2009.

Gemischter Satz, Neuburger, Furhroter Veltliner, and Scheurebe each experienced double-digit percentage acreage loss. The international varieties (Weissburgunder, the Austrian name for Chardonnay, Riesling, and Sauvignon Blanc) were the only varieties to increase acreage and value.

Of the Austrian wine regions Weinbaugebeit is perhaps the most famous and produces the finest white wines. It is located on the east shore of the Neusiedlersee, a large shallow inland lake. The lake produces morning fogs in the autumn that promote the development of *Botrytis cinerea*. Additional air drying before pressing is permitted.

BULGARIA

The ancient Greeks prized the wines of Thrace, modern Bulgaria. Wine production here began at least 4,500 years ago. Bulgaria was the leader in modernizing the Eastern European wine industry after the fall of communism. There are five recognized wine regions in Bulgaria. These regions and principal wine-producing communities are identified in Figure 7.7. About 30 percent of Bulgaria's vineyards are located in the Danube river plain (northern plain) and another 30 percent are found on the Black Sea coast (eastern region). The southern region, the Thracian Valley, accounts for 35 percent of the vineyards. The Valley of the Roses (sub-Balkan region) and the Struma Valley (southwestern region) contain the remaining 5 percent of the nation's vineyards.

Bulgaria was a significant exporter of wine to the West in the years after World War II. During the Cold War Bulgarian wine was exported to Eastern

Europe and Russia. Once the Iron Curtain fell and Bulgaria joined the European Union, Bulgarian products began moving west once more. Most Bulgarian wines are made in the old style and lack fruitiness.

Several wine operations have received significant funding by the European Union. The Minkov Brothers winery is an example. The property has hundreds of acres of young vines and an immaculate new winery building complete with guest center.

ROMANIA

What is Romania was once ancient Thrace to the Greeks and Dacia to the Romans. Grape growing began here before 3000 BCE. Legend has it that Dionysus was born here. In the first century BCE, King Burebista ordered the eradication of the vine in this famed region to discourage raiding pillagers who descended on the country to steal the annual harvest. His order did not take. Less than 200 years later Romans conquered the region. Romania was occupied and named for the Roman legionnaires who settled among the conquered people. They had a natural strong interest in and liking for wine.

Between 1500 and 1700 much of Romania was under the political dominance of the Ottomans. Wine culture suffered as a result. After the Hapsburgs drove the Ottomans out, wine making revived. During the Communist era Romanian vintners focused on quantity, with all exports shipped to Russia. Vineyard acreage has declined in recent years as quality has increased. In 1999 there were 255,030 hectares of vineyards; by 2002 the area had dropped to 242,700 hectares. The Moldova and Murtenia regions each contained about a third of Romania's vineyards. The location of Romania's vineyards is shown in Figure 7.7.

Romania boasts four indigenous varieties: *Feteasca Neagra* (black maiden), *Feteasca Alba* (white maiden), *Feteasca Regala* (royal maiden), and *Tamaioasa Romaneasca* (Romanian incensed maiden).

HUNGARY

The Romans introduced vines to Hungary, but Hungarian wine history really begins in 896 CE when the ancestors of the Hungarians migrated over the Carpathian Mountains and onto the Hungarian plain. When they arrived the land was sparsely populated by agriculturalists practicing viticulture. St. Stephen, 100 years later, was responsible for spreading Christianity across Hungary. Wine-making monasteries flourished for the next 500 years, until Turks invaded. In 1686 the Hapsburgs reestablished Christian rule and the wine industry was reestablished. By 1880 some vineyards were under attack by phylloxera, but not the vineyards on the sandy soil of the plain. War and communism practically destroyed the Hungarian wine industry.

Communist leaders blocked efforts toward quality to maximize quantity. The industry has rebounded very well over the past 25 years, but has yet to regain the fame of 150 years ago.

The loess soils of the Hungarian plain provide the basis for a tremendously productive phylloxera-free red wine region. The limestone hills on the edge of the plain produce the world's oldest, rarest, and finest white wine, Tokay. Figure 7.9 shows the distribution of Hungarian wine production. Typical of the bold reds of the plain is the famous wine of Eger.

Eger

The fame of Eger wine began in 1552 when the defenders of the citadel drove off an invading Ottoman army. The Ottoman troops ran from the

Figure 7.9: Wine regions of Hungary.

battle after a rumor started that the Eger defenders were berserk from drinking bull's blood. In actuality their beards were matted with strong dense red wine given to them to increase their aggressiveness. Later that year the same Eger defenders were defeated and the town was sacked.

Eger wine was traditionally made from the Kadarka and Kekfrancos grapes, but during the 1800s other varieties were introduced to the blend. The Communist takeover of Hungary led to mass production of low-quality wines between 1960 and 1990. To increase production Zweigelt vines were planted. Bikaver growers have replaced the Zweigelt with Kadarka since the collapse of communism. They are also growing Cabernet Sauvignon and Merlot to produce the desired wine. In recent years, Bull's Blood of Eger has become a popular Halloween beverage.

Tokaj

Tokay (Tokai, Tokaj) is known as the wine of kings because of its unsurpassed sweetness, body, density of flavor, and longevity. Historically, only nobility could afford this liquid treasure. It is so treasured that it has its own line in the Hungarian national anthem.

Noble rot was first discovered in Tokaj, as a result of a Turkish invasion in 1630 that delayed harvesting until after noble rot had formed. Thereafter, they encouraged its growth. This discovery predates all other purposeful growing of the mold on Furmint, Harslevelu, and yellow Muscat varieties. By 1700 Tokay established the world's first vineyard classification system. In the late 1800s when the Hungarian Empire was at its peak, this wine was in great demand.

The producers of Tokay employ a special maturation process. They store the wine in caverns beneath the hills where a mold encrusts the bottles. The mold creates a seal over the cork, completely preventing oxidation. Located in these caverns are wines bottle aging for more than 300 years. Tokay's sweetness comes from a combination of fermenting ripe grapes, then adding mature overripe, noble rotted, partially dried grapes paste to start a second fermentation.

Tokay quality is measured by the amount of the dried grape mixture added to the fermenting wine. The standard unit of measure is a *Puttoyo,* which is about 35 liters in volume. Most Tokay is treated with between three and six *Puttoyos.* Tokay production is limited to 13,500 acres of land along a twenty-one-mile-long stretch of south-facing hills.

SUMMARY

Throughout Europe the wine vine grows. In most regions wine making goes back at least 2,000 years and in some places as far back as 10,000 years.

After centuries of fine-tuning variety to vineyard Europeans have learned how to make an amazing assortment of wines. Each geographically isolated region drifted into its own techniques and wine styles. Since Columbus, European vineyards have set the standards to which New World regions aspire.

European vineyards declined from more than 7 to less than 4 million hectares between 1961 and 2009. The most marginal lands were those eradicated. They could not compete with the rising quality of New World wines. Today, these 3 million hectares are largely abandoned and lie fallow.

Chapter 8

Wines of the Southern Hemisphere

The vine is not native to the Southern Hemisphere. Grapes were transplanted there by European explorers. The Portuguese took the vine to tropical Brazil in the 1500s where it did poorly. Spaniards took the vine (Mission variety) to Mexico by 1525, then to Peru before diffusing it to Argentina and Chile before 1600. The Dutch took the vine to South Africa in 1653 where it did well. The English took the vine to Australia in 1788 and New Zealand in 1811. Wine production quickly increased in Australia but not in New Zealand.

Transport costs, mercantilism, and the unpredictability and rarity of transport largely prevented the international trade of colonial Southern Hemisphere wines before powered vessels. Opening the Suez Canal in 1869 created a new trade route that saved each ship 9,000 miles. England, however, promoted wine production in its colonies (having no domestic industry), and reducing transit time in larger ships meant lower transport costs. Spain and Portugal forbade the colonial production of wine to support their domestic industries. French colonies, with the exception of North Africa, were outside the vine's habitat.

Low post–World War II transportation costs and tariff reductions made wine trade with Europe and North America economically feasible. While they all made wine, for centuries Southern Hemisphere countries have appeared on global shelves like the proverbial fifty-year-old overnight success. Since 1990, first Australia, then Chile, then South Africa, then New Zealand, and most recently Argentina each entered the international market with a signature variety. Australia developed the model for accessing the international market: make a quality product at low cost using the best agricultural, vinification, and marketing methods available at an industrial scale.

AUSTRALIA

Australian Wine History

When Captain Arthur Phillip established the convict settlement of Port Jackson (now Sydney) in 1788, one of his first acts was to plant the vines he brought with him from South Africa. The vines died fairly quickly. In 1831, James Busby, the father of Australian wine, planted a number of vines he brought from France. He reputedly used the same location as Captain Phillip. Several of Busby's varieties succeeded.

Immigrants from Germany and Italy between 1840 and 1870 formed the knowledge pool necessary to select prime locations, grow grapes, and make wine. For the most part these immigrants settled in South Australia in the hills around Adelaide. As was common in most wine regions before 1950, Australian wines were fortified to survive shipment.

Australia initiated a wine quality program in the 1970s. The Australians envisioned wine as a major component of their economic future and made it happen. The real boom in Australian wine exports occurred between 1992 and 2002 when exports expanded eightfold. Australia exported more than $2.1B in wine in 2008. The number of wineries peaked at 2,573 in 2014 and has since decreased.

Consumers see Australian wine as high value, high quality, and innovative. The Australians are the masters of blending wine in unconventional ways. In 2005 Australia declared for the screw cap as the national closure of choice.

Australian Wine Production

Australian wine production occurs on a large scale. The average winery holds 50 hectares of vines. Treasury Wine Estates farms 9,133 hectares and seven others have more than 1,000 hectares each. Two companies, Casella and Accolade, have facilities capable of processing more than 200,000 tonnes of grapes, and another four are capable of processing more than 100,000 tonnes.

Australia produced 1,670,000 tonnes of grapes on 132,436 hectares in 2015, very near the eight-year average of 1,700,000 tonnes. Unfortunately, this is too much wine. Poddar reported the following to the Australian Senate Standing Committee on Rural and Regional Affairs and Transportation Reference's Inquiry into the Australian Grape and Wine Industry:

> The industry enjoyed a prolonged boom from 1991 to 2007, during which exports soared, grape prices escalated, and the area planted to vines increased dramatically. . . . An expert review of the Australian wine industry by Centaurus Partners, commissioned by the Winemakers' Federation of Australia in 2913 reported that the industry had been hit

> by a "perfect storm" of events including: The global financial crisis in 2007; Falling demand for Australian wine in key markets, especially the US, UK, and Canada; A steady rising $A from 80 US cents in 2004 to virtual parity by 2010; . . . competition and choice from other exporters . . . ; "Damage to Brand Australia" by a range of factors from exports of low quality and brand proliferation.[1]

Australia has, since 2008, responded to overproduction by restructuring the extent and contents of its vineyards. Half the continent's vineyard area is in South Australia and a quarter in New South Wales. Victoria has 17 percent and Western Australia 5 percent of the area. Australia has undergone a restructuring of its vineyards in recent years. Most visible is the reduction in Chardonnay from 32,151 hectares in 2007 to 21,441 hectares in 2015. Shiraz declined by 3,500 hectares. During the same period Colombard, Riesling, Sémillon, Cabernet Sauvignon, and Merlot areas declined by 1,000 to 2,000 hectares. Pinot Noir and Pinot Gris are the only varieties to increase in area by more than 500 hectares. Australians are intensifying their reliance on major varieties. Vines classed as "Other White" dropped from 4,740 hectares to 841 and "Other Red" declined from 5,334 to 914 in the last eight years.

Australian Wine Regions

Australia has more than fifty defined wine regions at the substate level. A region's growers may specify the wine's source on the label. According to the Australian Wine and Brandy Corporation,

> A Geographical Indication (GI) is an official description of an Australian wine zone, region or sub-region. The use of Geographical Indications in Australia commenced in 1993. Australian Geographic Indication can be likened to the Appellation system used in Europe (e.g. Bordeaux, Burgundy) but is much less restrictive in terms of viticultural and winemaking practices. In fact the only restriction is that wine which carries the regional name must consist of a minimum of 85% of fruit from that region. This protects the integrity of the label and minimally safeguards the consumer.[2]

The single largest Australian GI is Southeastern Australia. This mega-GI encompasses the states of South Australia, New South Wales, and Victoria. Wines identified as Southeastern Australia may be blended from grapes grown hundreds of miles apart. These wines are blended for consistency of taste.

Each Australian state defines the next level of GIs. Within states GIs are recognized and protected named quality regions. Australians grow most of their grapes along the southern edge of the continent where Mediterranean-type climates prevail. Individual GIs vary greatly in elevation, degree days, precipitation, and soil, resulting in terroirs appropriate for nearly all grape varieties.

Total Australian vineyard area is on the decline, dropping from 172,676 hectares in 2007 to 148,509 hectares in 2013. Marginal vineyards, planted during the boom years, are the obvious targets for eradication. All Australian states produce wine, although the Northern Territory quantity is negligible. In 2013 South Australia devoted 71,310 hectares to vines, almost double New South Wales' 39,097 hectares. Victoria, which had the largest vineyard extent in the 1880s, now is third in area with 25,409 hectares. Western Australia has 13,431 hectares of vines, all quite near the Indian Ocean.

New South Wales

New South Wales is Australia's most populous state. About 30 percent of Australian production originates here. The Mudgee region specializes in red wine production. In the interior the Mudgee River flows southeast parallel to the mountains. The Orange district rests at 3,000 feet above sea level and produces white wines of distinction. Riverina, in the interior, produces everyday wines.

Hunter was the first successfully pioneered area outside the original Botany Bay colony. It is about 100 miles north of Sydney. It is far enough north and inland to provide a hospitable climate for grape production. Hunter is surrounded by mountains while the valley floor undulates. It is hot and humid, making rot a serious problem. Regular daily afternoon cloudiness keeps Hunter Valley from overheating. February, the month before harvest, is the wettest month. Bloated berries and rot are annual problems. The Hunter Valley is well known for its Semillon, Shiraz, and Chardonnay.

South Australia

Vineyards in South Australia are between Gulf St. Vincent to the west and New South Wales to the east. South Australia produces half the nation's grapes. This large state includes the Barossa, Fleurieu, and Limestone Coast areas. The most serious problem facing South Australian vineyards is suburban sprawl from Adelaide.

The Barossa Valley is located in South Australia east of Adelaide. Shiploads of German immigrants settled here in the 1840s and the vine soon took root. According to Australian wine historian John Beeston, "In 1843, a Silesian family, the Aldenhovens . . . are believed to have planted grapes. By 1847, local wines were being exhibited. . . . In the same year, Johann Gramp planted near Jacob's creek and Joseph Gilbert also began his vineyard at Pewsey vale."[3] Between 1850 and 1870 settlement and grape growing intensified.

Barossa Pearl, a sparkling white wine, sold millions of bottles in the 1960s and brought the valley its initial fame. The Australian national drive for quality wines led the Barossa to reinvent itself in the 1970s. Because of its early success, Barossa was the training ground for many current winemakers and grape growers. Australia's largest wineries are here. Barossa trainees spread outward to the adjacent regions of Coonawarra and Padthaway where austere wines are produced. The Eden Valley subregion contains the Hill of Grace Shiraz vineyard that has vines more than 140 years old.

Fleurieu is south of Adelaide and includes Kangaroo Island and McLaren Vale. Both produce a range of varieties. The aptly named Limestone Coast is quite cool and produces quality wines. The Coonawarra area yields excellent Cabernet Sauvignon, and Padthaway is known for its whites. The Clare Valley to the north of Barossa is famed for its dry Riesling.

Victoria

Victoria is the smallest Australian state. Victoria's numerous small wineries produce about 15 percent of Australian wine. Victorian subregions include Goulburn Valley, Rutherglen, Geelong, Mornington Peninsula, and Yarra Valley. Rutherglen is among the hottest Australian wine regions. The region is known for its sweet fortified wines. Goulburn Valley produces full-bodied Shiraz. The soils of Heathcote are different from the adjacent Goulburn Valley and yield elegant reds. Port Phillip Bay separates Geelong and Mornington Peninsula. The cool maritime climate supports the Burgundian varieties, Pinot Noir and Chardonnay. Lindeman's is located in the Murray River region, which yields value-driven wines.

Western Australia

This region produces highly aromatic Rieslings. Perth, the region's capital, is the world's windiest city, yet coastal areas of Western Australia are temperate and river valleys fertile. In Australia's southwest corner the Swan and Margaret river valleys provide homes for a number of varieties. Wine production began along the Swan River early. Olive Farm was the site of the region's first winery in 1829, followed by Houghton in 1836. It is the hottest wine region in Australia. The Margaret River received its first vines in 1967. The growing area in 2012 was only five hectares divided among 215 wineries. Great Southern is the largest non-state-level wine region in the country. It encompasses a 20,000-kilometer rectangle, 23 of which hold vines. With 48 wineries, the average size is 120 acres. The first vines were planted here in 1859 with commercial production beginning in the 1930s.

Australian Wine Varieties

Australians produce more Shiraz than any other variety. In 2004 the 36,500 hectares of Shiraz yielded 438,000 tons of grapes. There were 28,400 bearing hectares of Cabernet Sauvignon that yielded 305,400 tons of grapes. The Chardonnay yield was 312,700 tons from 22,500 hectares.

According to the Australian Wine and Brandy Corporation, the number of wineries in Australia reached 2,146 in 2007, up from 1,625 in 2003. Australian wine production first exceeded one billion tonnes in 2000. The 2007 crush produced 955 million tonnes, down from the 2003 level of 1,038 million tonnes, but a much higher percentage of the 2007 production was exported than in 2003; 785 million in 2007 versus 524 million in 2003. By 2013 the volume increased to 1,748 million tonnes yielding 1,245 million liters.

Between 2007 and 2013 yields fell for every tracked red variety. Three white grapes increased yield. Pinot Gris tonnage doubled, Sauvignon Blanc saw a 50 percent increase, and Colombard saw a 10 percent increase. During the period 1995–1999 Cabernet Sauvignon produced more wine than any other variety in Australia. Shiraz exceeded Cabernet Sauvignon yields in 2000. Since then Shiraz and Chardonnay have traded first place multiple times. In 2013, Shiraz yielded 432,000 tonnes and Chardonnay 397,000 tonnes. Cabernet Sauvignon production varies greatly; it jumped from 207,000 to 250,000 tonnes between the 2012 and 2013 vintages.

Australian Wine Industry

The Australians provide the modern model for developing a national wine industry dedicated to ever-increasing quality and competition. The most significant success factor has been the Australian Wine and Brandy Corporation (AWBC). The other Southern Hemisphere countries adopted a similar agency and programs after observing the AWBC's success. The AWBC provides technical, marketing, and exporting support for the Australian wine industry. It is the source for wine and grape growing information for growers, winemakers, and the public. It assists growers and winemakers with promotional support and technical assistance.

The boom years for Australia are over. Competition from other Southern Hemisphere countries eroded the Australian hold on the market. Exports over the past seven years averaged 722,778,000 liters. In 2014 exports were 683,935,000 liters, the smallest value in the series. A third of Australian exports go to the United Kingdom and a quarter comes to the United States. Canada receives 8 percent of Australian exports. China, Germany, and New Zealand each receive about 5 percent of the exports.

NEW ZEALAND

New Zealand was discovered by the Maori people about 1000 CE. Eight hundred years later it was rediscovered by Europeans. Reverend Samuel Marsden, an early colonist, attempted wine production in the northernmost part of North Island in 1819, and failed. The initial planting on North Island failed because the vines were planted on flat, poorly drained rich soils. The terroir of that vicinity encouraged vine-killing mold and rot.

Vines were successfully planted in the Hawkes Bay region at Mission Estate in 1851, which is still in operation. Auckland saw its first planting in 1916, and Gisborne in 1921. Marlborough, acclaimed for its Sauvignon Blancs, experienced its first vine plantings in 1973.

Following on the success of the Australian wine export program, New Zealand began its own quality export program by placing all aspects of the wine industry into a single government agency. Consistency of message and lack of regulatory competition helped New Zealanders make the splash they did on the international market.

Wine represents about 2 percent of New Zealand's exports, generating $1,424,000 NZ in 2015. Three-fourths of the exported wine goes to the United States, Australia, and the United Kingdom. In 1995 New Zealand exported about eight million liters of wine. Ten years later in 2005 exports exceeded 209 million liters. Sauvignon Blanc accounts for 86.5 percent of the export value. In 2015, there were 673 wineries and 762 growers, operating on 35,859 hectares. Six of 530 wineries each sold over four million liters in 2006. The number of four million liter producing wineries jumped to 17 in 2015.[4]

The vineyard changes in varietal plantings reflect changing consumer preferences. Total 2015 Sauvignon Blanc production, 216,078 tonnes, exceeds Chardonnay and Pinot Noir yields by a factor of eight. All other varieties yielded less than 20,000 tonnes in 2015. Muller-Thurgau vineyard area decreased from 712 hectares in 1996 to 117 hectares in 2006 to 2 hectares in 2015. Semillon is also on the decrease. The recent popularity of Pinot Gris is mirrored by the increase in acreage from 21 hectares in 1996 to 2,456 in 2015. To accommodate the growth the number of wineries grew from 530 in 2006 to 673 in 2015 (see Table 8.1).

Wine Regions of New Zealand

The wine regions of New Zealand were largely pioneered from north to south. There are ten recognized wine regions. Six are on North Island: Northland, Waikato, Auckland, Gisborne, Hawkes Bay, and Wellington. Four are on South Island: Nelson, Canterbury, Otago, and Marlborough. The regions and the locations of vineyards can be seen in Figure 8.1.

North Island

All North Island wine regions are on the east coast. The best regions on North Island, Gisborne and Hawke's Bay, are on the eastern slopes of rich volcanic alluvial river valleys.

Table 8.1: New Zealand Vineyard Area and Wine Production by Select Variety, 1996, 2006, and 2015 Compared

Variety	Production area (hectares)			Production (tonnes)		
	1996	2006	2015	1996	2006	2015
Sauvignon Blanc	1,140	8,860	20,266	12,354	96,686	216,078
Chardonnay	1,466	3,779	3,363	13,870	26,944	27,095
Riesling	276	853	777	2,877	6,745	4,535
Muller-Thurgau	712	116	2	13,838	1,573	0
Semillon	186	229	80	2,342	2,664	425
Pinot Gris	21	762	2,456	102	3,675	19,707
Cabernet Sauvignon	499	531	300	4,169	2,659	1,376
Pinot Noir	431	4,063	5,564	4,617	22,062	25,763
Merlot	302	1,420	1,320	2,857	11,206	9,397
Total	6,610	22,616	35,859	73,340	182,885	326,000

Source: HortResearch. *Fresh Facts: New Zealand Horticulture*. Auckland, NZ: Horticulture and Food Research Institute of New Zealand Limited. 2006. http://www.hortresearch.co.nz/files/aboutus/factsandfigs/ff2006.pdf, p. 5.

Auckland produces Bordeaux varieties and some Chardonnay. Waikato is the smallest wine region with only twenty-five hectares divided among a dozen wineries. Hawke's Bay, New Zealand's Merlot capital, is the sunniest region and produces other Bordeaux varieties as well. Wellington, across the strait from Marlborough, produces Riesling and Pinot Noir.

South Island

South Island is well suited to cool-climate grapes. Pinot Noir, Chardonnay, and Riesling are common outside Marlborough. Marlborough is considered *the* region for Sauvignon Blanc. Throughout South Island vineyard acreage is increasing.

At 45 degrees south latitude, Otago is the world's southernmost wine region. It is also one of the newest with 1987 as its first vintage. Pinot Noir thrives in this interior valley where the summer days are long, hot, dry, and sunny.

The Nelson region is the oldest on South Island with its first winery opening in 1868. Seifried Estate produced the region's first modern wines in 1976. Chardonnay is most widely planted with lesser amounts of Sauvignon Blanc, Pinot Noir, and Riesling.

Vines were first planted in Canterbury, which surrounds Christchurch, in 1977. Pinot Noir and Chardonnay are the most common varieties. Waipara, in the region's north, is achieving international acclaim for its Riesling.

Figure 8.1: Wine regions of New Zealand.

Table 8.2: New Zealand Vineyard Area by Region in Hectares (1 hectare = 2.5 acres; 1 ton = 1,000 kg = 2,200 lbs)

Region	Production Area (hectares)			Production (tonnes)		
	1996	**2006**	**2015**	**1996**	**2006**	**2015**
Auckland/Northland	193	504	422	1,610	1,553	1,027
Waikato/Bay of Plenty	117	150	24	761	261	nd
Gisborne	1,165	1,913	1,914	22,330	18,049	17,280
Hawke's Bay	1,794	4,346	4,773	21,172	33,287	36,057
Waipapa	174	777	1,006	1,072	3,008	3,559
Nelson	97	695	1,139	761	5,623	3,559
Marlborough	2,155	11,488	23,203	24,192	113,436	233,182
Waipara	213	925	1,488	1,059	3,051	5,395
Otago	92	1,253	1,951	376	4,612	8,951
Not assigned to region	610	565	0	7	5	159
Total	6,610	22,616	35,859	73,340	182,885	312,387

Source: HortResearch. *Fresh Facts: New Zealand Horticulture.*

Montana Vineyards, established 1971, was the first commercial winery in the Marlborough district. In an amazing growth spurt 11,500 hectares were planted by 2007. The spurt was spurred by the flavorful Sauvignon Blanc produced there. Situated in the Wairau River valley, the Marlborough district is protected on the east and west by mountains. The soil is largely volcanic alluvium.

The period between 1996 and 2006 saw significant vineyard expansion in most regions. As the vineyard area increased, the yield per hectare decreased through quality control efforts. The last decade experienced substantial vineyard expansion and increasing yields. The figures presented in Table 8.2 show Marlborough doubling its yield between 2006 and 2015. The importance of Marlborough increased from 32 percent of yield in 1996 to 62 percent in 2006 and 75 percent in 2015.

PERU

Wine making in Peru began in the first half of the sixteenth century shortly after Pizarro defeated the Incan Empire and subjugated highland South America. The honor of being the first to grow grapes and make wine in South America goes to Bartolomeu de Terrazas in 1540. His Mission grape vineyard was near Cuzco.

Figure 8.2: Wine regions of South America.

According to geographer Harm De Blij, Spanish explorations quickly discovered: "The desert coastal zone of Peru is interrupted by a string of some 40 oases sustained by stream water from the interior mountains. The oases have deeply drained alluvial soils of volcanic origin which resembled the terroir of Spain. De Carbantes planted vines in one of them, the oasis of Ica."[5] Vine production remains centered in the province of Ica on the Pacific coast.

During the colonial era vine growing, under the watchful eyes of missionaries, was conducted by Christianized and enslaved Indians wherever Andean runoff met the coastal plain. Most of these runoff-fed rivers never make it across the narrow desert coastal plain to the Pacific Ocean. Their waters now, as in Incan times, disappear into a network of irrigation channels.

Today, Peru grows about 6,000 acres of Muscat of Alexandria mostly in southern coastal river valleys in Ica Province. From these grapes, Peruvians distill Pisco. Pisco is Peruvian brandy and is named for the Pisco River valley in Peru. After distillation, Pisco retains the Muscat's floral perfume. Nearly all Peruvian wineries have adjacent stills for making Pisco.

CHILE

Like southern Peru, northern Chile is punctuated by riparian oases flowing out of the Andes. River valleys with Mediterranean climates in central Chile, near the capital city of Santiago, proved best for growing grapes. It is recorded that Francisco de Aguirre planted vines in 1550 at Copiapo and Diego de Caceres planted vines near Santiago a few years later. Both regions still produce wine. The Mission (here named Pais) grape made plentiful wine, which made the colonists happy. Unfortunately, it hurt Spanish wineries' ability to sell wine in the colonies. In 1595, in one of the earliest examples of mercantilism, "the king issued a decree to restrict planting . . . and further expansion. But Chile lay far away from the source of the edict, and . . . failed to halt the vine's diffusion."[6]

Chile has the most extensive and renowned vineyards in South America. The reputation for quality of Chilean grapes dates back to the 1820s when Chile first became independent from Spain. The Chilean wine industry actually benefited from phylloxera! In 1851, Don Silvestre Ochagavia Echazarreta, a wealthy Basque landowner, hired a French winemaker named Monsieur Bertrand. Bertrand brought with him more than 200 vine cuttings of the Bordeaux standards Cabernet Sauvignon, Merlot, Malbec, Semillon, Sauvignon Blanc, and Riesling, and all grew exceptionally well. The arrival of these cuttings in Chile predates the arrival of phylloxera in Europe by fifteen years.

The pre-phylloxera transit was incredibly fortuitous, for it saved several varieties from extinction. Today, Chile is the only country in the world totally untouched by the dual plagues of phylloxera and downy mildew. Chile is the only major wine-producing country in the world that does not have to graft its vines onto North American root stock. Thanks to strict government controls and quarantines dealing with plant material coming into the country, Chile may remain phylloxera free for many years to come.

Once phylloxera devastated Europe, Chilean exports soared until the 1930s when greed and politics practically eliminated exports. Greed, because wineries shifted their focus from quality to quantity. Between 1938 and 1974 the Chilean government "capped production at 60 liters (15.8 gallons) per capita and impos[ed] heavy fines on anyone caught planting vines."[7]

Modern Chilean wine history began in 1990 when exportation recommenced. In the early 1990s the Chilean wine industry was in trouble. Vineyard area fell to 53,000 hectares in 1994 with production less than 400 million liters. An influx of French immigrants and money helped reinvent Chilean wines. In 2005, there were 116,793 hectares planted in wine grapes and in 2015 there were 130,000 hectares. In 2014, Chile became the fourth-largest wine exporter. Total 2015 wine production exceeded 1.18 billion liters with about 0.8 billion liters exported.

Despite its narrowness, Chile has four distinct north-south zones: the coast, the coastal range, *Entre Cordilleras* (between the mountains), and the Andes. On the coast, fog forms routinely, moves inland, and blankets the grapes. This reduces direct solar radiation and slows ripening. The *Entre Cordilleras*, home to the majority of vineyards, has very diverse terrain and some areas of limestone. The vineyards are planted at the upper ends of the valleys where the alluvial volcanic soil is stony and deep. Fruit, flowers, and vegetable crops grow on the valley floors. Vines planted in the Andes foothills commonly experience diurnal temperature shifts of 40°F. The cool nights are coupled with daytime katabatic winds, which limits the need for antifungal treatments.

More than 75 percent of Chilean plantings are red grape varieties. Cabernet Sauvignon covers the largest area with 40,789 hectares in 2005. The Cabernet Sauvignon area has changed little, adding only 48 hectares by 2015. Merlot covered 13,368 hectares in 2005, dropping to 11,432 in 2015. Carmenere plantings, in 2015, covered 10,040 hectares and Syrah 7,393 hectares. For the whites, Sauvignon Blanc and Chardonnay have similar extents, 8,697 and 8,548 hectares, respectively, in 2005. Areas for both grapes grew significantly in the following decade, reaching 13,992 hectares and 10,970 hectares, respectively. The Muscat of Alexandria plantings cover about 6,000 hectares and are concentrated in the more northerly regions. Most of the Muscat is distilled into Chilean Pisco.[8]

There are fourteen designated Chilean wine regions. Many are listed in Figure 8.2. Each region is centered on a river valley ranging between 32 and 39 degrees south latitude. The regions can be grouped into three zones: the

North, Center, and South. The North includes the river valleys of Elqui, Limari, Choapa, Aconcagua, Casablanca, and San Antonio where Mission and Muscat of Alexandria provide the base for Pisco. The Southern regions include, from south to north, Malleco, Bio-Bio, and Itata river valleys. Southern regions are beginning to produce Pinot Noir and Chardonnay, replacing Muscat and Mission.

The core of Chilean wine is the *Entre Cordilleras* and includes, from north to south, the Maipo, Rapel, Curico, and Maule regions. These four regions produce more than 80 percent of Chilean wine. Concha y Toro, Santa Rita, and Vina San Pedro Tarapaca, the nation's three largest companies, are headquartered here. Master Sommelier Evan Goldstein, author of *Wines of South America,* describes these companies:

> These three companies are formidable. They own wineries of all shapes and sizes in every wine region of Chile. This consolidated approach was very important to Chile in the early 1990s, when offering quality wine at a good price distinguished Chile from many other exporters.[9]

Maipo is the chief Chilean wine region. Maipo Cabernet Sauvignon is among the nation's best. Production is evenly split between reds and whites. The Rapel region produces Bordeaux-style whites and reds. Maule is cooler and produces Merlot, Sauvignon Blanc, and lower-end Cabernet Sauvignon. Maule is home to many centenarian Mission Carignan, Cabernet Franc, and Cinsault vines.

ARGENTINA

Argentina is a turbulent country that has had several dictators and revolutions over the past 100 years. Their wine industry suffered through these disruptive events and the resulting financial crisis following each event. Fluctuation in transportation, price, costs, labor availability, and government debt made getting wine to market consistently very difficult. The last coup d'etat occurred in 1976 when the military took control. In 2002, for example, the peso dropped from parity with the U.S. dollar to 3.15:1 in four months. Former president Cristina Fernandez de Kirchner (2007–2015) was charged in May 2016 with misappropriating about $5.2 billion in government funds. Despite the corruption, political stability seems likely. The Argentine wine industry is on track to remain a steady high-volume wine exporter.

The Andean foothills are home to most Argentine vineyards. Receiving little precipitation, the vineyards in the foothills are irrigated by Andean meltwater runoff.

There are three primary and five lesser wine regions in Argentina. Mendoza is the largest with about 150,000 hectares and produces two-thirds of Argentine wines. San Juan is second largest with 47,000 hectares, and La Rioja, the oldest, has 7,000 hectares. The lesser regions are Catamarca (2,540 hectares), Salta (2,500 hectares), Rio Negro (1,733 hectares), La Pampa (211 hectares), and Tucuman (78 hectares). Refer to Figure 8.2.

The first vineyards were planted in Mendoza in the mid-1500s. Goldstein describes the difficulty of their situation: "transport on the backs of mules to the urban centers was expensive and slow, and much of the wine spoiled in transit."[10] Conditions changed radically in 1885 when a rail line connecting Mendoza with thirsty Buenos Aires was completed.

Salta, to the north, supports a diverse range of crops. Deep well-watered valleys produce bananas while the higher and dryer slopes support Malbec and Torrontes with increasing acreage devoted to Cabernet Sauvignon. Catamarca, opposite the Andes from the Atacama Desert, is largely planted in Mission grapes and little is exported. La Rioja vineyards are a mile high with summer temperatures commonly exceeding 100°F. Limited water access prevents additional La Rioja plantings. San Juan is home to twenty subregions with the single area Tulum producing about 90 percent of the region's harvest.

Twenty years ago the most widely planted grape in Argentina was the Pedro Gimenez, but no longer. Malbec and Torrontes Riojano replaced the Pedro Gimenez vines. Malbec is the traditional red of Argentina, producing 348,618 tonnes in 2014. Torrontes Riojano is the traditional white. Riojano and its two siblings Sanjuanino and Mendocino each resulted from separate crossings of Muscat of Alexandria and the Mission grape. Pedro Gimenez is still the most widely grown white grape with 162,683 tonnes produced in 2014. Torrontes Riojano is second with 139,443 tonnes and Chardonnay a distant third with 69,721 tonnes produced. Among red grapes, Malbec produced 33 percent, Bonarda 23 percent, Shiraz 14 percent, and Cabernet Sauvignon 12 percent of the 2014 harvest.

Argentine wines were rarely exported outside South America before 1995. What wine Argentines did not consume went to neighboring Brazil. Based on Chilean success, individual Argentine wineries began exploring new markets in the mid-1990s. Wines of Argentina became the national brand identifier in 1998. Shortly thereafter Argentina's *Instituto Nacional de Vitivinicultura* enacted rules governing the use of appellation designations. Promotions, tastings, and lectures hit the United States and United Kingdom in 1999. The advertising campaign worked very well. Malbec infatuated the U.S. market. Malbec sales grew rapidly through 2009 when a poor growing season dropped yields 22 percent from the previous year. That year Argentine exports plummeted from 414 million liters to 283 million liters, with subsequent improved harvests exports rebounding to 360 million liters in 2013. Total Argentine yield was 2,870,000 tonnes in 2015, with 265.6 million liters exported—half of which went to the United States. Argentine wine will remain a standard on U.S. and U.K. shelves for years to come.[11]

BRAZIL

The first, and unsuccessful, attempt to grow grapes in Brazil occurred in 1532. Other attempts over the next 200 years also failed. The vineyards failed because the vines would not grow in the tropical sugar cane growing region then occupied by the Portuguese. When Rio de Janeiro and São Paulo were settled in the early 1700s, the Portuguese found locations where grapes would grow. Following the tenets of mercantilism, the Portuguese king banned Brazilian wine production in 1789 under pressure from growers at home. The measure was repealed in 1808 when the Portuguese royal family, fleeing Napoleon, moved to Brazil and wanted to drink wine.

Production remained limited until Italian immigrants came to southernmost Brazil. The earliest documentation of wine production was in 1817, when Manoel Macedo began producing approximately 45 barrels annually. By the late 1800s immigrants transformed the area around Porto Alegre into a model of Italy. The Italian wine-making tradition was augmented in the 1920s by wine-making German immigrants.

The immigrants from Italy and Germany spread throughout the southern states of Rio Grande du Sul and Santa Catarina. These two provinces are home to five of Brazil's six wine-producing regions. The average vineyard holding is about five acres, hand harvested and family owned. Rio Grande du Sul contains about 40,000 of Brazil's 78,000 hectares of vineyards. Rio Grande du Sul maintains Brazil's highest standard of living.

Each of the five Rio Grande du Sul growing areas has temperate climates and altitudes ranging from 700 feet in Campanha to 3,800 feet in Planalto Catarinese. The Campanha and adjacent Serra Sudeste regions are inland on the border with Uruguay and Argentina. The soils derive from granite and limestone. Rolling hills and wide valleys cover the landscape where a wide range of varieties grow, including Gewürztraminer, Tannat, Touriga Nacional, and Tempranillo alongside the noble varieties of France.

The Serra Gaucha wine region is Brazil's largest and most important as it produces 85 percent of Brazil's crop. The soils are of basaltic origin. Specialty varieties include Tannat, Ancellota, Malvasia, and Prosecco. Within this region are Brazil's only two protected regions, the Vale dos Vinhedos Denomination of Origin and Geographic Indication Pinto Bandeira. The Vale dos Vinhedos (Valley of the Vineyards) is about 100 miles inland from the city of Porto Alegre. The Vale dos Vinhedos covers 28 square miles and Pinto Bandeira 31.5 square miles. Vale dos Vinhedos wines must contain 60 percent Merlot for reds and 60 percent Chardonnay for whites. Permissible reds are Cabernet Sauvignon, Cabernet Franc, and Tannat. Italian Riesling, Chardonnay, and Pinot Noir may be blended to make sparkling wines. Pinto Bandeira produces sparkling wine from blends of Chardonnay, Pinot Noir, Italic Riesling, and Viognier. The Planalto Catarinense sits on the basaltic Santa Catarina plateau. The high altitudes result in colder temperatures, longer hang times, and later harvest.

The Vale do Sao Francisco, in contrast, is Brazil's hottest wine region. It is also among the world's most unique. Temperatures over 90°F are coupled with high humidity. Located between eight and nine degrees south latitude, it is nearer the equator than any other commercially viable wine region. Because of its equatorial location there are two harvests per year. After harvest the vines are placed in an induced resting period of 30 to 60 days between 120- to 130-day growing periods. The soils are sedimentary river deposits. The near desert conditions make this the only Brazilian wine region requiring irrigation. In these conditions, heat-tolerant varieties such as Syrah, Tempranillo, Malvasia Bianca, Muscat, and Chenin Blanc grow.

In 2012 Brazil exported $6.8 million in wine, an increase over 2009 of $4.5 million. Nearly a quarter of Brazil's wine exports went to Russia. The volume of wine exported was more than 6.1 million liters.[12] Not much in the total world economy, but the quantity and quality are increasing rapidly.

SOUTH AFRICA

The first vines in South Africa were planted in the 1650s by Dutch colonist Jan van Riebeeck who described his first harvest in 1659 this way, "Today, praise be to God, wine was made for the first time from Cape Grapes."[13] From then on passing ships stocked up on wine on their way to and from Europe and the Orient. The first Cape Town colony governor

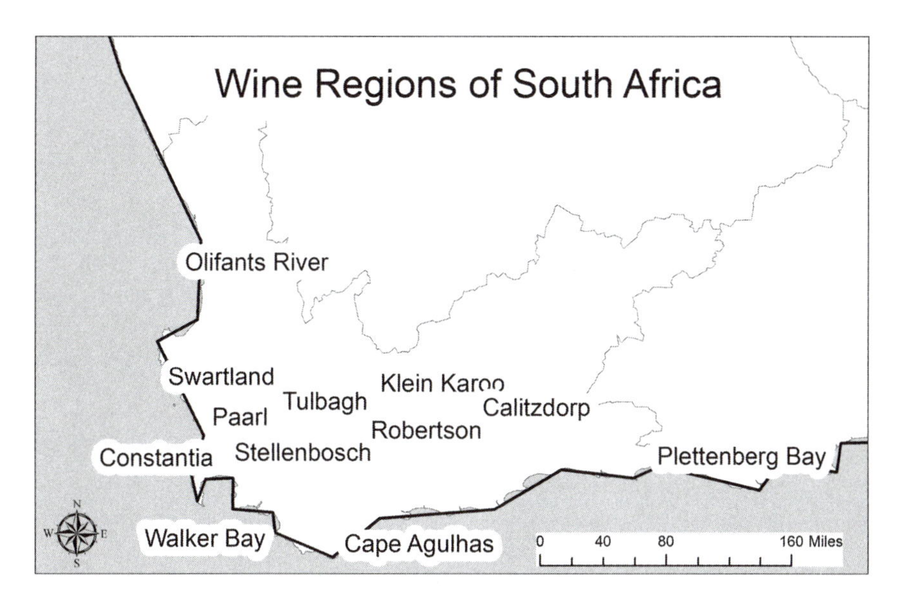

Figure 8.3: Wine regions of South Africa.

established Stellenbosch in 1679 and founded the Constantia estate with the planting of 100,000 vines. After falling into disrepair, Constantia was resurrected in 1788 by Hendrik Cloete. His descendants produced a wine that drew international acclaim. Made from the Orange Muscat grape, it was an exquisite, smoky, sweet, and racy dessert wine. It quickly became a favorite of European elites. Phylloxera bankrupted the Cloetes and the estate was abandoned a second time. In 1980 the Klein Constantia estate was revived and production of the exquisite dessert wine resumed.

During the first half of the nineteenth century South Africa transited from a Dutch to British colony. The colony became a significant source for British wine imports during the Napoleonic era when war meant a British boycott of French wine. South African wine continued to partially satisfy the British demand until phylloxera arrived, sometime before 1870. Waiting for young grafted vines to mature, production and exports diminished. Export terminations during the Boer War and again during World War I led cellars to overflow. In 1918 the *Ko-operatieve Wijnbouwers Vereniging van Zuid-Afrika* (KWV) was created to deal with the surplus wine. KWV set quotas and prices for grapes from existing vineyards. The KWV also prevented exploration of new regions until 1994. "No longer restricted in the choice of where to grow, what to plant, and how much to produce, quality minded vintners began exploring new growing areas."[14]

White rule in South Africa led to a near global boycott of their products. As the international outcry grew, export sales of South African wine plummeted. When majority rule began in 1995 South African wine became socially acceptable and exports grew. About then Wines of South Africa (WOSA) was established to promote quality and sustainability at home and exports abroad. Within WOSA continuous pressure to make better-quality wines is coupled with strong emphasis on protecting their unique biome. At home, WOSA has taken significant steps toward racially integrating the wine industry. Internationally, WOSA produces some of the most plentiful and attractive promotional materials.

Perhaps the most interesting promotion is the Sustainable Wine South Africa program. The program requires growers to produce grapes in an environmentally and socially sustainable way. Growers and wineries that meet the requirements, and about 85 percent now do, can place a certification seal on their wines. Consumers can visit www.swsa.co.za, enter the code numbers from the seal, and get verification of origin information.

Between 2012 and 2014 South Africa production averaged 920 million liters on 99,500 hectares. About two-thirds of the area is planted in white grapes. South African exports are on a steady upswing. South Africa is now deeply integrated in the global wine trade. Seventeen countries received more than one million liters in 2013. Exports grew from 236 million liters in 2003 to 309 million liters in 2007 and climbed to 414 million liters in 2014. The Germans and British imported 77 and 71 million liters respectively in 2013.

Russia and France each received more than 33 million liters. Relatively cheap land and labor combine to make wine expansion economically viable.

Climatically, South Africa ranges from Mediterranean types along the coast with savannah fading into desert to the north. Mountains introduce intricate meso-climatic patterns to the overall trend. In the area surrounding Cape Town there is adequate precipitation to support grapes. Inland and northward, vineyards must be irrigated.

Wine Regions of South Africa

South Africa's vineyards cluster around Cape Town. Experimental plantings further along the coast in both directions developed into the Northern Cape, Olifants River Region, Walker Bay, Cape Agulhas, and Plettenberg Bay. Interior vineyards experience desert conditions. In the interior the vineyards of Breedes Valley grow on lime-rich alluvial fans. The Klein Karoo valley is sheltered by the Swartberg mountains to the north and the coastal Langeberg Outeniqua Mountain range.

There are nine recognized wine regions in South Africa: Northern Cape, Olifants River, Swartland, Klein Karoo, Paarl, Robertson, Stellenbosch, Worcester, and Breedekloof. Figure 8.3 shows their locations.

The town of Stellenbosch, thirty-seven miles east of Cape Town, was founded by Simon van der Stel in 1679. The Stellenbosch wine region is South Africa's largest with more than 16,000 hectares planted. The 2013 yield was 151,878 tonnes. The four most widely planted varieties, in order, were Sauvignon Blanc (29,000 tonnes), Cabernet Sauvignon (27,000 tonnes), Syrah (22,000 tonnes), and Merlot (21,000 tonnes).

Paarl was settled by the Dutch in 1688. The town is situated in the narrow Berg River valley with Paarl Mountain to the west and the escarpment to the east. Huguenots arrived in 1689 and planted vines on decomposed granite alluvium at the base of Mount Paarl. The city sits on the eastern slopes of Mount Paarl. Vineyards surround the city in the other three directions. In 2015, Paarl, with slightly less than 16,000 hectares, was the second largest South African wine region. Twenty percent of the 2013 harvest of 162,000 tonnes was Chenin Blanc. The Cabernet Sauvignon harvest was 23,000 tonnes.

The Northern Cape region had 4,659 hectares of wine grapes in 2013. Grape production is only possible in this portion of the Kalahari Desert because the Orange River offers irrigation water. Irrigation projects are expensive, so small wineries are nonexistent. Of the 2013 yield, 55.7 percent was Colombard (73,000 tonnes) and 22.6 percent was Chenin Blanc (30,000 tonnes). Red wine yield was a paltry 3,800 tonnes.

The Klein Karoo valley is the largest wine region in South Africa, but it has the smallest vineyard area, 2,660 hectares. The region's climate is similar to that of the Douro Valley of Portugal. Diurnal temperatures changes are great. Hot days alternate with cool nighttime ocean breezes. In the Klein

Karoo, Colombard and Chenin Blanc together yielded 28,000 tonnes of the region's 44,000 tonnes of grapes in 2013. No other variety yielded more than 2,000 tonnes. Fortified wines are the region's specialty.

Bush-trained vines of Chenin Blanc and Syrah dominate desert-dry Swartland vineyards. Moorreesburg is near the center of the Swartland region. With no reliable source of water, irrigation is uncommon. The undulating hills grow vast quantities of grain in addition to producing 144,000 tonnes of grapes, 20 percent of which was Chenin Blanc. Syrah, Cabernet Sauvignon, and Pinotage each yielded more than 15,000 tonnes.

South Africa's citrus and 10 percent of its wine industry are centered on the Olifants River valley. The Olifants valley is reminiscent of Peruvian river valleys. The Olifants River provides essential irrigation as precipitation averages less than seven inches annually. Cooler conditions prevail near the coast. Inland heat and diurnal variation increase and the land is more suited to Cabernet Sauvignon and Syrah. Topographic variation modifies the climatic trend. River sediments form the soils in the lower reaches and valley floor. Along the lower reaches the slopes are gravelly, and in the upper reaches there are sandstone-based soils.

The Olifants Co-operatives controls the majority of the Olifants region, 7,000 hectares of white grapes and 3,000 hectares of red grapes. The co-operatives makes large amounts of export-quality wine. Boutique wineries making premium wines are located along the valley periphery. Chenin and Sauvignon Blanc vines prevail in the cooler lower reaches and Cabernet Sauvignon and Syrah in the upper.

The Breede River enters the Breedekloof through a water gap and flows between two linear mountains. The valley terminates near Worcester. The Breedekloof region produces more wine than any other South African region, 250,000 tonnes divided among numerous varieties. Like Robertson and Worcester, Chenin Blanc and Colombard combine for about 40 percent of the total yield.

The Worcester district is sandwiched between Breedekloof and Robertson. Vineyards and olives are irrigated by Hex- and Breede-extracted water. KWV Cellars houses the world's largest brandy still here. Worcester produced two-thirds of the nation's 32,000 tonnes of table grapes grown in 2013. Colombard and Chenin Blanc produced 70,000 tonnes in 2013, 38.1 percent of the regional yield.

Robertson, twenty-five miles east of Worcester, is best known internationally for its Syrah, Chardonnay, and Sauvignon Blanc. Soils here are limestone derived.

Popular South African Varieties

South Africa produces wines from a broad range of varieties. Noble varieties, destined for export, were widely planted in 2014. Cabernet Sauvignon

(11.407 hectares), Shiraz (10,410 hectares), and Merlot (6,098 hectares) combine for half the red acreage. Pinotage covers another 20 percent (7,357 hectares). Pinotage resulted from a cross between Pinot Noir and Cinsault performed in 1925 in South Africa and became commercially available in 1961.

Among white grapes Chenin Blanc covers 17,934 hectares, more than any other variety. Colombard is second in area covering 11,907 hectares, followed by Sauvignon Blanc with 9,224 hectares and Chardonnay with 7,356 hectares.[15]

SUMMARY

Europeans found favorable conditions around the Southern Hemisphere for grapes. In Spanish colonies the Spaniards were quick to establish vineyards. English and Dutch colonists did not know viticulture and their initial efforts usually failed. The introduction of French Huguenots into these colonies brought about successful viticulture and vinification. Oddly, French colonies were either tropical or in Canada, therefore French colonies did not compete with French wines. Yet, the Southern Hemisphere wines are French varieties made in the French style.

After 1980 global wine demand shifted away from Europe and into the Southern Hemisphere, which produced wines of equal quality at much lower prices. In the United States, wines from Australia, then Chile became popular during the 1980s. South African wines emerged in the mid-1990s after the end of apartheid. We were introduced to New Zealand Sauvignon Blanc about the same time. Argentine Malbec and Torrontes were first marketed in the United States in 2000. What will be next and where will it come from?

Chapter 9

Wine in North America

When we consider the Latitude and convenient Situation of Carolina, . . . being placed in that Girdle of the World which affords Wine, Oil, Fruit, Grain, and Silk, . . . sweet Air, moderate Climate, and fertile Soil; these are the Blessings . . . that, renders the Possessors the happiest Race of Men upon Earth.[1]

—John Lawson

Wine in California is still in the experimental stage; and when you taste a vintage, grave economical questions are involved. The beginning of a vine-planting is like the beginning of mining for the precious metals; the wine-grower also "Prospects." One corner of land after another is tried with one kind of grape after another. So bit by bit they grope about for . . . those pockets of earth . . . where the soil has sublimated under sun and stars to something finer, and the wine is bottled poetry.[2]

—Robert Louis Stevenson

Fortunately, the tide has turned. Since 2010 there have been more wineries outside California than in it.[3]

—Richard G. Leahy

HISTORICAL GEOGRAPHY OF WINE IN THE UNITED STATES

North American historical wine geography can be categorized into six periods. The first period begins with wine made near L'Anse aux Meadows in

Canada about the year 1000 CE, continues with the success of Spanish colonists making wine near Santa Fe (in modern New Mexico) in 1580, and ends with the California planting of Mission vineyards. The second period coincides with the English colonial period when colonists planted vines from Canada to Florida. Their vineyards consistently failed, mostly through abandonment. There were faster and easier ways to make a living. The third period began about 1800 when viticulture became successful in the Ohio Valley and ends in the 1880s when the Hayes presidency served only lemonade at the White House. The anti-alcohol movement of the late nineteenth and early twentieth centuries and Prohibition comprise the fourth period. From 1880 to 1930, bowing to political pressure (which resulted in a lack of sales), wineries went out of business. In the fifth period, 1920 to 1980, Prohibition and post-Prohibition social upheavals kept most Americans drinking distilled spirits and fortified wines. The sixth and current period, 1980 to the present, began with the combination of world-class California wines and the passage of state farm-wine laws that permitted direct winery to consumer sales. These laws opened business opportunities, leading to the current continuing expansion of the wine industry and wine consumption.

Period 1: Exploration and Visitation Attempts

After the Vikings visited North America, we know that Columbus drank wine every day of his voyages, and he performed a special drinking ceremony when he claimed the New World for Spain. Magellan spent more on wine than armaments for his globetrotting adventure. The first wine made from vinifera grapes was made in Mexico under conditions previously discussed. Spaniards in Florida also made wine very early. The first successful long-term wine-making region in the United States was in New Mexico, as agricultural historian Rick Hendricks explains:

> In the 1630s the fruit of the vines from this mission [San Antonio de Senecu, south of present-day Socorro, New Mexico] yielded the home-produced wine for all other missions in New Mexico. However, it did not produce enough for all of the celebrations of Mass. Every three years more than one hundred liters of wine arrived from Spain to meet their needs, but at a price too high for most of the population.[4]

In addition to early English attempts to grow vinifera grapes there were continuous efforts to make beverages from native grapes (and everything else). With no chance of getting better, native wines were either enjoyed or tolerated. Thus we find a smattering of anecdotal references to native grape wine production that only hints at the considerable volume of the wine made, such as this statement reported by colonial historian Albert Myers, "Captain Rapel (the Frenchman) with he made good Wine of the grapes (of the country) last Year [1684]."[5] The U.S. Census Bureau reported, "In 1769, the French settlers in the Illinois River Valley made upwards of 100 hogsheads of strong wine from the American wild grape."[6] The illiterate and hardscrabble farmers obviously kept no records and they were the most likely to press the local grapes.

In 1769, missionaries planted the first California vineyard at Mission San Diego. Missionaries then enslaved the local Native Americans and forced them to tend the vineyards. Subsequent and more northerly California missions repeated the process. The Mission grape dominated California for the next 100 years.

Period 2: Wine Making and Drinking Habits in the Eastern Colonies

Every group of immigrants arriving in North America planned to grow grapes and make wine. The exploits of the Virginia colonists provide an excellent example of the failure of people throughout the colonial period to make wine. North American grapes had a "foxy" flavor that satisfied the rough colonists and early Americans but met with disdain among hopeful investors back in Europe.

Attempts at wine making failed on the Atlantic coast for several reasons. Most significant was that the colonists were abandoning their wine-making plans for more profitable enterprises. Other factors included relocation, Indian attacks, poor site selection, infestation, and disease. Many growers moved inland after a few years, abandoning their coastal vineyards. Promising advances were made in Virginia before the 1624 Indian uprising but reduced population led to the abandonment of the project. Several early vineyards were planted in coastal areas and suffered from saltwater intrusions in dry years. Fungal diseases attacked the vines during wet summers.

Colonial grape-growing attempts failed for one of six reasons. These six reasons can be placed into one of two categories, human and environmental. On the human side, the primary cause of failure was ignorance. The English, Dutch, and Swedish colonists had no experience growing grapes. To alleviate the problem they recruited Huguenots (Protestants) from France, who were mostly happy to escape from Catholic-dominated France.

The second factor again deals with France. There is documented proof that France bribed British colonial leaders not to plant grapes, or to abandon existing attempts. Huguenots settling in South Carolina were successfully making wine in 1768. Huguenot colonist Pierre Legaux accepted an offer to return to France (he was wanted) with a reward for destroying his South Carolina vineyards.[7] American wine historian Thomas Pinney relates the events of 1772 when their leader

> took wine from South Carolina with him to London and submitted it to Lord Hillsborough, then secretary for the American colonies, but

> received no encouragement. It was later reported that Hillsborough was paid 250,000 pounds by the French to dampen the enterprise, since the French were terrified lest South Carolina take away the American wine trade.[8]

Economic considerations are third on our list. Growing grapes takes several years before a profitable harvest occurs. Colonists, and their backers, did not have time to wait. Virginia colonists, for example, abandoned their vineyards after only a year or two and switched to the more profitable tobacco farming.

The environmental conditions for vinifera viticulture were even more inhospitable. Fourth, throughout the eastern United States and especially in coastal areas, excessive humidity is conducive to bacteria, molds, and mildews. Rotten vines, leaves, and fruit were reported from several colonial vineyards. Fifth, coastal areas are also prone to sandy soils, which are not good for vines. The rich and fertile forested lands of the interior were hard to clear and caused excess vine vigor before the vines succumbed to disease, rot, or pests. Sixth and finally, the vineyards were subjected to infestations of insects and worms unknown in Europe. The vinifera vines, having no resistance to these pests, quickly succumbed to their devouring jaws. Interestingly, phylloxera probably played no part in the demise of early vineyards because phylloxera dislikes the sandy soils of the coastal plain.[9]

Growing grapes and making wine is difficult, demanding work. Most colonists dreamed of vineyards but had neither the knowledge nor temperament to wait several years for a crop. Not when any number of other products, tobacco being foremost at the time, could make them rich enough to buy all the Madeira wine they could drink. Better still, rum cost less and was more potent, an important consideration across trails and rough roads. Robert Fuller recounts the need for beer and wine among the Pilgrims:

> Beginning with the Pilgrims at the Plymouth Colony, alcohol was a staple in the early American diet. In fact, when the *Mayflower* arrived off the coast of Massachusetts, the beer supply was already nearly consumed. . . . The settlers "were hastened ashore and made to drink water" so that "the seamen might have more beer." . . . Future settlements came better prepared. Thus, for example, when the *Arbella* set sail for Boston in 1630 with a load of Puritans . . . it carried 10,000 gallons of wine and three times more beer than water.[10]

Eastern colonial attempts to produce wine were generally failures because of intervening economic opportunity. Almost any economic activity, colonists discovered, paid better than making wine. The few successes were novelties. Virginians first made native grape wine in 1608. Virginia was the scene of many attempts to establish vineyards during the colonial era. Most failed, but Robert Beverly kept a vineyard and winery operating for more than twenty years between 1720 and 1743. When he died his descendants abandoned the vineyard. Yet, assurances of impending success persisted for more than 100 years. Writing 110 years ago, Samuel Purchas recounted this statement by Richard Hakluyt:

> Our Frenchmen assure us that no Countrie in the World is more proper for Vines, Silke, Olives, Rice, *riser.* &c. then Virginia, and that it cxcelleth their owne Countrey. The Vines beeing in abundance naturally **over** all the Countrey: a taste of which Wine they have alreadie sent us, with hope the next yeere to send us a **good** quantitie. There bee Mulberie *trees* in wonderful ~ ~ ~ abundance, and much excelling both in goodnesse and greatnesse those of their Countrey of Languedocke. To the full perfecting of both which rich Commodities **of** Wine and Silke, there wanteth nothing but hands. And of the Mulberies ma bee made also wholesome Wine for the people there. And of a certaine Plumme in the Countrey, they have made good drinke.[11]

Ben Franklin's assertion that wine was "the good creature of God" prevailed from the dawn of civilization until 1800. The colonials and their ancestors in Europe were heavy drinkers. In Franklin's day, alcohol was the only commonly available painkiller. Many adults started the day with a drink to drive off the morning chill and receive a burst of energy. Colonial-era accounts make it difficult to decide whether clergymen or sailors drank more.

Wine drinking was largely limited to the elite who could afford it. The average person might have had wine at celebrations. In the vast majority of instances, however, colonists drank anything they could get their lips around including cider, perry, mead, and other fermented fruits. Distilled spirits were preferred because they were potent, cheap, transportable, and durable. Publication of Benjamin Rush's "An Inquiry into the Effects of Spirityous Liquors on the Human Body and Mind" in 1790 lit the spark that became Prohibition. Besides health considerations, he identified several other reasons for reducing the alcohol consumption of others. Early industrial leaders wanted to reduce consumption to improve productivity and increase profits (they were less interested in accidental death or injury occurring to drunken workers). Religious leaders began identifying alcohol as the source of all society's ills. Social leaders and reformers formed clubs and fraternal societies of men pledging to remain alcohol free. Political leaders wanted to discourage hard-drinking immigrants from coming to their states by making alcohol difficult to acquire.

Period 3: Wine Making and Drinking Habits in the Early Republic

Thomas Jefferson was a wine expert of great renown. He traveled extensively in France. There, he established connections in the best wine regions and with top winemakers. Through these connections he purchased thousands of bottles of wine throughout his long life. The following quote taken from Marie Kimball's "Some Genial Old Drinking Customs" provides an example of Jefferson's appetite:

> From a list of "Wine Provided at Washington." Preserved among the Jefferson papers, we learn that in 1801 the President bought "5 pipes of Brazil Madeira, a pipe of Pedro Ximenes Mountain (126 gallons, 424 bottles of it sent to Monticello), a quarter cask of Tent; a keg of Pacharetti doux; fifteen dozen claret; 15 dozen Sauterne; 2 pipes of Brazil Madeira; 220 bottles of claret; 2 pipes of Brazil Madeira, 30 doz. bottles of Sauterne."[12]

The record for the following year is no less magnificent.

Quality wine became harder to come by during the early republic. It was an expensive European import that few could afford. Alice Morse Earle recounts the availability and price of wine during the early republic period:

> At the United States Hotel in Philadelphia . . . in 1840 . . . Mumm's champagne was two dollars and a half a quart; Ruinard and Cliquot two dollars; the best Sauternes a dollar a quart; Rudesheimer 1811, and Hochheimer, two dollars; clarets were higher priced, and Burgundies. [*sic*] Madeiras were many in number and high priced; Constantia (twenty years in glass) and Diploma (forty years in wood) were six dollars a bottle. At Barnum's Hotel there were Madeiras at ten dollars a bottle, sherries at five, hock at six.[13]

During the early republic era, when they could get it, Americans' favorite wine was Madeira. Rum was ubiquitous in the New World colonies; gin was the European counterpart. The people of the frontier made whiskey and apple brandy in their homemade stills.

Every group attempting to establish a wine colony before 1800 failed. There were a variety of reasons for their failure, most notably intervening opportunity. This changed in the early 1800s when Swiss immigrants produced wine along the Ohio River. The Swiss, led by John James Dufour, first attempted wine production in Kentucky, then successfully in Vevry, Indiana, along the northern shore of the Ohio River. They grew a variety known to them as the Cape grape, believing it to have come from vinifera cuttings taken from South Africa. More likely it was a cross between *V. labrusca* and some other species, perhaps *V. vinifera*.

The real success came with the discovery of the Catawba grape. Historians are unsure of the origin of the grape, as there are two competing stories. In both Major John Adlum received cuttings from a widow living in eastern

Maryland before 1820. Adlum popularized the grape by distributing cuttings to other growers. Adlum was very proud of his discovery, saying, "In bringing the Catawba grape into public notice with my vineyards, I have rendered my country a greater service than I could have done had I paid off the National Debt."[14]

In 1842 Nicholas Longworth invented sparkling Pink Catawba wine, which quickly became a global success. The Catawba, a cultivar of *V. labrusca,* has a copper-colored skin that readily produces a pink wine. Ohio historian Alan Rocke describes Longworth's discovery this way:

> Longworth began growing grapes seriously about 1820, and he put his heart, and his considerable fortune into the effort. . . . He had modest success with the Catawba grape, producing a dry or off-dry wine that was popular among the city's large German immigrant community. . . . In 1842 a batch of wine inadvertently . . . [became] . . . sparkling wine. . . . The result was a lovely, tasty, semi-dry pink sparkling Catawba . . . this wine exploded commercially.[15]

Longworth made Ohio the nation's great antebellum wine producer. With 1,200 acres on the Ohio River near Cincinnati, Longworth made his fortune. Longworth's success led many others to emulate him throughout Ohio. Table 9.1 shows the rapid growth and decline of Ohio-produced wine.

Catawba wine gained vast notoriety and was widely celebrated, as evidenced by the following poem by Longfellow in commemoration of Longworth's products.

Ode to Catawba Wine

This song of mine
Is a Song of the Vine,
To be sung by the glowing embers
Of wayside inns,
When the rain begins
To darken the drear Novembers.

It is not a song
Of the Scuppernong,
From warm Carolinian valleys,
Nor the Isabel
And the Muscatel
That bask in our garden alleys.

Nor the red Mustang,
Whose clusters hang

O'er the waves of the Colorado,
And the fiery flood
Of whose purple blood
Has a dash of Spanish bravado.

For richest and best
Is the wine of the West,
That grows by the Beautiful River;
Whose sweet perfume
Fills all the room
With a benison on the giver.

And as hollow trees
Are the haunts of bees,
For ever going and coming;
So this crystal hive
Is all alive
With a swarming and buzzing and humming.

Very good in its way
Is the Verzenay,
Or the Sillery soft and creamy;
But Catawba wine
Has a taste more divine,
More dulcet, delicious, and dreamy.

There grows no vine
By the haunted Rhine,
By Danube or Guadalquivir,
Nor on island or cape,
That bears such a grape
As grows by the Beautiful River.

Drugged is their juice
For foreign use,
When shipped o'er the reeling Atlantic,
To rack our brains
With the fever pains,
That have driven the Old World frantic.

To the sewers and sinks
With all such drinks,
And after them tumble the mixer;
For a poison malign

Is such Borgia wine,
Or at best but a Devil's Elixir.

While pure as a spring
Is the wine I sing,
And to praise it, one needs but name it;
For Catawba wine
Has need of no sign,
No tavern-bush to proclaim it.

And this Song of the Vine,
This greeting of mine,
The winds and the birds shall deliver
To the Queen of the West,
In her garlands dressed,
On the banks of the Beautiful River.

Henry Wadsworth Longfellow, 1854[16]

During this era, wine communes were often religiously inspired. The examples of Amana in Iowa and Harmony and Economy, Pennsylvania, related by Robert Fuller, author of *Religion and Wine,* are typical. "At Economy or Amana or Zoar the people receive either beer or wine daily, and especially in harvest-time, when they think these more wholesome than water. At Economy they have very large, substantially built wine cellars, where some excellent wine is stored."[17] During the early republic period this wine drinking and cultish behavior helped inspire the wave of Prohibition sentiment that swept the North in the 1850s.

Winemaking from native grapes continued unabated on local farms right through the Civil War and Prohibition. During the mid-1800s, independent reports of wine-making success abound. The focus of each is on the gallons produced in total. Vineyard owners won awards for maximizing gallons per acre. Widely cited was the Hermann, Missouri, grower who coaxed 400 gallons of wine per acre from his vines.

According to the Olmstead brothers Alan and Paul, "In 1849, 300 acres of vineyards were to be found within twelve miles of Cincinnati, which yielded over 50,000 gallons of wine."[18] By 1850, according to the U.S. Census, Ohio's output was double that of California. The official counts did not include the wines made from wild grapes on farms across the northeast. The Ohio and upper Mississippi river valleys were America's wine center.

The Ohio wines were a hit in the eastern United States, but disliked in Europe because they have a "foxy" taste. The origin of the term "foxy" has baffled millions for generations. Campbell provides a reasonable origin for the term, but does not provide a date when she says, "European visitors who tried such beverages could never forgive any product of the labrusca grape's

Table 9.1: U.S. Census Reported Wine Yields in Gallons, 1840 to 1910 for Selected States

State	1840	1850	1860	1870	1880	1890	1900	1910
California	nd	58,055	246,518	1,814,656	13,557,155	14,626,000	5,492,216	16,005,519
Ohio	11,524	48,207	568,617	212,912	1,632,073	1,934,833	350,615	264,213
Pennsylvania	14,328	25,590	nd	97,165	114,535	nd	194,610	106,756
Indiana	10,266	14,055	102,895	25,000*	99,566	224,500	126,730	130,976
N. Carolina	28,752	11,058	50,000*	50,000*	334,701	388,833	146,699	205,152
Missouri	22	10,563	nd	326,173	1,824,207	1,250,000	122,382	245,656
New York	6,799	9,172	50,000*	82,607	584,148	2,528,250	290,365	346,973
Kentucky	2,209	8.093	179,948	nd	81,170	nd	51,668	45,138
Connecticut	2,666	4,260	50,000*	nd	5,336	nd	26,589	30,572
Illinois	474	2,997	50,000*	111,882	1,047.875	250,000	223,819	247,951
U.S. Total		221,249	1,627,192	3,092,330	23,458,827	24,306,905	8,246,344	18,636,225

*approximation

Source: Alan Olmstead and Paul Rhode. "Quantitative Indices on the Early Growth of the California Wine Industry." Davis, CA: Robert Mondavi Institute, Center for Wine Economics. http://vinecon.ucdavis.edu/publications/cwe0901.pdf. Accessed May 8, 2009.

curious animally undertaste. 'Musky' they called it—'foxy'—*le gout de renard*. The polite called the taste *framboise*, 'raspberry.' The more direct called it *passat de renard*—'fox piss'."[19] From "fox piss" to "foxy" seems a simple puritanical transformation.

The French wine judges in 1855, however, did not have problems with vinifera wines produced in California. In blind tasting, the French judges surprised themselves by how much they liked the California products.

In 1850, Ohio was America's wine center. The dream of wine production drew many immigrant religious communities to settle in the Midwest, the Amana colonies of Iowa, for example.

During the 1850s, Longworth's winery was producing 100,000 gallons a year. His neighbors produced another 500,000 gallons. Then disaster struck: Blackspot and downy mildew funguses invaded the vineyards. Fungal attacks were followed by the Civil War, which drew away the labor supply. After the war, growers migrated northward and planted their vineyards near the Lake Erie shore. South Bass Island and Conneaut became and remain focal points for Ohio wine production.

Missouri also had, during the late 1800s, a thriving wine industry. The region was settled in the 1840s by Rhinelanders familiar with growing grapes. Noted geographer Carl Sauer explains why vineyards were successful here:

> An important step toward the success of the German settlements was the introduction of wine-growing at Hermann by immigrants from the Rhine. In 1845 there were 50,000 vines at this place; in 1846, 150,000; in 1848, 500,000; and in 1849, 700,000. In the last mentioned year it was predicted that the wine crop of a few townships in Gasconade County would be of greater value than the hemp crop of the state. The success of [g]rapes at Hermann led to the extensive planting of vineyards at Ste. Genevieve, at Booneville, and in Franklin, Warren and St. Charles Counties. The vineyards were located on loess hillsides which afforded warm soil, excellent drainage, and protection from unseasonable frosts. They were supposed by vintners of the time to benefit by their nearness to a large stream. The climate was said to be better than in the Rhine country because of the sunny fall weather, which permitted the grapes to ripen with high flavor. Previous to the introduction of the grape Hermann, with its mediocre farmland, [had] been losing by emigration.[20]

The favorable environmental conditions Sauer describes, coupled with the knowledgeable immigrants led to their continued success. William Glasgow is credited with making the first sparkling wine in St. Louis in 1854.[21] L. U. Reavis in his 1853 book, *St. Louis, the Future Great City*, relates the scale of Missouri wine production:

> During the last five years [1848–1853] the increase has been at the rate of about 300 acres per year. Within the period last named, several companies have been formed for producing wine on a large scale. The Cliff Cave Wine Company, in the south part of St. Louis County, has about twenty-five acres of vines. . . . The Augusta Wine Company of St. Charles County, has 22,775 vines, and made last year 8,000 gallons of wine. The Bluffton Wine Company, of Montgomery county has 59,834 vines, and made last year . . . 13,490 gallons of wine . . . the American Wine Company of St. Louis . . . claims to have made last year over 100,000 gallons of still wines, and half a million bottles of champagne. The vineyards of the town of Hermann yielded last year over 150,000 gallons of wine.[22]

Fifty years later Missouri was still a major wine-producing state. In 1904 Missouri, according to William Bek, "shipped a twelfth of the wine placed on the market by all states. . . . the surplus number of gallons of wine Missouri produced is 3,068,780 gallons."[23] Prohibition ended the wine industry in Hermann, Missouri.[24] One hundred years later, the Missouri wine industry is beginning to recover.

Period 4: Post–Civil War: The Rise of California

It was the transcontinental railroad, completed in 1869, that brought California to the forefront of American wine production. Vast quantities of wine could be safely shipped east for the first time. California's vast lands in the Mediterranean climatic zone are ideal for growing wine grapes. Much of the wine comes from lackluster vineyards in the central valley. The best wines come from the coastal hills and valleys. In the coastal areas ocean breezes keep the temperatures low and produce regular morning fogs to dampen the leaves. Sumner and colleagues tell us of the difficulties associated with the first century of California wine production:

> During the 17th and early 18th Centuries vines and wine spread throughout what is now Mexico and the southwest of the United States with Spanish soldiers and missionaries. Like much else in the history of the State, the wine industry in California was created by the establishment and spread of the missions. . . . Beginning in 1790 in San Diego, over the next four decades, Spanish missions moved up the Coast bringing vines and winemaking abilities with them. The local grapes were of no use for wine making, but a European-based variety known as "Mission" did well, and became the basis of California wine production for many years. . . . The mission vineyards and winemaking facilities fell into disuse and disarray when the Spanish were forced out after Mexico (which included California) became independent in the 1830s.[25]

Between 1830 and 1895 the California wine industry grew rapidly and imports declined. Agoston Haraszthy, friend of the last Mexican governor of California, was a major early promoter of California wine. In 1861 he was sent to Europe to obtain cuttings for experimental plantings in California. When he returned to California with more than 100,000 cuttings from numerous varieties, the backers who sent him refused to pay for his expenses. He kept and planted the cuttings on his own property. Haraszthy did much to popularize the Zinfandel grape, but he did not introduce it to California. While running a rum distillery in Nicaragua he fell in a river and was eaten by alligators.

The 1880s were an important period of growth and expansion for California's wine industry. During that decade Sonoma and Napa were first put under vines. Thousands of immigrants, a high proportion of them Chinese, were brought to the valleys to work the vineyards.

Among the visitors to California at this time was Robert Louis Stevenson. While there he wrote *The Silverado Squatters*. The following excerpt from that book discusses what was occurring in California at that time. His words could apply to any newly pioneered wine region:

> Wine in California is still in the experimental stage; and when you taste a vintage, grave economical questions are involved. The beginning of vine-planting is like the beginning of mining for the precious metals: the wine-grower also "Prospects." One corner of land after another is tried with one kind of grape after another. This is a failure; that is better; a third best. So, bit by bit, they grope about for their Clos Vougeot and Lafite. Those lodes and pockets of earth, more precious than the precious ores, that yield inimitable fragrance and soft fire; those virtuous Bonanzas, where the soil has sublimated under sun and stars to something finer, and the wine is bottled poetry: these still lie undiscovered; chaparral conceals, thicket embowers them; the miner chips the rock and wanders farther, and the grizzly muses undisturbed. But there they bide their hour, awaiting their Columbus; and nature nurses and prepares them. The smack of Californian earth shall linger on the palate of your grandson.[26]

Three tragedies struck the California wine industry over the period 1895 to 1920. Phylloxera destroyed the capacity to grow grapes and make wine, the 1906 earthquake destroyed the capacity to make and store wine, and Prohibition destroyed the capacity to consume wine.

The first was the arrival of phylloxera. While phylloxera may have arrived in California by 1880, it took time for the population to build sufficiently to reach plague proportions. Table 9.1 clearly shows the devastating impact of the arrival of phylloxera in California in the late 1890s. Production plummeted from more than 14 million gallons in 1890 to 5.5 million in 1900. Production in Napa County alone dropped from 3 million to 400,000. California quickly rebounded, producing 16 million gallons in 1910. California growers had two advantages over the French infestation: phylloxera did not strike everywhere at once, and the French grafting cure was well known by then.

The second tragedy was the 1906 San Francisco earthquake, which devastated the California wine industry's stored wine. The impressive block-sized California Wine Association Building containing more than thirty million gallons of wine was destroyed in the earthquake. Much of that wine was stored in large cement underground vats, which ruptured during the earthquake. In outlying smaller facilities falling debris smashed barrels and cracked wooden tanks. The subsequent fires burned what was left. Prohibition fever slowed efforts to re-construct the wineries and wine storage facilities.

The third tragedy was Prohibition, which eradicated the legal supply chain. Most wineries simply closed their doors and shifted to other crops, a few were licensed to make sacramental wines, and a few went into bootlegging.

Period 5: Wine Drinking after Prohibition to 1980

When Prohibition was repealed, most drinkers were used to low-quality, highly alcoholic products at high prices. Prohibition's end brought prices down but did not change drinking preferences. Their key common characteristic was intoxication. This trend continued as drinking remained socially unacceptable in many regions. During and after World War II, liquor was the preferred beverage. Only recently has beer outperformed liquor. Most wine was fortified and sweetened. Wine drinking was treated with disdain. Gallo wines like Thunderbird, Bartles & Jaymes, Night Train Express, and Boone's Farm became cultural icons of poverty and a cheap drunk.

New York state wines prior to 1970 were generally sweet and fortified. New York at that time was the home of Mogen David, Manischewitz, and Kings. It is from these strong, sweet wines that Americans derived the false impression that all kosher wines were sweet and strong.

This trend continued as wine drinking remained socially unacceptable in many regions. During and after World War II, liquor was the preferred beverage.

Beginning about 1970 winemakers around the world began remaking the industry. The new focus was and remains on producing quality products of character that commanded high prices. The wine program at UC Davis is most responsible for advancing viticulture and wine making in the United States. Coupled with marketing efforts, wine consumption has slowly increased around the world. Now wines that do not meet the minimum quality standards do not sell.

For the pioneers of quality the early years were difficult. This was the era of the big winery. Gallo, Almaden, and Italian Swiss Colony dominated the wine industry in 1970. Like the proverbial snowball, demand for quality wine started slowly and has grown to dominate the market. To satisfy the demand for diverse choices and yet keep their dominance, the big wineries created multiple brands.

On the forefront of those seeking quality wine were California's newly enriched. They became enamored of Napa and Sonoma wines. Several bought land, built big new winery buildings, and planted vines only to discover wine making was hard work and wine selling harder still. Many failed, leaving the developed land ready for the big wineries to gobble up. The standard joke around the Napa Valley was, "How do you turn a large fortune into a small one? Buy a winery!"

The 1980s was a decade of corporate vineyard consumption. Wineries were bought, sold, and sold again. Distilling, tobacco, and soft drink companies were among the leading purchasers. Agglomeration continues today. A newly proved winery is ripe for corporate buyout like any successful startup business.

Period 6: Wine in the United States Today

Since 1970 we have seen the expansion of wineries into every state. States with more than 1,000 acres of bearing grapes include California, Washington, New York, Oregon, Michigan, and Pennsylvania. States with less than 100 bearing acres are North Dakota, Wyoming, Montana, Maine, Nevada, Utah, South Dakota, New Hampshire, Rhode Island, Delaware, Hawaii, Alaska, and Louisiana. These low-acreage states are either at climatic extremes or are very small.

AMERICAN VITICULTURAL AREAS

A significant event in American wine history was the establishment of American Viticultural Areas (AVA) by the U.S. government. The structure of the AVA system is patterned after the French Appellation Control system, which allows the nesting of regions. The first AVA approved was the historic Augusta area of Missouri. To receive AVA status, petitioners (mostly winemakers and grape growers of the area) submit a proposal to the government. In the proposal they must provide evidence that the name of the proposed area is locally (or better, nationally) known as referring to the area. Evidence that environmental conditions including climate, soil, geology, and topography are distinctive is also required. The proposers must also include a boundary line and justify the boundaries identified based on historic or current evidence. The cost of proposing, reviewing, and approving each AVA is highly variable and may be a function of the number of people involved. The San Antonio Valley AVA of Monterey County, California obtained their approval after about 100 staff hours and $2,500 in fees and land survey. The process took three years. Lodi, California spent $30,000, expended 1,500 hours of effort, and took five years.

To include an AVA indicator on the label 85 percent of the grapes must come from within the designated area. Refer back to the Generalized EU geographic classification system discussion in chapter 7 and Figure 7.1.

CALIFORNIA

California is the source for upward of 90 percent of U.S. produced wine. Its vast lands in the Mediterranean climatic zone are ideal for growing wine grapes. Much of the wine comes from lackluster vineyards in the Central Valley. The best wines come from smaller wineries in the coastal hills and valleys. In the coastal areas ocean breezes keep the temperatures low and produce regular morning fogs to dampen the leaves.

In California wine is big business. In 2006, there were nearly 5,000 wine growers employing more than 207,000 people. The wine industry produced more than 532 million gallons of wine in 2005, of which 441 million gallons were sold domestically. Wine sales added almost $40B to the California economy in 2004. In addition, wine tourism added another $1.3B to the state's economy. Winemakers and growers paid $1.9B in taxes to California and another $3.7B in federal taxes.

California wine grape plantings covered 864,831 acres in 2012. Grape acreage changes between 2000 and 2005 indicate subtle changes in varietal preference as seen in Table 9.2. Chardonnay lost the most land, more than 5,000 acres. Cabernet Sauvignon gained more than 4,000 acres. Chenin Blanc and French Colombard, both prime components of blended table wines, lost more than 5 percent of their land area. Sauvignon Blanc and Syrah acreage expanded by more than 6 percent.

California Climatic Variation

Two factors define macro-scale climate in California. First is California's geographic location, which places it squarely in the latitudes prone to wet winter–dry summer climates. Except in January, when it pours, widespread rain is uncommon south of San Francisco. Through April the native vegetation flourishes. From May onward the landscape browns. Wildfires, fed by this herbaceous bracken, begin flaring up in October.

Second, the California current, part of the great North Pacific vortex, brings cold water southward from Alaska. The oceanic influences are most pronounced along the coast and diminish inland. Microclimatic variation is intense in California.

Table 9.2: California Wine Grape Bearing Acreage by Variety, 2000–2014

Variety	2000	2005	2009	2014	% change 2000–2014
Cabernet Sauvignon	48,285	74,970	73,797	86,258	79
Chardonnay	89,272	92,089	90,434	97,970	10
Chenin Blanc	19,127	10,618	7,628	5,967	–69
French Colombard	42,058	27,202	24,772	22,273	–47
Merlot	42,070	52,185	45,871	45,296	8
Pinot Gris	939	5,094	8,676	15,009	15
Pinot Noir	11,769	23,323	30,339	41,301	251
Rubired	10,841	10,801	11,257	12,511	15
Sauvignon Blanc	10,808	14,082	14,670	15,444	43
Syrah	6,553	17,481	18,890	19,019	190
Zinfandel	47,152	48,969	48,807	48,638	3
TOTAL	**403,287**	**445,141**	**448,957**	**495,951**	**23**

Source: California Department of Food and Agriculture 2009 and 2015. "California Grape Acreage Report 2009" and "California Grape Acreage Report 2014." Sacramento, CA. www.nass.usda.gov/ca.

California Geomorphologic Variation

California is geologically active. Beyond the catastrophic effects of earthquakes, the steady shaking means the mountains are young and rugged. The valleys between sharp peaks are choked with thick alluvial beds, a near perfect soil for vines.

California Terroir

Winter rains and dry sunny summers coupled with thick alluvial soil create a near ideal terroir for successful viticulture. Matching variety to terroir is a young science in California. As a result, California vineyards are a patchwork of producing varietal blocks. This diversity is unlike the varietal uniformity of European vineyards.

The interplay of atmospheric and geological phenomena creates a myriad of individual terroirs from the coast to almost fifty miles inland. The San Francisco and Monterey bays create the largest opportunities for land–sea interactions that create large diurnal temperature variations often accompanied by cooling fogs. Inland is the great Central Valley. Fruits and vegetables grow up and down its length. Table grapes and wine grapes destined for jug wines thrive near the valley's northern end. Further eastward is the Sierra Foothills district where vines grow on the alluvial slopes.

The California American Viticultural Areas

There are more than one hundred wine grape growing regions, designated and defined by governmental decree, within California. These regions overlap. Almost any wine can claim three levels of designation. First, if the grapes are from California, the winemaker will proclaim that designation on the label. Second, wines may bear a multicounty regional designation; such as North Coast, Central Coast, Sierra Foothills, or South Coast. Third, wines not blended from grapes across county lines are probably eligible for a county-level AVA designation. These are the familiar designations of Napa, Sonoma, Monterey, or Amador. Nested within the county designations are smaller AVA designations. The smallest AVA, which is in Mendocino County, California, is the Cole Ranch AVA at only sixty-two acres.

NEW YORK

There are six New York AVAs. Each exists in close relationship with a large body of water providing spring and fall lake effects. Successful wine making came to New York long after Ohio. The delay was caused by the Iroquois Indians' occupancy of upstate New York and past Hudson Valley viticultural failures. New York growers produced table and fortified wines from Native American grape varieties. In 2012 there were 36,919 bearing acres in New York.

New York is the great sweet fortified wine center of America. These wines remain popular with many Americans. The efforts to build a quality wine industry in New York keep getting in the way of profits. From New York vintners Americans obtained the mistaken idea that all kosher wine is sweet, port should cost less than ten dollars a gallon, sherry is for cooking, and you can make Concord grapes taste like most anything.

Grafted vinifera vine plantings were nonexistent before innovator Dr. Konstantin Frank planted Riesling in his Finger Lake vineyard fifty years ago. The passage of the New York winery act in 1980 spurred the planting of Riesling and hybrid varieties in the six New York AVAs.

The Finger Lakes were long known for their native and hybrid wines. The Mogen David, Taylor, Bully Hill, Manischewitz, and many other familiar American wine companies began in the Finger Lakes region.

New York's first vineyard was planted on Long Island by Moses Fournier when it was still New Amsterdam. This vineyard failed, as did subsequent attempts to grow vinifera grapes. By 1670 (and now under British control) New York production shifted to native grapes. Long Island remained the focal point of production throughout the colonial era and has emerged in modern times as a prime location for Merlot. Upstate New York was largely unavailable because the Iroquois lived there before the Revolutionary War.

In 1818, Elijah Fay came up the St. Lawrence and selected a homestead site along Lake Erie. There he planted a number of native grape varieties. Joseph, Elijah's son, blended Catawba and Isabella to make sparkling wine. Joseph's wine company, Great Western, won prizes and fame in Europe. Since then the reputation of the Lake Erie wine region has grown steadily. Great Western's success inspired winemakers for generations and ultimately led to the establishment of the entire Great Lakes wine region.

The Finger Lakes region, the heart of the old Iroquois Confederation, saw its first vineyard in 1829 when Reverend William Bostwick planted Catawba and Isabella grapes. His success led his neighbors to emulate him. Soon the Finger Lakes district became a major wine producer specializing in fortified and domestic grape wines.

Expansion of vinifera grape territory in New York proceeded slowly, but thanks to Dr. Konstantin Frank they are now commonly available. In the 1950s Dr. Frank grafted Riesling tops to rootstocks from Quebec successfully grown in New York. Detractors told him the vines would freeze and die during the cold New York winters. They did not. Largely through his efforts, a sizable vinifera harvest is achieved each year in the East.

The Finger Lakes region is now the real commercial wine district; the remainder are mostly producers of juice and homes of tourist wineries. The Finger Lakes region has the coldest winters and is renowned for dry Riesling. More than seventy years ago Frank Schoonmaker described the Lake Erie–Chautauqua grape belt:

> [It] begins in Chautauqua County, NY, and continues westward, through the "neck" of Pennsylvania which reaches to the lake, some miles into Ohio. A high escarpment rises a few miles back of the lake and runs parallel to the shore for a hundred miles or more. The vines are planted between the escarpment and the water's edge, on what is actually the ancient lake beach. Nearly all the vines are Concord.[27]

The region has changed little since then, aside from a few hundred acres planted in vinifera. The continued predominance of the Concord grape is assured in this region because of the influence of the Welch's Corporation, which purchases the larger proportion for juice.

WASHINGTON

The first attempt to plant grapes in Washington occurred at Fort Vancouver in 1825. By the time of Prohibition many areas of Washington were growing grapes. The first commercial plantings by Columbia Winery and Chateau Ste. Michelle occurred in the 1960s. The northern location of Washington was misunderstood by growers in the 1970s. Thinking that Washington was a cool-climate region, growers initially planted Riesling and Gewürztraminer. While appropriate for western Washington, the hot dry climate of eastern Washington proved unsuitable for Riesling and more suited to Bordeaux and Rhone varietals.

In 1981 there were only eleven wineries in Washington. In 1988 Riesling covered 26 percent of Washington vineyards and Cabernet Sauvignon only 10 percent. Total vineyard acreage grew to 11,100 in 1993. In 2001 there were 900 wineries on more than 50,000 acres, which crushed 220,000 tons of grapes. In 2010 Washington had 40,000 acres under the vine yielding 160,000 tons. In 2012 bearing area was 67,180 acres. By 2013 tonnage soared to 210,000 tons. Making 5 percent of U.S. wine, Washington has been the nation's second largest wine producer for more than a decade. Compare the historic production statistics, presented above, with 2013 statistics quoted from the USDA's "Washington Wine Grape Release, 2013" to evaluate how the Washington wine industry grew and how varietal selection has shifted:

> Cabernet Sauvignon was the top producing variety grown in the State at 42,600 tons or 20 percent of the total. Chardonnay was ranked second, at 40,500 tons or 19 percent of the total. In third place was White Riesling with 40,100 tons or 19 percent of the total. Merlot ranked fourth with 36,000 tons or 17 percent of the total. The average Cabernet Sauvignon price increased $103 per ton from last year to $1,440 and the average Chardonnay price was up $12 per ton from 2012 to $916. Growers received an average of $796 per ton for White Riesling, $13 more than 2012. The average Merlot price per ton increased $82 per ton from last year to $1,186.[28]

Washington experiences a number of climatic regimes. The coastal region of Washington is too cool and wet for wine production. The Puget Sound and Seattle area provide environments suitable for Riesling, Scheurebe, Pinot Noir, and other cool-climate varieties. The Cascade Mountains separate Western from Eastern Washington and create a rain shadow in Eastern Washington. Within the rain shadow the days are sunny and given the state's northern latitude, the summer days are two hours longer than in San Francisco.

The fifteen Washington AVAs are summarized in Table 9.3. The Columbia Valley AVA in Eastern Washington is the state's largest. This vast region contains all other Washington AVAs except Puget Sound and Columbia Gorge; 42,236 acres of the 50,316 acres are inside one of the smaller AVAs. This vast region is a near desert with deep soils created by volcanic eruptions, glaciers, the Missoula floods, and windblown sediments.

The Yakima Valley AVA, inside the Columbia Valley AVA, houses five smaller AVAs: Snipe's Mountain, Rattlesnake Hills, Red Mountain, Horse Heaven Hills, and Naches Heights. The Puget Sound region, surrounding Seattle and on the San Juan Islands, produces about 1 percent of Washington

Table 9.3: Washington State AVA Characteristics

AVA Region	Year established	Total acres	Planted acres	Soil	Precipitation	% Red
Ancient Lakes	2012	162,762	1,608	Dusty alkaline	6”	< 50
Columbia Gorge	2004	191,000	336	Varied	10–36”	36
Columbia Valley	1984	11,313,300	50,316	Varied	6–8”	58
Horse Heaven Hills	2005	579,000	12,956	Sandy loam	6–8”	66
Lake Chelan	2009	24,040	261	Sandy glacial	6–8”	51
Lewis and Clark Valley	2016	306,658	81	Stony	11–22”	> 50
Naches Heights	2011	13,254	39	Loess	10–13”	40
Puget Sound	1995		200		15–30”	61
Rattlesnake Hills	2008	68,500	1,747	Silty loam	6–12”	56
Red Mountain	2001	4,040	1,647	Sandy loam	6–8”	> 50
The Rocks of Milton-Freewater	2015	3,770	250	Rocks & loess	12”	> 60
Snipes Mountain	2009	4,145	807	Loess & gravel	6–8”	
Yakima Valley	1983	600,000	13,215	Silty loam	8”	
Wahluke Slope	2006	81,000	8,491	Sandy alluvium	6”	> 50
Walla Walla Valley	1984	322,794	1.466	Sandy loess	12”	96

Source: Washington State Wine, https://www.washingtonwine.org/wine/facts-and-stats/regions-and-avas, Seattle, WA, accessed October 11, 2016.

wine. The Columbia River Gorge region is the most downstream AVA in the Columbia Valley. Washington shares the Walla Walla, Columbia Valley, Columbia Gorge, and Lewis and Clark Valley AVAs with Oregon or Idaho.

OREGON

Oregon is similar to Washington in terms of geography. Indeed, Washington and Oregon share the Columbia Gorge, The Rocks, and Columbia Valley AVAs. The coastal mountains protect the interior valleys of Willamette Umpqua, Rogue, and Applegate Valley. To the east of this chain of valleys are the Cascades. Oregon's eastern border is dry, although the Blue Mountains in northeastern Oregon receive enough precipitation for forests.

In 2010 there were 419 wineries farming 16,900 acres. The majority were planted in Pinot Noir (12,406 acres). The area planted by major variety in 2010 was Pinot Gris on 2,747 acres, Chardonnay on 950 acres, Riesling on 798 acres, and Cabernet Sauvignon on 639 acres, which rounds out the top five Oregon varietals. In total Oregon produced 31,200 tons of grapes and sold 1,930,763 cases of wine.[29] In 2012 Oregon contained 17,884 acres of bearing vines.

Pinot Noir is Oregon's adopted variety. Pinot Noir is a low producer so it is not surprising that it represented 58 percent (45,239 tons) of Oregon's tonnage and occupied 63 percent (17,146 acres) of the acreage in 2014. Pinot Noir is also a high-value variety and 2014's yield was 67 percent ($113,879,000) of Oregon's total gross value ($168,552,000). Pinot Gris, the second most widely grown variety in Oregon, is planted on about 3,500 acres. Chardonnay is the only other variety covering more than 1,000 acres.[30]

Oregon's wine industry is still expanding. Total Oregon production jumped 39 percent between the 2013 and 2014 harvests. Sixty percent of the increase came from the maturation of young vineyards.

There are seventeen AVAs in Oregon. Between the Coastal and Cascade Mountains lie the Willamette, Umpqua, Rogue and Applegate valleys. The Willamette AVA contains Chehalem Mountains, Yamhill-Carlton District, Ribbon Ridge, Dundee Hills, McMinnville, and Eola-Amity Hills AVAs. Sixty-five percent of all vineyard acreage is in the Willamette AVA, with 90 percent of that concentrated in the valley's northern end where the six sub-AVAs reside. The Southern Oregon AVA includes Umpqua Valley, Red Hills Douglas County, Rogue Valley, and Applegate Valley AVAs. Oregon shares the Snake River AVA with Idaho.

TEXAS

Central and west Texas have all the characteristics of great wine regions. Much of Texas is underlain by limestone with thick alluvial soils near the surface. The climate is similar to the Mediterranean with wet winters and dry summers. There are eight AVAs in Texas. Bell Mountain and Fredericksburg AVAs are within Texas Hill Country AVA. The Texas High Plains AVA, surrounding Lubbock and covering parts of twenty-four counties, is the second largest in the state. A late spring freeze in 2013 killed most of the High Plains vines. The Texoma region, shared with Oklahoma, is along the Red River, which separates Texas from Oklahoma. Texas shares the Mesilla Valley AVA with New Mexico. The Escondido Valley is along Interstate 10, east of Fort Stockton, Texas. Texas Davis Mountains, Bell Mountain, and Escondido Valley AVAs contain no wineries.

Texans are experiencing a wine boom. The number of wineries jumped from 54 in 2003 to 273 in 2013. The vineyard area has not kept pace. In 2012 there were 4,052 bearing acres and 3,500 acres of vines too young to harvest. The yield is inadequate to satisfy demand so Texans import grapes, juice, and wine from California to blend with Texas wine. In 2007 Texas bottled 2,381,469 gallons, presumably from 2,900 acres.[31] That is an impossible average of 821 gallons per acre (typical range is 120–180 gallons per acre), except by legal deception.

Texas has one of the world's most liberal Geographic Indicator laws. Wine can be labeled "Texan" if more than 25 percent of the grapes were grown in Texas. This conflicts with federal law, which requires a 75 percent minimum from the state of origin for state labeling. Wines with less than 75 percent state content are labeled "American" in the other forty-nine states. Wines with less than 75 percent Texas grapes must bear the statement "For Sale in Texas Only."

OHIO

Five AVAs have been designated within Ohio. The Ohio River Valley AVA includes the site of Longworth's Catawba vineyards. The Ohio River Valley AVA extends into Kentucky, Indiana, and West Virginia. Three AVAs benefit from the Lake Erie lake effect: Lake Erie, Isle St. George, and Grand River Valley. The Lake Erie AVA extends along the southern shore of Lake Erie across Ohio, Pennsylvania, and New York. The Lake Erie AVA contains 42,000 acres of vines, a very large percentage of which is Concord slated for juice. The Grand River AVA is contained within the Lake Erie AVA. Isle St. George AVA is located on North Bass Island in Lake Erie. Loramie Creek AVA in Shelby County has no wineries at this time.

The Ohio Wine Producers Association was formed in 1975 and had 181 member wineries in 2016. Ohio winemakers concentrate their activity on Cabernet Franc, Chardonnay, Pinot Gris, Riesling, and the hybrid Vidal Blanc. Bearing acres in 2012 were 1,648. Combined, Ohio wineries produced an estimated 800,000 cases in 2015.[32]

VIRGINIA

In 1950 Virginia had only fifteen acres of commercial grapes. In 2010 there were 210 wineries and 2,700 bearing acres. In 2012 there were 3,733 bearing acres. Today, there are more than 275 wineries on 3,144 vineyard acres, which yielded 8,038 tons of grapes in 2014. Hybrid grapes grew on 490 acres. Chardonnay is the top variety, occupying 441 acres, which yielded 1,104 tons. Cabernet Franc and Merlot both yielded more than 950 tons. To satisfy the growing demand, Virginians will need to plant more grapes.[33] The humid continental and maritime climates require extensive pesticide, herbicide, and fungicide use to bring a crop to maturity.

In a flurry of activity five Virginia AVAs were approved in the early 1980s: Monticello, Shenandoah Valley, Northern Neck George Washington Birthplace, North Fork of the Roanoke, and Rocky Knob. Of these only Monticello can be considered a success with twenty-nine wineries. Rocky Knob and North Fork of the Roanoke are each home to a solitary winery. Virginia's Eastern Shore AVA, approved in 1991, has three wineries. Middleburg AVA was approved in 2002 and already contains twenty-three wineries.

Tourism is a major component of Virginia's wine industry. Most wineries cater to Civil War buffs who tour the region each year. Across the diverse terroirs of Virginia a wide range of varieties can be grown. Norton, a robust red hybrid, was very popular before Prohibition. Revived by Horton Vineyards, Norton is a Virginia specialty. Petit Verdot, a Bordeaux variety, is gaining in popularity and acclaim.[34]

PENNSYLVANIA

The first commercial winery in Pennsylvania after the repeal of Prohibition was Penn Shore Winery. Its opening was shortly followed by the opening of Presque Isle Winery in 1970. Both opened shortly after Pennsylvania made it legal to make and sell wine in 1968. Pennsylvania has five AVAs including the shared Lake Erie AVA. The remaining four are Central Delaware Valley AVA, Cumberland Valley AVA, Lancaster Valley AVA, and Lehigh Valley AVA. The Cumberland Valley AVA is largely unsuited for grape production and contains about 100 acres of vines within its three-quarters of a million acres and no wineries. The Central Delaware Valley AVA has no wineries.

In 2008, 123 wineries produced 971,191 gallons of wine.[35] Yield increased to 1.25 million gallons from 160 wineries in 2011 and 11,779 bearing acres in 2012.[36]

Pennsylvania has lagged in transiting from native to vinifera grapes. "More than 70 percent of Pennsylvania wine grapes are Native American with 67 percent being Concord, and about 4 percent being Niagara."[37] The

Lehigh Valley AVA with twelve wineries on about 220 vineyard acres has declared Chambourcin the region's signature grape.

MICHIGAN

Grapes were first planted in Michigan near Detroit in 1702 by French colonists. The first winery was constructed in 1870 near Monroe. Within two years of Prohibition's repeal, ten wineries opened. Four AVAs reside in Michigan; all approved in the 1980s.. Fennville and Lake Michigan Shore (eighteen wineries) AVAs are in the southwest corner of the state where the growing season is 160 days. Old Mission Peninsula (seven wineries) and Leelanau Peninsula (twenty-six wineries) AVAs are located on the northwest corner of Lower Michigan. Forty-seven Michigan wineries are outside of the four AVAs. The growing season on the Peninsulas is 145 days. All four areas experience lake effect conditions from Lake Michigan.

Michigan is home to 13,917 bearing acres of grapes, 2,850 acres of which are wine grapes, the remainder table and juice grapes. Riesling is the most widely grown variety, covering 670 acres in 2014, followed by Pinot Noir at 245 acres, Chardonnay at 230 acres, and Pinot Gris at 230 acres. Vidal Blanc, Cabernet Franc, and Merlot occupy more than 100 acres each. In addition, Michigan produces substantial quantities of fruit wine. Michigan wineries produce sizable quantities of ice wine.[38]

OTHER STATES

In 1999, Delaware, Montana, North Dakota, Nebraska, Nevada, South Dakota, and Wyoming had no wineries. Today, all fifty states have at least one winery. Delaware, the last state without a winery, saw the opening of four wineries in 2013. In 2014 there were seventeen states with more than 100 wineries, and six states with less than ten wineries.[39] A large proportion of all wineries, whether in California or Delaware, are profitable because of tourists.

CANADIAN WINE

In the early 1600s, Samuel de Champlain wrote of the short-lived success of vinifera vines he planted in New France.[40] During French occupation, the number of vines remained small because the French, fearing competition, discouraged cultivation, and because the Acadian colonists needed to expend all their energy to grow enough food to live.

Today, Canadians grow wine grapes in a few select locations where there is adequate warmth. As conditions warm the area, types of Canadian wine production will shift northward and Canada will become a more significant player in global trade. Grape varieties suited for Winkler Region One (less than 2,000 degree days) grow on the Niagara Peninsula in Ontario, British Columbia, and the Prince Edward Island sound of Nova Scotia. Several companies own vineyards in both Ontario and British Columbia.

The Vintner's Quality Alliance (VQA) is Canada's wine regulatory agency. The VQA guarantees appellation of origin. VQA certifies wines under testing to ensure quality standards are met. The VQA organization is active in British Columbia and Ontario only.

Wines of Eastern Canada

Ontario Wines

Wine production in Ontario is expanding at a rapid pace in Ontario's three recognized appellations. Glacial soils and lakes dominate Ontario. The oldest, the Niagara Peninsula, is a limestone escarpment jutting between Lakes Erie and Ontario. The Lake Erie North Shore is similarly situated with Lake St. Clair to the north, Detroit River to the west, and Lake Erie to the south. Pelee Island is in Lake Erie. Prince Edward County, Ontario (not to be confused with Prince Edward Island) juts into Lake Ontario from the north. Surrounded by water, this area feels the lake effect in full force for all appellations. These regions exist on grainy glacial soils above limestone bedrock.

The Niagara Peninsula is eastern Canada's premier wine region and home to the largest concentration of vineyards in Ontario. The first winery to open on the Niagara Peninsula was Inniskillin in 1973. Chateau des Charmes, established in 1978, was the first winery in the region to plant only vinifera grapes. Four varieties, Riesling, Chardonnay, Cabernet Franc, and Merlot, combine to yield 53 percent of the 2014 harvest. Anthony Shaw of Brock University, expert on environmental conditions on the Niagara Peninsula, quantified the terroir characteristics of the region and recognized ten subappellations. The results of his work are available at http://www.vqaontario.ca/Resources/Library. Of particular importance are the Beamsville Bench and Twenty Mile Bench where crumbling limestone alluvium lies on the edge of the escarpment.

Wineries on the Ontario Peninsula produce most of the nation's ice wine. Winemakers produced ice wine from traditional varieties of Vidal Blanc (588,175 liters) and Riesling (68,428 liters) in 2014. Winemakers in Ontario also produce ice wines from nontraditional varieties including Cabernet Franc (210,571 liters), Cabernet Sauvignon (52,010 liters), and Chardonnay (5,111 liters).[41]

The Lake Erie North Shore with its glacially ground soils offers excellent agricultural opportunities. The North Shore of Lake Erie has been among Ontario's most productive fruit- and tobacco-growing areas. In 2015 it was home to fourteen wineries that cumulatively produced 19,309 cases.

Prince Edward County is located on the north shore of Lake Ontario east of Toronto. A summertime paradise for Toronto residents, this is Ontario's youngest wine region. Emerging from an experimental vineyard planted in 1993, the Waupoos Estate Winery, the appellation's first, opened in 2001. For the most part, the thirty-one wineries grow their own grapes. The county produced 26,186 cases in 2015.

Wines of British Columbia

There were 254 wineries in British Columbia in 2014 clustered in five recognized wine regions: the Gulf Islands, Vancouver Island, Fraser Valley, Similkameen Valley, and Okanagan Valley. Together they produced 8.5 million liters in 2005. British Columbia produced 21.4 million liters on 10,260 acres in 2015. Cool-climate varieties grow at higher elevations while Bordeaux varieties are found in coastal areas

Coastal Wineries

The Vancouver Island, Gulf Islands, and Fraser Valley regions all lie in close proximity to the gulf of Georgia (the southeastern side of Vancouver Island). Long known as a fruit-growing region, this area began wine making in earnest with the rise of Prohibition in the United States. The winters are very wet, but the summers are dry, creating a challenge for the grapes. Winkler region one grapes are most suited to these three districts.

Okanagan Valley

The Okanagan and adjacent Similkameen valleys are desert river valleys that drain the center of the Canadian section of the Columbia Plateau. Lake Okanagan is 155 miles long, but rarely more than three miles wide. The summers are sunny and warm. The soils are mostly sandy in the south and contain more clay as one explores northward. As a result, there is a great diversity of grape varieties grown in this long linear valley. The growing conditions here are not significantly different from that of parts of the Rhone and Rhine river areas. Hillside plantings facing the lake are the rule. Expansion over the crest of the ridge nearest the lake is occurring in response to demand. Global warming and continued high prices will undoubtedly lead to further expansion throughout Canada.

SUMMARY

Wine is a growing concern in North America. The number of wineries increases annually as more people enter the industry. Most wineries are small and rely on tourists and direct sales. Determining what variety to grow and how to grow it in each region is better understood each year. Nearly every North American vineyard was planted with the dream of joyous harvests, but many failed through ignorance, intervening opportunities, troubling times, and disease (human and plant). The first person to plant vines in an area was usually ridiculed, then emulated by neighbors when they proved successful.

Chapter 10

Frontiers

In the twenty-first century the grape is finding its way to new regions and returning to some old ones. This chapter reviews the wine industry in select nations where pioneering efforts are underway. Viticulture and wine making among the Islamic Mediterranean countries will remain curiosities and tourist destinations. Israeli producers are doing well supporting their niche market. Thai and Yunnan Chinese wineries are marvels producing two crops annually. China and India are the two nations with enough available land to become major producers. China has already planted land enough to rival Europe's most prolific countries.

When a pioneer succeeds in a region, the international wine industry swoops in—a sure sign of burgeoning profits. Both Chinese and Indian wineries are recipients of their expertise and funding, helping these industries mature.

CHINA

The people between the Caucasus and the Pacific consumed alcoholic beverages derived from fruit and milk. Many of these beverages are strange to the modern world. Products like fermented mare's milk was the most common intoxicant on the grassy plains of Asia's interior. Asian people fermented nonvinifera grape species wherever they were found. Plums and lychees, with nearly as much fermentable sugar concentrations as grapes, were also commonly fermented. These products date back at least 5,000 years, according to archaeologist Patrick McGovern.[1]

Vinifera wines were first introduced to China during the Qin Dynasty (221–207 BC) from Persia. Ancient Greek, Persian, and other invaders settled in western China where their descendants still make wine. Chinese records indicate that grape wine, imported from the West and available at unpredictable intervals at court, was a rare novelty. Li Bai, a famous Chinese poet of the seventh century, earned the title "Immortal of Wine" because of his love of alcohol; nearly 20 percent of his poems are about drinking.

The growing affluence of the Chinese is creating a huge demand. Over the last decade plantings grew by 20 percent each year. The growth in wine making is not keeping up with the growing demand and imports are also increasing annually. Since 2013 the Chinese have annually consumed more red wine than any other nation.

In China the most common drink was (and remains) either a rice or millet beer/wine, but wine is increasingly popular. Chinese vineyards are expanding rapidly. Vineyards now cover two million acres across several appellations. Yunnan in the south has the potential for two harvests annually. Desert Xinjiang became a wine-producing region in the fourth century CE when Greeks settled there. The remaining Chinese wine regions are located in the north central and east of the country. Wuwei, Helan Mountains, and Shacheng District are in the cool hills where tea grew formerly. Bohia Bay is in a protected coastal area climatically similar to the eastern United States. The northeast wine region is along the North Korea border and is a major producer of ice wine.

Partnerships with Western wine companies are leading to a transformation of the industry. The first company to venture into China was Remy Martin, which co-founded Dynasty wines. In 2012 the value of Chinese wine production exceeded seven billion dollars. The Chinese wine industry, between 2007 and 2012, grew by 21.3 percent each year.[2] Look to northerly arid provinces like Ningxia for continued vineyard expansion.

JAPAN

Grape production began in Japan in the eighth century. The wine industry is located in Yamanashi Prefecture where there are about eighty wineries. Production is based on the Koshu variety, which provides both table and wine grapes. Aaron Kingsbury, in his analysis of Japanese wine, revealed the historic opinion of Koshu drinkers, "My grandfather always said, 'I want to drink Japanese sake. But we are poor, so I drink this terrible *budoshu* [wine]. It was sour and oxidized and good for nothing but getting totally hammered.'"[3]

Modernization and an emerging pride in Japanese wine have led to a revitalization of the industry and increased consumer popularity within Japan.

THAILAND

On the Malay Peninsula the Hua Hin Hills Vineyard and GranMonte Estates are Thailand's pioneers. Tropical Thai vineyards are harvested two times a year. GranMonte boasts forty acres of French varieties. Siam Winery, the nation's largest, produced 300,000 bottles in 2010.[4]

INDIA

In 2001, policy changes and reduced licensing fees created a boom in vineyard development. The Nashik region of India is home to a developing cluster of vineyards and wineries. Located northeast of Mumbai, at elevations around 2,000 feet, are the wineries that are key to the region's success. Sula winery, one of India's most popular wineries, is located here with about fifty other wineries. These wineries are well positioned for a massive increase in domestic wine consumption, if they can plant enough land. Production increased by 1.6 million liters between 2009 and 2014. Imports, however, grew by the same amount.[5]

"In the next 10 years there will be 300 million upwardly mobile Indians who can afford wine and for [whom] it will be a lifestyle choice," said Ranjit Dhuru, owner of Chateau d'Ori winery.[6] Dhuru's winery produced 300,000 bottles in 2008. Further south there are select sites suitable for vineyards in Karnataka.

ISRAEL

The first modern winery, in what was then Palestine, was established by Rabbi Mordechai Avraham Shor in 1848. Since independence the Israelis have aggressively expanded vineyard acreage. Israel, in 2012, had 4,800 hectares under vine, double the area in 1995. Cabernet Sauvignon represents 27 percent of the total production, Merlot 21 percent, Carignan 17 percent, and Shiraz 7 percent. White grapes represent only 23 percent of total production. Israel's largest winery, Carmel, produces ten million bottles annually.

Israel has five uncontrolled voluntarily defined wine regions. Galilee is the northernmost, largest, and most productive region in the country. The Golan Heights, captured from Syria in 1967, produces more than 600,000 cases of wine annually. The Shomrom region is on the coast south of Haifa and includes the Carmel Mountains. Samson is on the coastal plain south of Tel Aviv. Samson plantings represent 27 percent of Israel's total plantings, but produce mostly inexpensive wines. Surrounding Jerusalem are the approximately thirty Judean Hills wineries. Wineries in this region are mostly kibbutz operated. The Negev region contains a few irrigated vineyards.

MAGHREB

The Maghreb is the northwest coastal area of Africa encompassing the countries of Morocco, Algeria, and Tunisia. Climatically the coastal region is much like Mediterranean France. All three countries became French colonies in the late 1800s as part of the "scramble for Africa." Algeria was colonized in 1830 and at first colonists were a trickle. Following the arrival of phylloxera, France needed the farmland. Jobless and wineless, nearly a million French migrated across the Mediterranean and attempted to put down roots.

The native population was poorly treated and many were simply evicted from their property. The French Foreign Legion was created to protect French immigrants and their vines from the ousted inhabitants in the same way the U.S. cavalry protected settlers from Native Americans. The Algerian desert was the scene of epic battles between the French Foreign Legion and Arabic tribal groups who wanted their land back. While colonized, the Maghreb colonies annually produced and exported more wine than the French homeland. Most French colonists returned to France during the 1960s after these nations gained independence

Since independence, vineyards in these countries have had divergent histories. The Maghreb governments generally support wine producers because they make exportable products. The Muslim prohibition against alcohol is not legally enforced in these countries, but strong social pressure is exerted within the community. Tunisia, according to Wine Institute statistics, has the highest per capita wine consumption of any Muslim nation.

ALGERIA

Algeria is the largest of the Maghreb countries. Its northern sections have a Mediterranean climate. South of the Atlas Mountains are large areas of alternating sandy and rocky deserts interspersed by artesian oases. Algeria produced wines long before the Romans conquered the territory more than 2,000 years ago. From 700 to 1830 Algeria was nominally part of the Ottoman Empire. Algeria was annexed by the French in 1830. During French colonial occupation, nearly a million French migrated to Algeria, escaping the phylloxera that devastated vineyards back home. In an economic history of Algerian wine, Giulia Meloni and Johan Swinnen reported, "Between 1880 and 1900, the area under vines increased from 20,000 to 150,000 hectares."[7] The French immigrants forced out the occupants and took the best lands where they planted wine grapes

During the colonial period wine produced in Algeria was classified as French. Production peaked in 1938 at 22 million hectoliters, making it the fourth ranked wine producer in the world at that time. Meloni and Swinnen also state, "In 1960 Algeria was the largest exporter of wine in the world. . . . it exported twice as much wine as the other three major exporters (France, Italy, and Spain) combined."[8]

With independence, in 1962, the Algerian government retained wine production for export. Algeria remained the world's largest exporter of wine from 1961 to 1974 according to the United Nations FAO.[9] Today, there are less than 40,000 hectares of vines; they are mostly in Oran Province, which produced 420,000 liters in 2001. Mascara is the most famous wine region in Oran.

MOROCCO

The Phoenicians, then Greeks were the first to bring wine culture to Morocco more than 3,000 years ago. Moroccan wine exports peaked in 1964 with 194,570 tonnes. Morocco became independent of France in 1962, and in 1967 the EEC (now EU) limited imports. The 1968 exports were only 67,413 tonnes. Since 1985 in only one year have exports exceeded 10,000 tonnes. In 2008 exports were only 5,987 tonnes. The wine from the 2008 grapes sold for more than $11,000,000 (UN FAO, 2010).

Approximately 75 percent of the wine produced is red with Carignan dominating, about 20 percent rosé, and 5 percent white. There are five designated Moroccan wine regions, containing among them fourteen quality regions. Generally, the higher the vineyard's altitude, the higher the wine quality.

TUNISIA

In ancient times Tunisia was called Carthage. Also an ex-French colony, Tunisia produced large volumes of wine before World War II. With independence and the resumption of land control by native Arabic peoples, wine production plummeted. In 2000 Tunisia produced about 40,000 tonnes, decreasing to 23,000 in 2011. National export value was less than $3 million. Grenache, Muscat of Alexandria, and Carignan are the most widely grown varieties.

Chapter 11

Modern Wine Containers

Throughout history, transporting wine was troublesome. An ancient winery discovered in Armenia contains heavy walled clay vessels. Ceramic containers for transport came slightly later. The terracotta amphora became the standard vessel for storage and transport. The oldest amphorae known (4800 BCE) come from north central China from a site called Banpo. Amphorae came into general use in the Mediterranean about 3500 BCE. Varying greatly in size, volume, and contents, amphorae were the plastic bottles of their day, except that they were larger. During the Roman era the wine amphora had a standard volume of about ten gallons and a quart (39 liters).[1]

The wine bottle was commonly used during the Roman era as a serving and mixing vessel, in the way that a wine carafe is used at restaurants today. The bottle began to be used as a vessel for selling wine shortly after England and Portugal signed the Treaty of Methuen about 300 years ago.

Some of the best examples of old wine bottles are in the Berry Brothers collection. These bottles held port (see Figure 11.1). Each bottle has defined parts dictated by history and technology. For example, bottles need to narrow at the opening. This reduces sloshing and allows the cork to seal its entire length. By 1700, bottle technology was rapidly replacing barrels for quality products. Since then, the bottle has become integral to wine preservation and consumption.

PARTS OF THE BOTTLE

The basic wine bottle has seven named features (see Figure 11.2). Crowning and sealing each bottle is the capsule, made of materials ranging from heavy

Figure 11.1: The evolution of the port bottle. (H. Warner Allen, Faber & Faber)

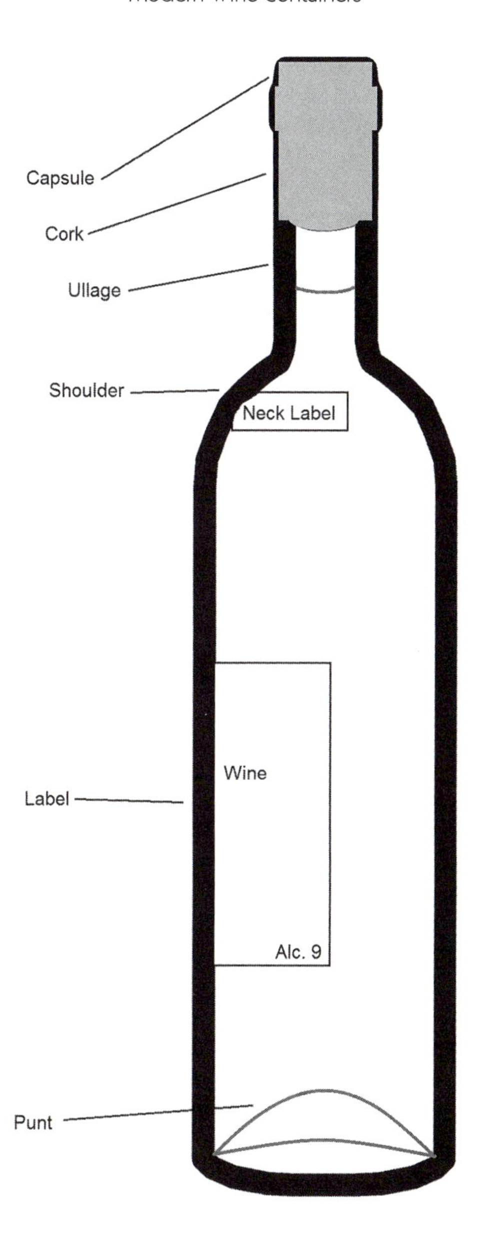

Figure 11.2: Parts of a wine bottle.

foil to shrink wrap. Underneath the capsule and inserted into the bottle is a cork. Between the bottom of the cork and the top of the wine is a small airspace called the ullage. Vintners concerned with quality fill the ullage with nitrogen or argon to prevent oxidation. An enlarged ullage in aged wine indicates imperfect bottle sealing.

The neck is the top segment of the bottle. Ideally it is straight sided throughout its length so it can hold the cork for its entire length to achieve a complete seal. From the neck the bottle widens out into the shoulder. Shoulders vary by region in Europe and are a good clue to the wine source. Wineries in non-European regions often use a bottle style that mimics the style of the region their wine emulates. Many wines wear a small label on their neck. The body of the bottle must have a front label and may have a back label (discussed further below). The depression at the bottom is called the punt.

BOTTLE SIZES

Older wine and liquor bottles were one-fifth of a gallon (25.6 fl oz, 0.72 l), but in 1979 the United States fixed bottle size at 750 ml. Table 11.1 lists names, sizes, and comparative capacity for wine bottles. Obviously, the larger the bottle the rarer its existence. Bottles like the Methuselah and larger appear at victory celebrations.

EVOLUTION OF THE BOTTLE

The first modern bottles were hand blown using rough molds made from partially hollowed-out tree trunks. Blown glass, especially in the early years, was inconsistent in volume and glass thickness. English glass made before 1615 was melted over wood fires. These bottles contained imperfections and air bubbles. Wood historian Joachim Radkau relates the story: "In 1615 James I issued a strict ban on the use of wood for glass production, since 'the great waste of timber in making glass is a matter of serious concern.'"[2] Glass makers were forced to switch to coal, whose higher heat output created purer and more free-flowing glass.

The first bottles were, squat, thick, and onion shaped. This shape, with its low center of gravity, proved ideal for shipboard use. During the ensuing one hundred years the bottle became taller as the glassmakers became more skilled. Figure 11.1 depicts how port bottle quality improved over time. The dissymmetry of the bottle necks from 1741 through 1770 demonstrates the individuality of hand-blown bottles and how a cork's seal was unlikely to be continuous. The interior of the necks of the earliest bottles are triangular. They held the cork only at the top of the bottle. The cork bottom was loose. During this period bottle makers strove to make the neck longer,

Table 11.1: Wine Bottle Sizes, Names, and Equivalences

Bottle Name	Metric Volume	U.S. Volume (qt)	Standard Bottle Equivalence
Split	187 ml	0,1976	¼ bottle
Half-bottle	375 ml	0.3963	½ bottle
Jennie	500 ml	0.5283	none
Bottle	750 ml	0.7925	1 bottle
Magnum	1.5 L	1.585	2 bottles
Double magnum	3 L	3.17	4 bottles
Jeroboam	4.5 L	4.755	6 bottles, half case
Imperial	6 L	6.34	9 bottles
Methuselah	6 L	6.34	9 Champagne only
Salmanazar	9 L	9.51	12 bottles, one case
Balthazar	12 L	12.68	16 bottles
Nebuchadnezzar	15 L	15.85	20 bottles

internally rounder, and stronger. Only then could the neck hold a cork through its whole length. The evolution of the modern wine bottle was essentially complete by 1770. Since 1770, regional variations in shoulder style and color evolved.

REGIONAL BOTTLE STYLES

As the European glass industry developed the character and shapes of the bottles, at first with individualistic shapes and multiple flaws, the bottles became regionally standardized. Identifying the source appellations for many Old World wines is simplified if one can read bottle shapes. Table 11.2 organizes the regional bottle shapes by their properties and characteristics.

Unfortunately, these traditions are breaking down. German wines today are often sold in Bordeaux bottles. Some New World winemakers adopt bottle shapes to conform to the wine style they imitate. As often as not, however, small wineries select a bottle for economy and convenience. They bottle Chardonnay, Riesling, and Merlot in the same style bottle. Large corporations or individual regions may select a unique shape for marketing purposes, but these efforts are generally short-lived.

The three most common bottle styles are Burgundy, Bordeaux, and German. In each case the bottle becomes associated with the wine it contains, fixing in the consumer's mind bottle style and color with the flavors of the wine. The German bottle is tall, slender, and without shoulders. In Germany brown bottles are used for Rhine wines, blue bottles for wines from the Nahe, and green bottles for the remainder of the country. Alsace winemakers

Table 11.2: Regional Bottle Styles

Region	Sides	Shoulder	Neck	Punt
Port	Straight	High	Bulbed	Pronounced
Sherry	Straight	High	Straight	Pronounced
Bordeaux	Straight	High	Straight	Pronounced
Burgundy	Straight	Sloping	Thin, slightly cone shaped	Small
Rhine, Mosel, Alsace	Slight slope	Sloping	Elongated	none
Franconia, Portugal	Oval, flattened	Rounded	Straight	none
Sparkling	Straight	Sloping	Conic, narrow	Pronounced
Old Chianti	Rounded	Rounded	Straight	none
New Chianti	Straight	High	Straight	Pronounced
Jura, Clavelin	Straight	Concave	Tapering	none
Cote du Rhone	Straight	Conic	Tapering	Slight

use green German-style bottles. The Bordeaux bottle has steep, high shoulders and is dark green in color. The Burgundy bottle is short, broad, and usually has a significant punt.

Other less common bottle styles are closely tied to their region of origin. The Provencal bottle has a bulge in the center and has a flared base. Common before 1960, the Chianti flask has largely vanished because consumers associate it with the cheap, poorly made Chianti of that era. Today's Chianti is mostly sold in Bordeaux bottles. The Loire and Sparkling wine bottles are similarly shaped, but Sparkling wine bottles have thick glass, narrow opening, and a deep punt. Port bottles are similar to Bordeaux bottles except they have a bulbous neck. Sherry bottles have the steepest shoulders and a long, thin neck. Through an EU decree in 1995, the goat-scrotum-shaped Boctesbutel may only be used by vintners in Franconia and Portugal.

SEALING THE BOTTLE

Cork

Cork is the traditional sealing media for wine bottles. Cork is the bark of the cork oak tree. This tree grows in Spain, Portugal, Morocco, and Algeria. This remarkable tree has an ever-growing bark that can be stripped off the tree every seven to ten years without harming the tree. The cork oak bark is elastic, nonconductive, watertight, and compressible. The cork is not completely airtight. Over long periods, oxygen penetrates the cork and slowly oxidizes the wine; this is the essence of aging. Often air leakage is not the fault of the cork but of irregularities in the bottle itself. As the demand for wine increases, the demand for cork has also grown. In response, some have started using composite corks made from cork chips glued together. Some benefits of cork include the following:

- Lightness—Cork is light and will float on water. Beneficial for buoys, floats, fishing rod handles
- Elasticity—The cellular membranes are flexible so that the cork can be fitted against the wall of a bottle under pressure (the gas in the cork cells is compressed, reducing volume) and when released maintains its original form.
- Impermeability—Cork does not rot due to the suberin, which makes the cork largely impermeable to gases, liquids, and bacteria. Suberin is a material consisting of lipids and phenolics that does not absorb water. It lines cork cell walls and prevents the passage of water.
- Low conductivity—The gaseous elements are sealed in tiny compartments in cork cells, insulated and separated from each other, providing low conductivity to heat, sound, and vibrations. Cork is one of the best natural insulators.
- Resistance to wear—Cork's honeycomb structure resists wear. It does not absorb dust, and it is fire resistant.

Synthetic Corks

Today, the demand for bottle sealing has exceeded the supply and vintners are searching for other sealers. Many have turned to a plastic plug. From the consumer's perspective, the synthetic cork acts like the real thing. The plastic plug has the advantage of being free from cork taint (usually associated with excess chlorine from processing), but it remains in a landfill forever. It has the disadvantage of not allowing cellared wine to slowly oxidize from gas exchange through the cork. Synthetic corks further lack the memory of natural cork. As a result, it is difficult to re-stopper a bottle with a synthetic cork.

Corkscrews

Once the cork and bottle became standard, a device for removing the cork from the bottle was required. Necessity caused the invention of the corkscrew. Andre Simon, an early twentieth-century wine expert, traced the origin of the corkscrew to 1732. According to Simon, the following was penned in 1732 and represents the first written record of the corkscrew:

The bottlescrew, whose worth, whose use,
All men confess, that love the juice.
Forgotten sleeps the man, to whom
We owe the invention; on his tomb,
No public honors grace his name,
No pious bard records his fame.[3]

The corkscrew was modeled on a tool called the bulletscrew. The bullet-screw extracts stuck bullets from gun barrels. The first corkscrew patent was issued in England in 1795. Initially called a bottlescrew or steelworm, the corkscrew was in common use by 1725.

There are several types of corkscrews and numerous patents have been issued for them around the world. William Rockwell Clough is credited with designing and manufacturing more than one billion corkscrews. In all, he holds forty-three corkscrew patents.[4] The basic corkscrew is shaped like the letter T with the down stroke being the screw and the horizontal stroke being the handle. A common variety of corkscrew employs one or two levers to assist in lifting the cork. Less common are cork expellers; most use forced air to lift the cork from the bottom. However, the wine captain, also known as the waiter's friend, is the single most versatile and common corkscrew. It includes a small knife for cutting the foil, a folding screw, and a folding lever. The lever includes an integrated bottle opener.

Many corkscrews were integrated into other kitchen hand tools. Novelty corkscrews shaped like people and animals abound. Donald Bull's Virtual Corkscrew Museum is a great place to see the range of these devices. Novelty or not, the importance of the corkscrew was absurdly overstated by W. C. Fields's character in the film *My Little Chickadee* when he says, "During one of my treks through Afghanistan, we lost our corkscrew. We were compelled to live on food and water for several days."[5]

Screw Caps

Long associated with inferior wine, fruit wine, and wine coolers, screw caps were disdained by consumers. Beginning with the 2005 vintage, Australia commenced a national program of screw cap closure for wine, culminating in 2009 when nearly all wines were exported with screw caps. The Australian national decision has been followed by a number of wineries in other regions and among the middle-level offerings of top wine companies. Industry changes have practically forced the elimination of the screw cap stigma.

Screw caps, like plastic plugs, form a complete seal preventing slow oxidation. Screw caps, however, are easily resealed, limiting oxidation. For wines designed for consumption within a year or two of bottling (as most are), this is not a concern, as the wine does not have time to mature over that short period. Even though they do not go "pop" they are the wave of the near future.

The Capsule

Whether made of metal foil, wax, or shrink-wrapped plastic, the capsule covers the cork. The capsule provides the final closure on the bottle and at one time protected the top of the cork from chewing rodents and insects capable of boring into it. Generally, the need for such precautions are no longer necessary. Finding and operating the capsule release cord is always difficult, and many times servers must resort to the little knife on their wine captain to get at the cork. Technically, the capsule does help in counterfeit reduction and tampering identification, but otherwise it is about as useful as a necktie.

THE WINE LABEL

The label is the most influential factor when selecting a wine. The label is what entices a buyer to select one wine over another. The label provides the necessary information to distinguish one wine from another. The name of the wine itself is often a key selling point. Whimsical or erotic terms are sure to have people reaching for bottles. Terms such as Chateau, Ridge, or Hills add to the retail price. Colorful, vibrant, and erotic labels attract customers as well.

Every wine-producing nation requires its commercial wines to specify the alcohol content, the nation in which the wine was produced, and the winery name. After that, national requirements vary. Experienced wine buyers know what key words to look for on the label when purchasing wine. Wine labels, like labels on other products, must be truthful and verifiable by the national government of the producing country. Further, governments restrict the use of certain terms, mostly specifying region of origin, to wines from those locations. International incidents result when one country tries to assume wine terminology associated with a region from another country.

In the United States, every wine container sold to the consumer must have pregnancy and sulfite warning. (Some organic wines possess sulfite levels below the notification-triggering threshold.) The label must identify the alcohol content and give the vintner identification. Any remaining information on the label is optional, but it must be true. To specify a varietal wine, the juice must be from grapes of that variety. To specify a vintage, the juice must be from grapes grown that year. To specify an appellation, the juice must be from that region.

Potentially, there are three wine bottle labels: the front label, the back label, and the neck banner. The neck banner is the least common and often contains little information. The back label generally provides information about the taste and character of the wine.

THE PUNT

The punt is an indentation at the bottom of the bottle. It serves several purposes. The greater surface area offered by the punt allows more rapid chilling than a flat-bottom bottle. The punt directs internal pressure from the bottom to the sides, reducing the likelihood of the bottom shearing. Both qualities are particularly significant for bottles designed to hold sparkling wine. It is probably serendipitous that the circle of contact (versus a flat bottom) more readily establishes a stable base for the bottle on rough surfaces. Aesthetically, the punt makes the bottle appear larger and may affect consumers who may think they are getting more.

BARRELS

The Celts, living in forested northern Europe, had highly developed wood technology. Our concern is one specific piece of that technology, the barrel (see Figure 11.3). The barrel's design is elegant in its simplicity and functionality. It has a large capacity and is lightweight. The barrel is fattest in its center, called the bilge. On its side, filled to a gross weight of 600 pounds, the barrel is easily rolled by one person. The extension of the staves beyond the head make it easily grabbed by a crane and hauled aboard a ship or wagon.

Since 500 CE the barrel has been the principal vessel for fermenting, aging, and storing wine—where it was available. Some of the most common

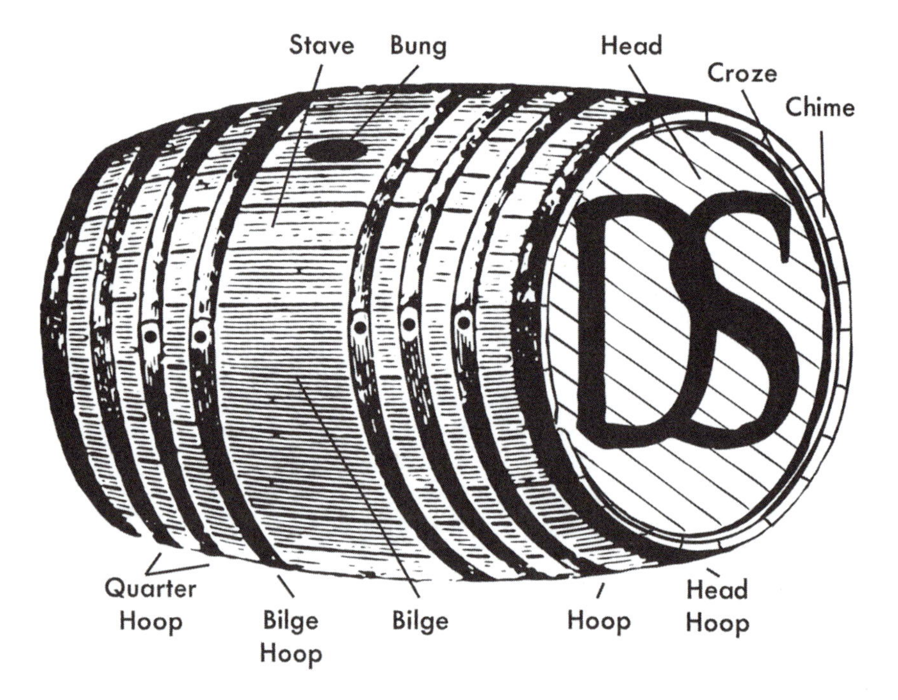

Figure 11.3: Parts of a barrel.

images of the Middle Ages show men working treadmills to raise and lower barrels of wine. Oak was most popular for barrels because of its strength. Later people began to enjoy and manipulate the taste oak wood imparts to wine. With oak there is the transference of tannins from the oak to the wine and the extraction of bitterness from the wine to the oak. The transfer is primarily manipulated by toasting the barrel's interior.

Coopers make wine barrels from several oak species. Each oak species contributes differently to the wine's ultimate flavor profile. European oak species include *Quercus petraea* (aka *sessiliflora,* sessile, and *rouvre*), a slow-growing, dense-grained wood, and the more common *Quercus rubar.* American white oak, *Quercus alba*, is the preferred North American oak species. Depending on the forest of origin, alternative North American species such as the swamp white oak, overcup oak, Durand oak, swamp chestnut oak, and chestnut oak may be substituted. *Quercus rubar* imparts a vanillin flavor while *Quercus alba* imparts the familiar oaky aroma. All oak species are rich in tyloses, which block the flow of liquids through the wood.

The most prized oak comes from the cool central highlands of France (the Massif Central). Napoleon originally had these forests planted for shipbuilding. There oak trees grow little during the short growing season. Trees are usually between 100 and 120 years old when harvested. Only the trunk below the first branch can be used for wine barrels. The wood grain is made perfectly straight by densely planting trees, forcing them to grow tall and branchless. The average tree produces enough wood for only two barrels. The rest is used in furniture making. *Wines and Vines* reporter Andrew Adams filed this report, "In June of 2013 the French coopers' trade association Tonneliers de France reported that its 49 members sold a total of 532,990 barrels. . . . the United States accounts for 60% of its members' sales."[6]

To make a French oak barrel, strips of wood, called staves, are cut to length and split to a thickness of half an inch to an inch depending on the size of the barrel. Splitting the staves, rather than sawing them, provides a subtle transfer of flavors between wood and wine. American oak is naturally more watertight; therefore staves may be sawn. The broad stave edge does not cross the grain or else leakage will occur. The staves are aged at least two years before construction begins. Steam-treating staves creates a lengthwise bend. The staves, averaging one inch wide, are assembled in a cylindrical form. Iron rings placed around the outside of the ring of wood keep the staves in place. A wooden disk called the head is placed at one end. At this point the barrel is closed at one end and flared at the other. The barrel is then heat treated, ideally with excess wood chips, while dampened with sponges. The heating softens the wood and allows the cooper to hammer hoops into position over the flared end.

The chemical composition of the barrel's interior is modified by toasting. The most important chemicals transferred are lactones. Lactones possess the aroma of coconut. Vanillin transfer peaks at low toasting levels. Guaiacol and eugenol adds smoky and clove aromas. Their concentration increases with toasting level. Furfural, derived from sugar and carbohydrates in the wood, transfers caramel and almond aromas to the wine. Tannins extracted from the wine are called ellagitannins. Ellagitannins produce the astringent character of young red wines. The above chemicals blend uniquely in each barrel of wine, producing a unique flavor and aroma profile.

The size of the barrel has a great influence on the speed of maturation. The larger the barrel, the slower the process because of reduced wine-to-surface-area access. Table 11.3 lists barrel names, characteristics, and sizes.

BLADDERS AND NEW-FANGLED INVENTIONS

The wine bladder originated thousands of years ago when our ancestors learned how to make watertight containers from animal hides. The use of animal hides for storage was popular in dry lands throughout the Old World. Spaniards used skins for wine storage until after World War I. The skins imparted an interesting flavor even before oxidation turned the wine to vinegar.

The use of plastic bladders for transporting and distributing fluids began in the late 1960s. Twenty years later, vintners began distributing their wines

Table 11.3: Summary of Barrel Types, Capacities, Oak Species, and Purpose

Barrel Type	Capacity (L)	Oak Source	Purpose
Gorda	700	American	Whiskey
Madeira drum	650	French	Madeira
Port pipe	522.79	European	Port
Butt	572.8	European	Sherry
Machine puncheon	500	American	Varied
Sherry puncheon	500	Spanish	Sherry
Barrique	300	Varied	Varied
Leager	681.9	Varied	Varied
Hogshead	286.4	American	Varied
American standard	200	American	Varied
Quarter cask, firkin	50	Varied	Varied
Bordeaux	225	French	Wine
Burgundy	228	French	Wine
Tun	982	Varied	Varied
Blood tub	40	Varied	Beer

Source: Russ Rowlett, *How Many? A Dictionary of Units of Measure*. Chapel Hill, NC: http://www.unc.edu/~rowlett/units/index.html, accessed October 11, 2016.

in large-capacity (5 L) units. The advantage of the bladder over bottles is its airtight nature, inexpensive packaging, and volume discounting. The wine can be consumed slowly without oxidation because of the one-way valve on the bladder. The bladder presents tolerably good wine at otherwise ridiculously low prices.

Bob Walker, vice president of marketing for the Wine Group, cited the advantages of the bladder bag this way:

> Our Winetap® package [bag in a box] lets consumers enjoy the same taste with the last glass as they enjoyed with the first, something that is rarely possible with bottled wines. . . . The Winetap® guards against oxidation and maintains the complex bouquet that makes wine as pleasing as a beverage, rather than letting it volatilize away.[7]

The most recent packaging trend is individual-sized wine containers, ranging from cans to sealed plastic stemware to personal-sized wine pouches. The individually sized containers are usually tolerably good, but consumers remain leery of this packaging. Undoubtedly, these containers will become more prevalent.

Chapter 12

Cultural Geography of Wine

When you talk about wine, you have to talk about politics.[1]

—Tyler Colman (aka Dr. Vino)

While making every effort to focus the following discussion on wine alone, some of the following passages speak to generic alcoholic beverage manufacture, sale, and consumption.

Wine, as identified by Donald Horton, satisfies four basic human needs: (1) hunger and thirst, (2) medicine, (3) religious ecstasy, and (4) social jollification.[2] Together these characteristics make wine an important factor in human physical and mental health. Past societies channeled alcohol consumption through festivals, games, and deities, and it occurred at specific locations, sporting venues, parades, temples, and taverns. Modern festivals continue the tradition in the form of Mardi Gras, New Year's Eve, St. Patrick's Day, and college homecoming weekends. Beer advertising supports professional sports teams, but not Little League (it is illegal).

Americans are wary of overconsumption for it often leads to raucous behavior, especially in large groups. Society, therefore, restricts locations, times, and permissible quantities. Pressure to conform is great, ranging from friends suggesting you've had enough to prohibition or incarceration. Social pressures regarding allowing children to see wine consumption are strong in the United States. The social pressures exerted usually result in laws controlling what can be sold, whom it can be sold to, who can sell it, where it can be consumed, and who can see consumption (not children).

If wine were discovered and released to the public today, it would be quickly banned as a dangerous drug. Wine, however, entered the mainstream of the Western tradition in the remote past and can be suppressed, as

Prohibition demonstrated, only by quashing individual rights, just as current attempts at suppression of intoxicants from other traditions (tobacco, marijuana, opium, peyote, betel, psilocybin, and coca) are and will always be failures as long as those traditions exist. Throughout history, every attempt at declaring a substance to be contraband leads to smuggling, increased demand, increased sales, disrespect for legal systems, and a search for a more potent alternative.[3]

WINE AND WESTERN CIVILIZATION

From the earliest origins of Western civilization comes the Epic of Gilgamesh. In this ancient tale Enkido (Ink e do) is a wild man, raised by animals without human contact. Wine, with bread and oil, is one of the three symbols of civilized life:

> Enkido, eat bread, it is the staff of life; drink the wine, it is the custom of the land. So he ate till he was full and drank strong wine, seven goblets. He became merry, his heart exulted and his face shone. He rubbed down the matted hair of his body and anointed himself with oil. Enkido had become a man.[4]

Even during the time of Gilgamesh wine was an important symbol of civilization's luxuries over the barbarian wildness. Most symbolic linkages for wine began long before Dionysus or Judaism. Edward Hyams says, "The mystique of wine as an attribute of ancient order and sophistication is as powerful as that of ancient tongues or the ancient rules of architecture."[5] Ever after, wine was heaven sent and for the best people first. "Lands where the humblest peasant daily drank a liter or more were the envy of beer drinking neighbors. That wine was the symbol of wealth, prestige, intoxication, and joy only added to its diffusion and the ardor with which adherents cling to it."[6]

As a plant that thrived where other food crops failed, the grape became a symbol of fertility in the dry lands of Mesopotamia and the Mediterranean basin. This was reinforced by the socially lubricating effects of wine. Several early civilizations were enraptured by wine. Wine and its transforming effects, coupled with the fecundity of the plant itself, became the basis of early religious cults. Hebrew and Christian ceremonies require wine. Throughout the Christian–Muslim wars of the past 1,200 years and the Catholic–Protestant wars of the past 500 years wine, and opinions about wine, were group-defining cultural traits.

The customs and traditions of many peoples residing outside wine-producing regions require wine consumption and reflect the traditions of others, long ago. Take, for example, the role it plays in even secular Polish weddings:

> The sharing of the bread, salt and wine is an old Polish tradition. At the wedding reception, the parents of the bride and groom greet the newly married couple with bread, which is lightly sprinkled with salt and a goblet of wine.
>
> With the bread, the parents are hoping that their children will never hunger or be in need. With the salt, they are reminding the couple that their life may be difficult at times, and they must learn to cope with life's struggles. With the wine, they are hoping that the couple will never thirst and wish that they have a life of good health, and good cheer and share the company of many good friends.
>
> The parents then kiss the newly married couple as a sign of welcome, unity, and love.[7]

As a defining trait for so many peoples and over such a great period of time, wine and the vine are diversely symbolic. A partial listing of religious and life-cycle symbolism associated with the vine and its products is shown in Table 12.1.

Table 12.1: Wine and Vine Symbolic References in Western Civilization

Trait	Symbolic Meaning
Bud burst	Link to fertility cycle, birth, rebirth
Bocksbeutel	Fertility (bottle shaped like goat scrotum)
Corkscrew	Sexual penetration
Cork popping sound	Sexual orgasm
Destruction of wine	Political statement, religious statement
Dionysus leaning on a broken pillar	Suggests the havoc that overindulgence can cause
Fermentation	Divine fluid that transforms itself (water to wine), takes the place of blood in ceremonies
Fermentation	Bubbling action = magical transformation of juice to wine
Flowering	Link to fertility cycle
Intoxication	Communion with . . .
Glass breaking	Finality, conviction, sacrifice, dedication
Grape	Holds the secret of rebirth (seeds)
Libation	Prayer, offering, remembrance
Leaf fall	Link to death
Pressing grapes	Sacrifice, flowing blood
Toasting	Ritual—ritual behavior, prayer
Vine	Death and rebirth (perennial)
Winemaker, process of wine making	Knowledge of God's ways, talking to God, making wine considered a religious activity
Wine, bread, oil	First agricultural products, symbols of civilization

(Continued)

Table 12.1: (Continued)

Trait	Symbolic Meaning
Wine (good)	Goodness, joy of life, communion with God, breaks down barriers, religious opposition to atheism, purity, clean
Wine (bad)	Revealing, destructive, excess

Table 12.2: Wine is . . .

Metaphor	Attribution
Bottled geography	Percy Dougherty
Bottled poetry	Robert Louis Stevenson
Sunlight held together with water	Galileo
Constant proof that God loves us and loves to see us happy	Benjamin Franklin
One of the most civilized things in the world	Ernest Hemingway
The most healthful and most hygienic of beverages	Louis Pasteur
Like the incarnation—both divine and human	Paul Tillich
Just a conversation waiting to happen	Jessica Altieri, *Kiss My Glass*

In addition, wine is a common metaphor for many of the good things in life. In trying to describe wine, we see further symbolization. Table 12.2 lists a few of these metaphors and their attributions.

WINE AND RELIGION

Beginning with writing, records of wine and its role in society abound. Daniel Stanislawski's analysis reveals that "[t]he earliest written records of viticulture and winemaking refer to the use of the beverage in temples . . . of the fourth millennium BC."[8] Historian Edward Hyams describes how one of the earliest civilizations deified wine:

> One of the oldest of the Sumerian divinities was a wine-goddess called either Gestin or Ama-Gestin, which apparently means "Mother Vine-stock"; there was a temple [to] her in Lagash." [Other Sumerian wine deities include] Pa-gestin-dug, which apparently means "the good vine-stock." His wife was a goddess with two names according to her function at the moment; Nin-kasi, "the lady of the inebriating fruit," and also Sa-bil, "she who causes burning." We may safely take this pair

> as wine-deities. A third name by which the consort of Pa-gestin-dug is sometimes known is Kastin-nam, "the intoxicating drink which degraded life."[9]

The wine goddess lived on the "mount of the taverner." Pa-gestin-dug and Nin-kasi had nine children, each representative of an aspect of intoxication. They and their associated symbolism are listed in Table 12.3. The females represent beer and wine and the males represent aspects of drunken behavior. The male figures are easily recognizable as states of inebriation and characterize individual behavior associated with excess consumption.

From the earliest times, wine, like other intoxicants, was associated with religion. Other cultures devised cosmological legends for the divine origin of wine. Religious historian Robert Fuller says:

> Several ancient religions developed mythical accounts of the creation of wine from the body of a primordial divine being. Iranian legend recounts how wine originated from the blood of a ritual sacrifice of a primordial bull. Each year's crushing of grapes thereby reenacts this sacrificial slaughter and the resulting wine is consequently thought capable of bestowing the bull's strength, energy, and vital force upon those who consume it.[10]

Drinking sessions routinely began with a prayer. A libation accompanied the prayer. A libation is the spilling of a few drops of wine from the first wine cup in honor of the gods or in memory of a hero. The libation dates to at least 5000 BCE and probably is much older, as it was practiced in Mesopotamia. Libations were common among ancient religions and could be water, milk, wine, oil, or the blood of a sacrificial animal. There is a

Table 12.3: God Children of the Sumerian Wine God and Goddesses

Child's Name	Gender	Symbolism—Translation of Name
Siris	Female	Wine goddess of the Hamrim mountain vineyards
Sin-kas	Female	Spiced barley beer
Sin-kas-gig	Female	Another type of barley beer
M-hus	Male	The drunken brawler
Me-azag	Male	He who makes drunkards speak the truth
Eme-te	Male	He with eloquence of tongue
Ki-dur-ka-zal	Male	He who brings mirth
Nu-silig-ga	Male	The braggart
Nin-ma-da	Male	Lord of the land (hubris)

Source: Edward Hyams, *Dionysus: A Social History of the Wine Vine*. London: Thames and Hudson, 1965.

bas-relief of King Assurbanipal (688–627 BCE) pouring a libation over four dead lions.

Modern social references to the libation abound even though Europeans generally replaced the practice with the toast centuries ago. The beverage poured out is in tribute to those not present, symbolically representing what the missing person would have consumed.

The rise of Dionysus and the spatial diffusion of his cult occurred slowly. In Mesopotamia, Dionystic worship began nine thousand years ago, although for most of that time the wine god was the lesser consort of the fertility goddess. The worship of Dionysus began as a subcult among the Mycenean Greeks. The euphoria resulting from actually consuming the god led to physical effects leading worshipers to believe they were truly in the presence of Dionysus. The ability of this god to touch humans in a unique way, an ability no other god had, expanded the cult. Geographer Daniel Stanislawski completes the story:

> Only in the early centuries of the first millennium BC did events conduce to the dramatic improvement of his status. At that time the internal affairs of several nations were changed by such innovations as new types of ships, transport of products previously not common to sea trade, and the release of trade from the shackles of ancient localized systems. This economic and social flux created further consequences, [one] of which was an enhancement of the Dionysian cult. Furthermore, the conjunction of dynamic commercial growth with the intensification of the cult symbiotically gave impetus to economic and territorial expansion attested not only by history but also by effects still visible in the Mediterranean region.[11]

Greek historian Thucydides delivered the Greek parallel to Enkido's transformation and extends it from the personal to the cultural level when he said, 2,500 years ago, "The peoples of the Mediterranean began to emerge from barbarism when they learnt to cultivate the olive and the vine"[12] In his mind, wine created the opportunity for Greek civilization to flower.

The Mesopotamian wine god was the only god transferred from the old female-dominated pantheon to the new male pantheon brought in by invading Indo-European herders. Eventually the wine god took the name Dionysus among the Greeks and Bacchus among the Romans. Neither group fully trusted the wine god. Both Greeks and Romans outlawed his ceremonies, fearing the excesses they encouraged.

Philosopher Bertrand Russell in his landmark book, *A History of Western Philosophy*, explains the deification of wine:

> In intoxication, physical or spiritual, the initiate recovers an intensity of feeling which prudence had destroyed; he finds the world full of

> delight and beauty, and his imagination is suddenly liberated from the prison of everyday preoccupations. The Bacchic ritual produced what was called "enthusiasm," which means etymologically having the god enter the worshipper, who believed that he became one with the god.[13]

Jesus co-opted the transformative properties identified by Thucydides in the Dionysian Communion ceremony and even older traditions of ceremonially substituting wine for blood. As the Catholic Church solidified its doctrines, wine represented the civilized attitudes it wished to foster. Without wine the Catholic sacramental ceremonies are incomplete.

The relationship between Muslims and wine is clear: It is not permitted. The enforcement of Islamic religious prohibition can be strong at some times and in some places, not so much at others. Muslims use alcohol destruction for political symbolism. During the 1979 Islamic revolution in Iran hotel wine cellars were raided and their contents smashed (the revolutionaries did not pay for the wine). Muslim reaction to wine is a distinction separating Christian and Islamic peoples, and at times has been a focal point for tensions between the religions.

Wine enters prominently in the schism between Roman Catholics and Protestants. According to long-standing Catholic doctrine the wine in the communion ceremony is transformed from wine into Christ's blood. Martin Luther, in 1517, challenged church doctrine, counter-claiming that the wine remains wine and merely represents Christ's blood. The distinction, no matter how small you might believe it to be, led directly to the deaths of millions between 1525 and 1800. The distinction is mocked in *Gulliver's Travels* when the Lilliputians warred with the Blefuscu over which end of an egg to break. In France the religious civil wars between 1562 and 1598 killed between two and four million. In the Thirty Years' War (1618–1648) estimates range from three to eleven million killed. Millions more died in the small wars, local uprisings, and retaliations, and from murder, disease, starvation, and exposure.

In the late 1800s several Protestant sects splintered over whether unfermented juice or fermented wine was the appropriate fluid for ceremonies, termed the "two wine theory," more than a dozen authors wrote books on the impossibility of Jesus Christ ever consuming an alcoholic beverage. Juice drinkers maintain that an alcoholic beverage is inappropriate for communing with God; the exact opposite was the belief of communicants before 1880 and many since. Oddly, this distinction has not led to religious civil war in the United States.

Wine is also symbolic of opposition to communism and atheism. In the 1947 film *The Fugitive* with Henry Fonda, Fonda plays a priest secretly performing Roman Catholic ceremonies in a godless country. His search for wine to perform ceremonies lands him in trouble with the authorities and he is forced to flee to an adjoining nation. After recuperating, he returns to the godless land in continued opposition. Here, wine symbolizes the desire for the prohibited, in this case both God and wine.

WINE AND INDIVIDUAL BEHAVIOR

The Mesopotamian anthropormorphism of consumption as children of the wine god (Table 12.3) is reversed in Greek as described to us by John Maxwell Edmonds's translation of the play *Semele or Dionysus* by Eubulus.

> The characters of the children of Pa-gestin-dug and Nin-kasi are mirrored as wine-bowls in this passage by the Greek playwright Eubulus c375 BCE. "Three bowls only do I serve for the temperate: one for health, which they empty first; the second for love and pleasure, and the third for sleep. When this bowl is drunk up, the wise go home. The fourth bowl is ours no longer, but belongs to violence; the fifth to uproar, the sixth to drunken revel, the seventh to black eyes, the eighth is the policeman's, the ninth belongs to biliousness, and the tenth to madness and hurling the furniture."[14]

Whether the units of measure are minor deities, bowls, bottles, or glasses, the reality is that wine modifies behavior. The form of that behavior modification has much to do with the frame of mind of the imbiber. Yet, as thousands of years of legends tell us, there are stereotypic and common behavior patterns.

Wine, as the ancients knew, had decidedly different effects, depending on rate of intake and co-consumption of food, as one drank more. For many people the general pattern of intoxication is as follows. First comes a warm glow and a cheery disposition. Inhibitions are relaxed and one desires to interact with others. With inhibitions reduced and intoxication looming, people are apt to say and do what they normally might not. They reveal their inner selves. The happy convivial level of intoxication enjoyed by the ancients is incompatible with the operation of machinery or keeping secrets. As one drinks more, coordination is lost and the mood often shifts from joyous to sour. The drinker may become garrulous, moody, and quarrelsome as noted by Eubulus above.

Because of the tendency for wine to lead people to reveal what they otherwise might keep private, there is a strong tendency to localize drinking parties in predefined locations. Thus we have the symposium, literally "drinking party," of the ancient Greeks, the tavern of the Middle Ages, and the restaurant/bar of today. The barroom brawl is a common element of popular literature, thus defining the culturally approved location for such activities. Too much wine and the consumers become like the male children of Pa-gestin-dug and Nin-kasi. During such events, property, friendships, skin, bone, teeth, and even lives are routinely destroyed.

Helping or ordering people to curb consumption is a long-standing goal of society; recall the Roman crushing of the Bacchic cult in 187 BCE. Intoxication laws (laws limiting where consumption may occur, who may be intoxicated, and to what level) have been widely employed. Gentle efforts are offered by support groups like nineteenth-century temperance societies that guided drinkers from excess and Alcoholics Anonymous guiding people from addiction. Draconian measures, such as Prohibition, can create a whole new set of socially disapproved behaviors, starting with possession and consumption.

ALCOHOL DRINKING AGES BY COUNTRY

Most countries have laws defining who can and cannot consume what alcoholic beverages at what age and where. Islamic countries maintain prohibition or highly restricted access. Political geographer Colin Flint reveals just how restrictive prohibition can be: "A teenager reports that his father was murdered because he consumed alcohol: 'The Hizbul Mujahidden had warned him about drinking but when he didn't care they killed him.'"[15]

In countries permitting the consumption of alcohol, most set the drinking age at eighteen, the generally recognized age of adulthood. Enforcement is highly variable. The general result is a long-standing pattern of underage persons in one country traveling to other countries with lower drinking ages. Many border communities derive significant revenue from this exchange. Tijuana, Juarez, Niagara, and Windsor on the U.S. borders with Mexico and Canada are focal points for servicing these customers.

The United States has the highest drinking age of any nation, twenty-one. The statute is rigorously enforced in most jurisdictions. This suggests that either Americans require greater social control than people in other countries, or that our government is more paternalistic. The federal government imposed the twenty-one years old drinking age in 1984 using a backdoor technique. States failing to implement this minimum drinking age were denied their share of federal highway money. The federal government spends millions each year promoting restraint until age twenty-one. They spend millions more on enforcement, millions more on punishment, and even millions more studying the problem. The result of these expenditures is hundreds of thousands of Americans with criminal records for underage drinking and twenty-year-olds, unfamiliar with alcohol's effects, let loose on their twenty-first birthday.[16]

Many believe slow, gradual introduction of wine to a child's diet generally leads to a long, healthy, and respectful relationship with wine. Anecdotal evidence suggests experienced young adults are not drawn toward binge drinking. Certainly, they are more familiar with the effects and their capacity than twenty-one-year-old first-time drinkers. In countries where alcohol is withheld past adulthood, those newly permitted enter a protracted period of binge experimentation. The Association of University Presidents proclaimed that the drinking age in the United States is too high. For them, it certainly is. With the drinking age at twenty-one, they are, unjustly, held personally responsible for every student excess.[17]

WINE AS A SOCIAL LUBRICANT

At the beginning of a social event people can be uptight and interaction might be limited. Social lubricants help individuals overcome their fear of looking foolish in front of strangers. Social lubricants such as humor, pets, or babies are known ice-breakers. Wine is among the most ancient of social lubricants. It relaxes inhibitions; people overcome their hesitancy to interact with strangers. Wine also melts the barriers between the sexes. Unfortunately, some people with social anxiety problems overindulge and their antics become a humorous social lubricant for those nearby.

For generations wine was the principal Indo-European procreative lubricant, only to be replaced by brandy in Europe and rum in the rest of the world (where permitted).

There is a perception that wine is effeminate. This is a residual from the time of the Greeks when they worshiped Dionysus. Dionysus is often depicted as a beautiful young man. The concept was reinforced when the beer-drinking barbarous tribes of northern Europe encountered the civilized wine-drinking residents of the Mediterranean basin and conquered them. Huns, Celts, and the Norse viewed civilization as something that softened and weakened men. The civilized wine drinkers, naturally, maintained equal disdain for beer-drinking barbarians. The expense of wine in colonies around the globe led colonists to disdain wine in favor of distilled spirits.

POLITICS AND WINE

The British used their lack of domestic wine production to their great advantage by playing suppliers off against one another to obtain favorable prices while furthering national policy. French and British conflicts have had detrimental effects on the French wine industry. During times of war French imports were outlawed, but even during peacetime British citizens desiring French wine paid extra for it. Prior to 1693, the tariff was 1s. 4d. (one shilling four pence) on all wine imported to England. The initial tariff rate differential was 5d. (five pence) per gallon. In 1697 the differential increased to 2s. 10d., culminating, at the end of the Napoleonic wars, in the 1813 differential of 10s. 9d. per gallon imported. The extra expense of French wines and patriotism ensured the popularity of port and sherry. As Jonathan Swift put it in this couplet,

> Bravely despise Champagne at Court,
> And choose to dine at home with Port.[18]

The gin craze 1689–1751 was politically motivated. When William of Orange became king, the English people switched from French brandy (which had a punitive tariff) to Holland gin and rapidly to English-made gin. The ready availability of nearly free gin and the widespread ruin it caused made Parliament act. At first Parliament sought to control access points by requiring a very high license fee. Dram shop owners simply did not purchase the license and operated clandestinely. The gin craze officially ended when Parliament passed the Gin Act of 1751. The Gin Act separated manufacturers and retailer licensing. Licensed manufacturers could only sell to licensed retailers, establishing a two-tier distribution system.

The use of tariffs to punish other countries continues. In 2013 a trade dispute between the European Union and China broke out when the EU accused China of dumping solar panels on EU countries. (Dumping means selling products below cost. It is a measure designed to drive competitors out of business.) The EU placed tariffs on the Chinese products. The Chinese retaliated with claims that the EU was dumping wine on China and raised its tariff on wine. The nations involved were still bickering over the issues in 2016.[19,20]

In the United States, wine is a touchy subject. Among the three type of alcohol (beer, wine, liquor), it is the most socially acceptable. Persons drinking beer or liquor are seen as drinking for fun. Persons drinking cheap wine are seen as alcoholics. Persons, however, who drink popular or expensive wine are seen as participating in a cultural event. A "wine and cheese party" is usually associated with conversation, a lecture, a musical performance; the opposite image of that associated with a keg party.

Since the 1840s, when Americans first began interfering with their neighbors' actions, (wine) drinking has been under attack. Often the antidrinking cause was linked to anti-immigrant sentiments. First were the attacks of the Know-Nothing political party that sought to keep Irish and German immigrants out of the United States. At the beginning of the twentieth century American opposition to immigrants from southern and eastern Europe help fuel the Prohibition movement.

With Prohibition's repeal, wine once again became a significant source of government revenue. Wine is taxed at ever-increasing rates by federal and state governments. There are wine consumption taxes on the consumer over and above sales taxes. There are also wine manufacturing taxes. Makers and sellers of wine must carefully track the volume of juice and wine made so that it might be properly taxed.

Political controversies surround the best way to make wine and other alcoholic beverages available to the public. Most states have an open system where retailers are licensed to collect and remit the taxes. Less than a dozen states now maintain a monopoly on wine sales. The main advantages of state monopolies include complete collection of taxes, enhanced control of access to minors, consistency of price over a wide area, and trained adult staffing. On the negative side one must make a special stop at a store with limited hours. Monopoly control states face continuous political pressure to convert to open retailing. The two largest groups that oppose monopoly control are retailers desiring to profit on wine sales and newspapers that do not get to sell advertisements. In open states newspapers derive considerable revenue from competing liquor advertisements.

Yet another aspect of wine in politics is the displeasure some have shown with other nations by destroying wine. When the Iranian Revolution occurred in 1979 thousands of bottles stored in hotel cellars were smashed. In 2004, Sandy Block reported, "The efforts of many consumers to register their displeasure with French opposition to the American-led war in Iraq by boycotting wine purchases is not a new chapter in the evolving story of inter-connections between wine, culture and politics."[21]

GOVERNMENTAL OVERSIGHT

The government is conflicted about alcohol. The direct taxes on alcohol and the taxes on worker income fill government coffers. At the same time the Department of Health and Human Services is doing all it can to prevent the consumption of alcohol.

Some potable fluids are "better" than other potable fluids, according to the government. What is "better" is determined by the U.S. Department of Health and Human Services and the Bureau of Alcohol, Tobacco, Firearms, and Explosives (BATFE) policies. This determination is not challengeable by facts. For example, the heart-healthy benefits of wine and oat bran were reported about the same time in the early 1990s by medical researchers. Quickly Quaker Oats and Cheerios developed heart symbols to place on their packages and emphasized the healthy benefits of eating their products. Winemakers were prevented from incorporating a similar symbol on their labels. Critic Lewis Purdue, author of *The Wrath of Grapes,* maintains, "The petty irrationality of the BATF even extended to threatening San Francisco wine retailer Jerome Draper with revocation of his license in 1993 because he was giving away red heart-shaped buttons with the words, 'Have you had your wine today?'"[22]

The Department of Health and Human Services and the Bureau of Alcohol, Tobacco, Firearms and Explosives have a demonstrated paternalistic attitude toward the welfare of the American people. During the 1990s, there were many similar incidents involving both agencies. The conflict between the Constitution and the proponents of governmental paternalism was addressed in the Supreme Court case *44 Liquormart, Inc. v. Rhode Island*. In 1995, the Route 44 Liquormart in Rhode Island put signs in its store windows advertising the prices of alcoholic beverages. This was a violation of state law wherein the public advertising of alcohol prices in flyers, newspapers, and posters in store windows was forbidden. The U.S. Supreme Court held that any merchant can advertise any truthful statement about their products. The Court decided that prohibiting alcohol price advertising was an unconstitutional suppression of free speech. The Court's action simultaneously struck down several other state laws and federal policies.

Prior to the court's decision, the BATFE justified its attitude and the suppression of facts relating to the benefits of moderate wine drinking because they believed that the public would act irrationally. In other words, they were afraid all Americans would go on a wine bender were they to learn of, for example, the "French paradox." The Court made it clear that such restrictions on truthful advertising were violations of the First Amendment, which took priority over the 21st Amendment right of the states to control alcohol distribution. To quote the U.S. Supreme Court,

> There is, of course, an alternative to this highly paternalistic approach. That alternative is to assume that this information is not in itself harmful, that people will perceive their own best interests if only they are well enough informed, and that the trend means to that end is to open the channels of communication rather than to close them. . . . It is precisely this kind of choice, between the dangers of suppressing information, and the dangers of its misuse if it is freely available that the First Amendment makes for us.[23]

The *44 Liquormart, Inc. v. Rhode Island* Supreme Court decision also opened the door for pharmaceutical companies to advertise prescription drugs. Drug manufacturers realized they could spark sales by truthfully informing the public of their drugs' benefits. This unanticipated consequence resulted in the explosion of prescription drug advertisements. Despite this decision the BATFE continues to censor winemakers by refusing to allow them to publicize linkages between wine and health. The BATFE also controls all text printed on wine labels, including proprietary wine names.

In another example, the Alabama Alcoholic Beverage Control Board banned the sale of Cycles Gladiator because the label showed the side view of a naked woman flying alongside her bicycle. According to the board the image posed the woman in an immoral/sensuous manner.

Agreeing with the above agencies, the U.S. Department of Health and Human Services seeks to curtail consumption. The agency defined alcohol consumption levels in these terms:

Moderate alcohol consumption: According to the Dietary Guidelines for Americans, moderate drinking is up to 1 drink per day for women and up to 2 drinks per day for men.

Binge Drinking: NIAAA defines binge drinking as a pattern of drinking that brings blood alcohol concentration (BAC) levels to 0.08 g/dL. This typically occurs after 4 drinks for women and 5 drinks for men—in about 2 hours.

Heavy Drinking: SAMHSA defines heavy drinking as drinking 5 or more drinks on the same occasion on each of 5 or more days in the past 30 days.[24]

TEETOTALLERS AND NEO-PROHIBITIONISTS TODAY

Alcoholism is a serious global problem. The National Council on Alcoholism and Drug Dependency reports more than 17.6 million alcoholics in the United States.[25] While many persons are genetically disposed to alcoholism, habituation also leads to dependence. For them, abstinence (teetotalling) is a solution. Governmental, private, religious, and medical facilities without number are available to help and treat them. The good they perform cannot be overemphasized. The most laudable, Alcoholics Anonymous, offers free support services to all in need while residential treatment programs may charge thousands. Comparisons between program effectiveness is near impossible because the industry (and it is an industry) lacks standardized measures. Consumers (alcoholics and their supporters) only see favorable statistics, if any at all, selected from a larger pool of unreported data.

There are some people who wish to see a return to Prohibition. To them, alcohol is a dangerous substance that cannot be allowed to exist. Not content to restrict their own behavior, they demand the moral right to dictate the behavior of others. As Richard Mendelson said of their nineteenth-century predecessors, "Few things [immigrants and foreign influences] could have propelled Americans . . . so decisively as the threat of urban decay and physical and moral degradation."[26]

Beginning on the date of repeal, these people have organized, seeking a return to Prohibition. The Prohibition Party, symbolized by a camel, ran candidates for president until 1980 when Earl Dodge (who had been running as their candidate since 1948) failed to participate. The Women's Christian Temperance Union still has thousands of members and seeks pledges in its continuing efforts.

In 1980 Candace Lightner created the nonprofit Mothers Against Drunk Drivers (MADD). Unlike the Prohibition Party, MADD did not seek to elect its own slate of politicians. MADD raises public awareness, gathers support, and forces legislative change. MADD's efforts prior to 1985 reduced highway death tolls, saving thousands of lives to date. With success MADD diversified its mission and its staff sought to justify retaining their positions as bureaucracies do. The mission became alcohol eradication. Later drug prevention and victim services were added to the mission.

David Hanson summarized the main points of neo-prohibition this way:

> Neo-prohibitionists tend to assume that
>
> - The substance of alcohol is, in itself, the cause of drinking problems,
> - The availability of alcohol leads people to drink,
> - The amount of alcohol consumed (rather than the speed with which it is consumed, the purpose for which it is consumed, the social environment in which it is consumed, etc.) determines the extent of drinking problems, and
> - Alcohol education and policy should focus on the problems that excessive alcohol consumption can cause and should promote abstinence.[27]

In addition, I maintain that neo-prohibitionists

- Think *they* know best about how you should live your life.
- Believe others need to be protected from themselves.
- Are certain their God does not approve of alcohol.
- Maintain they have the right to determine the availability of "undesirable" products.

Like their predecessors, the arguments of neo-prohibitionists are hard to put aside. The dead at the hands of a drunk driver, the wasted lives of derelicts, and the alcoholic's hungry children give them a moral high ground. Historically, they misused this advantage by exaggerating the threat. The worst of these fabrications involve quasi-scientific data mis-reporting by medical and social researchers. Many advocates and leaders fell out of favor when their fabricated horror stories were revealed as frauds. Take, for example, the statement by critic David Hanson regarding the Center on Addiction and Substance Abuse (CASA), a private think-tank funded by the Robert Wood Johnson Foundation:

> CASA has a long record of producing highly questionable papers about alcohol [that are] later discredited. For example, a researcher "examined some of the references in (a) CASA paper and found the conclusions in the articles to be shockingly different from the way CASA depicted them." Report after report has been exposed as lacking credibility, leading *The Washington Times* to observe that CASA has a "proven disdain for the facts." Understandably, scholars have a lot of

> negative things to say about the Center on Alcohol and Substance Abuse, "some of it unprintable" observed Christopher Shea in the *Chronicle of Higher Education*.[28]

There is also a swell of neo-prohibitionists at the international level. UN-funded studies conclude that alcohol and tobacco are "enemies of health."[29]

SOCIAL CONTROL AND WINE

There are places where drinking is acceptable and places where it is not. Mostly, drinking in the United States takes place behind closed doors. Drinking in public places, especially parks, is generally illegal. Why? Because Americans have a great fear that children may see adults behaving like adults. Oddly, this fear extends to the most natural of activities, such as breastfeeding an infant.

Anti-alcohol pressure on drinkers has all but eliminated the custom of treating. Treating, or buying drinks for another, is an ancient tradition relating back to offering a visiting friend a drink. Through the course of an evening treats spiraled through a group, bringing greater joy and camaraderie to all.

Those attending New York Philharmonic concerts in Central Park brought candelabras, picnics, tablecloths, and bottles of wine to accompany the music. The illegal practice went unquestioned for years. Then, during the Puerto Rico Day festivities on June 11, 2000, fifteen to twenty beer-drinking men attacked several women, bringing the consumption of alcohol in the park to public attention.[30] The whole issue has since been swept under the rug, and drinking at concerts is once again acceptable.

Wine reduces inhibition, and the public display of uninhibited behavior disturbs those subjected to it. Excess wine consumption is usually associated with uncontrolled expression of joy, fighting, loss of motor control, and vandalism. For these acts alcohol is blamed, not the individual, although individuals are punished for their acts while intoxicated. Blaming alcohol is easier than blaming the individual. Such riotous behavior often results in police intervention to restore stability and social control.

Prior to 1980 neighborhood taverns, most not selling wine, were widely distributed in small towns. Many of these rural and smaller establishments are closing. More and more, drinking establishments cluster in close proximity to one another in an entertainment district. More and more they are not independent small businesses; instead they are corporate franchises. Rather than a short trip to the corner bar of fifty years ago, people today travel longer distances to pleasure destinations consisting of multiple restaurants, theaters, and drinking establishments. Whether localized in a downtown or commercial area or a resort, the clustering provides for expanded pleasure opportunities from multiple vendors and enhanced social control by police.

TRENDING

Wine is trendy. Millennials (those age 21 to 38, about 79 million Americans) averaged two cases per person in 2015, according to the Wine Research Council (WRC). Other statistics from the WRC show a doubling of high-frequency wine drinkers (three days in seven) to 13 percent of the population.[31]

Online wine purchases continue to increase and are more prevalent among male millennials than any other segment of the population. Online purchasers tend to buy multiple bottles, often in case lots. These purchases often occur as part of a wine club, where once a month wine is shipped to subscribers. Clubs can be based on the products of a single winery or be a separate business that acquires, organizes, and ships wine to members.[32]

Wineries are trendy. Hundreds of new wineries are licensed each year. To make a go of it, most must open their vineyards and wineries to tourists and hold events. Winery weddings, anniversaries, concerts, and other celebrations are common affairs.

To reduce the confusion created by national wine quality terms (refer to Table 7.1), the European Union addressed the problem created by required memorization of locations of appellations and their synonymous grapes. The first step was allowing the inclusion of varietal names on labels where it was previously banned; this has improved sales in the United States. The second was to implement an EU-wide standardized wine categorization. As of 2008 there are three categories of wine: Wine, Protected Geographic Indication (PGI), and Protected Designation of Origin (PDO). Recategorization has had a more limited impact on consumers.[33]

Labels in general are undergoing a vast change. Staid old labels are being replaced by flashy new labels with proprietary wine names. German Rieslings now bear names like Twisted River, Fish Eye, and Clean Slate instead of Weingut and Schloss. Wineries in the United States are using catchy terms instead of the name of the winery owner. "The unconventionality seems to go mainly with inexpensive wines."[34] Proprietary names like Marilyn Merlot, 7 Deadly Zins, and Cardinal Zin have been superseded by names like Mommyjuice, Mommy's Time Out, and Mad Housewife. Vampire Merlot is a big hit around Halloween.

Changing morals have opened up the naming of wines. Words once forbidden have entered the common parlance. The acceptability of the word "bitch" is at the forefront of the transformation. Nancy Friedman's study of wine names revealed,

> In the 21 years before 2001, we find not one BITCH approval [by TTB, the Alcohol and Tobacco Tax and Trade Bureau]. By contrast, in the 7 years since 2001, we find not less than 65 BITCH approvals. We don't see any great need for the government to banish this term, on adult beverages, but we do wonder why it's so prevalent. . . . Where is the wine relating to cranky men?[35]

In addition to all the "bitch" wines there are also a number of "bastard" wines.

Talking about wine is trending. For decades morning TV programs, newspaper cuisine sections, and books were the only source of wine information available to most consumers. This was one-way communication from the expert to the masses. Blogging has allowed a multitude of writers to publish at practically no cost, greatly broadening the opportunity to express one's opinions. Wine writers abound as never before. Friends instantly share photos of wine bottles and recollections. The explosion of available information must be a major factor in the doubling of "high-frequency wine drinkers" over the past fifteen years.

Knowing about wine is trending. Classes at wine schools are filling regularly. The Society of Wine Educators and the Wine and Spirits Educational Trust train, test, and certify individuals who want to achieve the requisite knowledge and skills. Several wine databases, accessed via smartphone app, provide consumers with ready reviews and suggestions for selecting and serving wines. Other apps offer wine trivia quizzes.

Together the increased general wine knowledge from participation discussions, tasting events, courses, and ready information access has eradicated the stereotypes of wine drinkers. Today, it seems that people who drink wine are people.

SUMMARY

Major components of cultures around the globe deal with the consumption of alcoholic beverages. Wine was an innovation unlike any other, consumable, desirable, and a portal to another, happier world. At some times and in some places, however, alcoholic beverages seem to have exerted too much control over the populace. Social instability resulted wherever Bacchus reigned. Whether in ancient Rome, the English gin craze, or in American pre-Prohibition saloons, social instability and upheaval shook the roots of their societies.

Restoration of order required strong measures. The United States purged alcohol, then created a controlled system of access and personal limitations. Overseeing its continuation is the assigned responsibility of the Tax and Trade Bureau (TTB), the Bureau of Alcohol, Tobacco, Firearms, and Explosives, state liquor control boards, police at all levels, and health agencies, Each has a specific responsibility. The TTB gathers excise taxes, states oversee the sales and collect taxes, and the police keep individuals in line. The U.S. Department of Health and Human Services spends millions annually encouraging Americans to avoid alcoholic beverages.

The call for even greater alcohol restrictions continues by social groups and government agencies. Meanwhile, bending to broad public demand for greater access, politicians loosen laws (by increasing purchase points and hours of operation) to generate tax revenues while creating opportunities.

An informed, educated, experienced, and socially restrained (moderation in consumption) population can direct and manage the behaviors resulting from both occasional and excess consumption. The growing sophistication of Americans leads me to believe that may be a description of us.

Conclusion

Wine is as old as the thirst of [humanity], not the physical thirst . . . but the heaven sent thirst for what will still our fear–that our mind be at peace; and stir our sense and sensibility—that we shall not ignore nor abuse God's good gifts—wine not the least of them.[1]

—André Simon

Numerous species of grape vines grow naturally across temperate portions of the Northern Hemisphere. One grape species, *Vitis vinifera,* was selected by proto-Indo-European people living near the Caucasus Mountains. The grape plant itself is hardy and fecund, thriving where other plants wither. These people treasured the luscious fruit and its fermented juice. The effects of the fermented juice brought unparalleled ecstasy. Nine thousand years ago the mountain people began exporting wine to the plains below. On the Mesopotamian plains only the elite could afford wine, while the rest drank beer. Images of vines and grapes appear in prehistory as wine drinking became deeply integrated into religious ceremonies.

The desire for the intoxicating treasure found in the grape became a driving factor in the rise of Western civilization. Wine was a fundamental component of the Mediterranean diet by 3000 BCE. Wine mystified the ancients. Juice, to them, magically transformed into wine. As wine spread westward, it became symbolic of birth, death, rebirth, plenty, joy, and sometimes disharmony. Wine became integral to Greek, Phoenician, Egyptian, Hebrew, Etruscan, Carthaginian, and Roman religions. Wine retains a central role in Jewish and Christian ceremonies. After the collapse of Rome and the corresponding disintegration of the transportation network, people of the Dark and Middle Ages drank mostly local products. While this was tolerable for the locals, wine for the sacraments had to be the best. For many monasteries, wine became a significant source of revenue for the raising of armies to fight in the Crusades and later to build hospitals and universities.

Wine lost its preeminence in the Middle East before 900 CE. The Prophet Muhammad forbade wine drinking, and true believers have since abstained. Actual eradication of wine drinking was, however, a slow process. Enforcement of the ban varied geographically and temporally throughout the Islamic world. Omar Khayyam, a Muslim who lived in thirteenth-century Persia, wrote poetry about his love of wine.

As war and plagues traversed Europe between the time of Charlemagne and Martin Luther, the vineyards of monasteries were rarely disturbed. Monks had years to study the terroir influences on grape plants. For processing the local must into the best "blood of Christ" possible, monks made technological changes.

With the Renaissance and awakening capitalism came the rebirth of trade. No commodity was more important than wine. The necessity of the wine trade was one of the few points of agreement between the French, Dutch, and English empires. Then as now, wine was taxed. Taxation inspired smuggling, followed by the invention of distillation to concentrate the intoxicants, with the toxic part emphasized.

During the Age of Discovery, Christians brought the joy of wine to their conquests around the globe. When Europeans exploded out of their homelands they brought wine with them. Vineyards quickly proliferated in a few locations but it took hundreds of years before most places developed viable wine industries. Early European meetings with Native American, African, and Asian kingdoms were celebrated with wine in hand. By 1700 distilled alcohol (brandy and rum mostly) became the preferred political, economic, and social lubricant.

Wine's place among the world's commodities only began to change in the mid-1800s. After a series of calamities for the wine industry, including the near global phylloxera plagues of 1870 to 1890, the 1906 San Francisco quake, Prohibition, and the Prussian and two world wars, wine is resuming its place at the American dinner table after a 150-year absence. The European wars not only cut production during the wars but resulted in slow reestablishment from labor shortages afterward. The San Francisco quake and subsequent fire destroyed millions of gallons of wine stored in barrels, large wooden vats, and underground ceramic tanks. Prohibition was a major benefit for the home wine-making industry, but destroyed the professional winemakers. During Prohibition and during the wars, people wanted something strong and drank quickly, instead of leisurely enjoying a glass of wine.

From Prohibition's end until the 1970s the cocktail was king. Globally, wine drinkers suffered with homemade wines, low-cost bulk or fortified wine, and low availability of quality wine. Beginning in the 1960s Californian and Australian winemakers applied science to winemaking. They started in the field with informed varietal selection, and then they invented new methods of viticulture and vinification. Most important of all, they learned to keep the winery clean. Most of what was learned and communicated to growers was accomplished by researchers at the University of California–Davis and the Australian Wine Research Institute.

The Australian and Californian wine pioneers had their individual motivations, but as a group making quality wines was their passion. They succeeded in their efforts. Californians, through the "Judgment of Paris" in 1976, ended forever the myth that only the French could make quality wines. More importantly, they inspired the ongoing global race for quality wines by a diverse cross-section of humanity who make high-quality wine on six continents. Even more diverse are the billion people who find joy in wine drinking.

The bottle is currently the most popular unit for wine sales. This has not always been the situation, nor will it be for much longer. Other packaging will emerge in the near future that will make wine more transportable and increase shelf life. Certainly within the next twenty years the corked glass wine bottle will be uncommon.

Today, wine transport is second only to petroleum. More countries produce wine than petroleum. The value of wine exceeds that of all other food commodities. Over the course of the preceding chapters we explored where grapes grow, why grapes grow where they do, the characteristics of varieties, wine-making methods and technology, the characteristics of the people who grow them, and the global pattern of modern wine production, transportation, and consumption.

People in many countries are looking to wine for their personal and national futures. Vineyards and wineries are increasing in number in the Southern Hemisphere and North America while dwindling in Europe. These changes are inspired by demographics that disfavor aging Europe, and by the desire to establish experimental plots in places many would consider unlikely, such as the Sao Francisco vineyards of Brazil or those in Khao Yai, Thailand. Those experimenting in these far-flung locations are chasing the dream described by Robert Louis Stevenson in *The Silverado Squatters*, "So bit by bit, they grope about for their Clos Vougeot and Lafite . . . these still lie undiscovered . . . there they bide their hour, awaiting their Columbus."[2]

What will happen in the future? Environmentally, will climatic fluctuations mean that Syrah will grow in Norway? Geographically, the industry will diffuse into more countries and regions. Unknown locales will become famous for yet to be determined varieties. The trade network will become more complex yet more accessible as the Internet (high-end wines) and big-box retailers (everyday and inexpensive wines) break down old distribution systems. Demand for wine will also expand with improved standards of living.

Culturally, because of its high value and eminent taxability, wine will never go away, despite the wishes of busybodies who want *your* body to be *their* business. Wine will have its ups and downs, its denouncers and its glorifiers, through times and across space. We can speculate that the release of a study declaring wine is good for you will have consumers clearing the shelves, or that the next time French and American interests conflict we are likely to see French wine being poured into the sewers of New York. The only facts we can be sure of are that (1) wine (alcohol) will continue to be controversial in society, and (2) wine will be taxed.

Appendix A: Affirmation of Grape Variety Produced in Commercial Quantities by Country of Origin

Grape Variety	Argentina	Australia	Bulgaria	California	Chile	France	Georgia	Germany	Greece	Hungary	Israel	Italy	New Zealand	New York	Ohio	Oregon	Pennsylvania	Portugal	Romania	Slovenia	South Africa	Spain	United States	Washington
Alvarinho																		X						
Amarone												X												
Barbaresco												X												
Barbera												X												
Bardolino												X												
Barolo												X												
Bordeaux Blanc						X																		
Brunello												X												
Cabernet Franc				X													X							
Catawba														X			X							
Cayuga																	X							
Chambertin						X																		
Chambourcin																	X							
Chardonnay	X	X		X	X	X				X	X	X	X	X		X	X	X	X	X	X	X		X
Chenin Blanc				X																				

Chianti				X								X		X										
Concord											X						X						X	
Cabernet Sauvignon	X	X	X	X	X	X				X	X	X		X			X		X	X	X	X		X
Dolcetto												X												
Fortified				X										X				X				X	X	
French Colombard				X		X																		
Frascati												X												
Gamay				X		X																		
Gattinara												X												
Gavia												X												
Gewürztraminer				X		X																		X
Grenache				X																				
Heritage												X												
Hermitage						X																		
Lambrusco												X												
Macabeo																						X		
Madeira																		X						
Madeira																							X	

(Continued)

(*Continued*)

Grape Variety	Argentina	Australia	Bulgaria	California	Chile	France	Georgia	Germany	Greece	Hungary	Israel	Italy	New Zealand	New York	Ohio	Oregon	Pennsylvania	Portugal	Romania	Slovenia	South Africa	Spain	United States	Washington
Malbec	X																							
Marsala				X								X		X									X	
Merlot	X	X	X	X	X	X					X	X					X		X	X	X	X		X
Meursault						X																		
Montalcino												X												
Montepulciano												X												
Muscat Bianco												X												
Muscat						X						X											X	
Niagara														X			X							
Nobile																								
Orvieto												X												
Petit Sirah				X																				
Pinot Blanc												X												
Pinot Grigio				X								X					X							

Pinot Gris						X											X							
Pinot Noir		X		X		X		X				X	X			X	X		X	X				
Pinotage																					X			
Primitivo												X												
Prosecco												X												
Retsina									X															
Riesling				X		X		X						X		X	X			X				X
Salentino												X												
Sangiovese	X			X								X												
Sangria												X					X					X	X	
Sauterne														X										
Sauvignon Blanc				X	X	X							X								X			X
Semillon																								X
Seyval Blanc																	X							
Shiraz	X	X		X	X	X																		
Soave												X												
Tempranillo																						X		
Tokay																							X	

(Continued)

(*Continued*)

Grape Variety	Argentina	Australia	Bulgaria	California	Chile	France	Georgia	Germany	Greece	Hungary	Israel	Italy	New Zealand	New York	Ohio	Oregon	Pennsylvania	Portugal	Romania	Slovenia	South Africa	Spain	United States	Washington
Toscano												X												
Trebbiano												X												
Valpolicella												X												
Verdicchio												X												
Vernaccia												X												
Vidal Blanc																	X							
Vignoles																	X							
Viognier						X																		
Zinfandel				X							X													

Source: Data derived from Pennsylvania Liquor Control Board wine sales.

Appendix B: North American Wine History

Year	Event
985	Leif Ericson lands at L'Anse aux Meadows. Makes wine from local fruit. Called discovery Vineland in same spirit that his father named Greenland.
1180	Distillation introduced to Europe.
1269	First reference to distilled spirits containing juniper berries for flavor.
1495	First recipe for gin.
1524	Verrazzano sees vines growing around New York.
1554	Francisco Urdionla initiates wine industry near Parras, Mexico.
1565	French Protestants make wine in Florida near the mouth of St. John's River.
1568	Spanish on Parris Island, South Carolina plant grapes but never make wine.
1587	Roanoke Colony established, report wild grapes.
1608	Jamestown colonists send wine to England; made nearly 20 gallons from "hedge" grapes.
1609	Much drunkenness reported at the Jamestown colony; the source, local grapes.
1610	Ralph Hamor plants native vineyard near Henrico, Virginia of three to four acres.

(*Continued*)

(*Continued*)

Year	Event
1619	Eight French *vignerons* sent to Virginia. Jamestown's Acte 12 required each settler to plant 12 wine grapevines each year or be punished.
1620	10,000 vines arrive in Virginia for planting.
1621	King sends every household in Virginia a pamphlet on how to grow grapes. Vines die; colonists turn to growing tobacco.
1624	Dutch build first commercial distillery and make brandy.
1625	All vines dead (killed by fungus; phylloxera not present in Virginia then). French are blamed for failure.
1629	Spanish colonists plant grapes near Santa Fe, New Mexico.
1633	First Spanish harvest in New Mexico. Vineyard produces for 40 years before abandonment after Apache attack in 1675.
1637	Rum is made in Barbados by Pieter Blower (Dutch).
1648	English colonists make wine in Delaware.
1670	Long Island winery based on North American grapes.
1683	William Penn plants vineyard. Vines die, apparently after releasing pollen.
Ca. 1720	Jesuit priests establish vineyard at Kalkaska, Illinois.
Ca. 1730	English gin craze begins.
1740	Alexander hybrid discovered near William Penn's original abandoned vineyard.
1751	Official end of English gin craze, which began about 1730.
1760	Alexander vineyard destroyed by frost. Alexander variety extinct.
1769	Virginia Legislature passes the Act for the Encouragement of the Making of Wine in hopes of trying to develop a native wine industry.
1774	Phillip Mazzei comes to Virginia with 10 Tuscan wine growers and 10,000 vines to start a wine-making colony. Settles next to Monticello at Jefferson's insistence.
1774	Thirty-seven investors form the Wine Company. Among these ambitious investors are His Excellency Earl Dunmore, Royal Governor of Virginia, Thomas Jefferson, George Washington, Thomas Adams, and George Mason.
1776	Jefferson et al. abandon vineyard during the Revolution.
1790	Spanish mission at San Diego makes wine.
1793	Pennsylvania Vine Company founded to support John James Dufour.
1798	Kentucky Association for the Establishment of a Vineyard is formed to support Dufour and Swiss followers. The vines perish in two years.
1802	New Switzerland established on banks of Ohio near Cincinnati.
1806	800 gallons of wine produced at New Switzerland.
Ca. 1810	Harmony, Pennsylvania settled by vine planting Germans, who expand and establish Economy, Pennsylvania.
1818	Elijah Fay starts what becomes Great Western winery on shore of Lake Erie.
1829	First Finger Lakes vineyard established.

1832	Berlandier's botanical expedition to Mexico. Discovers *V. berlandieria*, which later proves phylloxera came from America.
1835	Virginian discovers the Norton grape variety and wine making prospers. Wardian case invented by Ward. Transoceanic botanical specimen survival ensured.
1841	Nicholas Longworth creates wine region on banks of Ohio River surrounding Cincinnati. The native Catawba is the basis for success.
1849	California gold rush creates thirsty miners.
1855	Asa Fitch identifies and describes insect later known as phylloxera.
1861	Agoston Haraszthy returns from European expedition with 100,000 vine cuttings.
1873	Monticello Wine Company wins gold medal in Vienna for its wine made from the Norton grape.
1878	Monticello Wine Company wins silver medal in Paris for its wine made from the Norton grape.
1900	U.S. Census reports a total of 240,864 grapevines in Virginia. During the next 30 years this number falls quickly.
1906	Great San Francisco earthquake. Thirty million gallons of wine destroyed.
1930	U.S. census reports a total of 5,016 grapevines in Virginia.
1955	Dr. Frank grows vinifera in New York's Finger Lakes region on Canadian rootstock.
1973	Piedmont Vineyards plants the first commercial post-Prohibition vineyard in Virginia.
1976	Judgment of Paris—California wine beats French wine in head-to-head tasting.
1979	Pennsylvania passes limited Winery Act (first in East). Penn Shore Winery in North East, Erie County, receives first license.
1980	Virginia Legislature passes the Virginia Farm Wineries Act to create tax incentives for small family wineries.
1983	BATFE establishes first American Viticultural Areas designations.
1984	Virginia Wine Marketing Office established to increase public awareness of the growing industry.
1985	Virginia Legislature forms the Virginia Winegrower's Advisory Board to help promote Virginia wine.
1993	A Virginia wine wins the "Best Wine East of the Rockies" award in a national competition.
2003	One hundred licensed wineries in Pennsylvania.
2014	Ravines Winery in the Finger Lakes region selected for *Wine Spectator* top 100.

Appendix C: Wine Glossary

Acetic	Describes a sour, vinegary odor referred to as volatile acidity, too much of which will make the wine undrinkable.
Acid	The sharp, tart effect of the green fruit of young wine on both the nose and tongue.
Aroma	The perfume of fresh fruit. It diminishes with fermentation and disappears with age to be replaced by the "bouquet."
Astringent	The rough, puckery taste sensation caused by an excess of tannin in especially young red wines. It diminishes with age in the bottle. Place an aspirin on your tongue to experience this sensation.
Baked	Quality of red wine made in a hot climate from ripe grapes. This flavor is characteristic of sherry and Madeira.
Balanced	Having all natural elements in good harmony.
Beery	The odor of stale beer from a white wine that is over the hill—usually in old Moselles.
Big	Full of body and flavor, high degree of alcohol, color, and acidity.
Bitter	Self-descriptive. Sign of ill-health caused by inferior treatment such as excessive stalks during crushing or even metal contamination.
Black currants	The slight smell and taste of black currants often found in Bordeaux wines.

Body	The weight and substance of the wine in the mouth; actually a degree of viscosity largely dependent on the percentage of alcohol and sugar content.
Bouquet	The fragrance a mature wine gives off once it is opened. It develops the two aspects of the olfactory sensations—aroma and bouquet.
Breed	Having the character, type, and qualities of its origin.
Brilliant	Bright and sparkling in appearance so that one can see the light through the wine. Opposite of dull and cloudy.
Broad	Full-bodied but lacking in acidity and therefore also lacking in finesse.
Character	Positive and distinctive taste characteristics giving definition to a wine.
Clean	A well-constructed wine with no offensive smells or tastes.
Clear	Transparent and luminous appearance. Any sediment rests on the bottom of the bottle.
Cloudy	Unsound condition of hazy, dull-looking wine. Not to be confused with the condition of a recently shaken old wine whose deposit hasn't yet settled.
Cloying	Too much sweetness and too little acidity.
Coarse	Rough texture; little breed or elegance.
Common	Adequate but ordinary.
Corky	Disagreeable odor and flat taste of rotten cork due to a defective cork in the bottle.
Depth	Rich, lasting flavor.
Dry	Completely lacking sweetness. Should not be confused with bitterness or sourness.
Dull	See *Cloudy.*
Earthy	What the French call *goût de terroir*. The peculiar taste that the soil of certain vineyards gives to their wine. Disagreeable when too noticeable.
Elegant	Well balanced, with finesse and breed.
Fat	Full-bodied but flabby, which in white wines is often due to too much residual sugar. When applied to red wines, it means softness and maturity.

Finesse	The breed and class that distinguish a great wine.
Finish	The taste that the wine leaves at the end, either pleasant or unpleasant.
Flabby	Too soft, almost limp, without structure. Lacking in acidity.
Flowery	The flowerlike bouquet that is as appealing to the nose as the fragrance of blossoms, as, for example, in a fine Muscat.
Foxy	A pronounced flavor found in wines made from native American grapes; the same smell as in grape jelly and soda.
Fruity	The aroma and flavor of fresh grapes found in fine young wines. It diminishes with age.
Full	Having body and color, often applied to wines high in alcohol, sugar, and extracts.
Geranium	Smelling of geraniums, an indication that the wine is faulty.
Grapy	The strong flavor that certain grape varieties, such as the Muscat, impart to certain wines.
Green	Harsh and unripe with an unbalanced acidity that causes disagreeable odor and a raw taste.
Hard	Tannic without softness or charm. It can mellow with age.
Harsh	Excessively hard and astringent. It can become softer with age.
Insipid	Lacking in character and acidity; dull.
Light	Lacking in body, color, or alcohol, but pleasant and agreeable.
Lively	Usually young and fruity, with acidity and a little carbon dioxide.
Long	Leaving a persistent flavor that lingers in the mouth. Sign of quality.
Luscious	Juicy and soft, filling the mouth without a trace of dry aftertaste. Usually attributed to sweet wine well balanced with acidity.
Maderized	Flat, oxidized smell and taste reminiscent of Madeira. Term is applied to wines that have passed their prime and have acquired a brown tinge.
Mellow	Softened with proper age.

Metallic The unpleasantly bitter taste a white wine can acquire from improper treatment that did not eliminate traces of the copper that was used to spray the vines.

Musty Disagreeable odor and stale flavor caused by storage in dirty casks in cellars; moldy.

Noble Superior and distinguished; not only possessing the right credentials but also having an impressive stature of its own.

Oxidized Having lost its freshness caused by contact with air.

Peppery The aromatic smell of certain young red wines, such as Syrah, from hot climates.

Pétillant Effervescent with a natural light sparkle.

Piquant Dry and crispy acid, prickling the palate with its tartness.

Powerful Usually applied to robust red wines of great substance, such as a Châteauneuf-du-Pape, or to white wines with full, assertive bouquet, such as a big white Burgundy.

Ripe Full; tasting of ripe fruit, without a trace of greenness.

Rounded Well balanced and complete. See *Balanced*.

Séve The sap of a great wine; the concentrated aromatic savor of a luscious and ripe sweet white wine of inherent quality.

Sharp Excessive acidity, a defect usually found in white wines.

Short Leaving no flavor in the mouth after the initial impact.

Smoky Self-descriptive for the particular bouquet of certain Loire wine, such as Pouilly-Fumé, made from the Sauvignon grape.

Smooth Of a silky texture that leaves no gritty, rough sensation on the palate.

Soft Suggests a mellow wine, usually low in acidity and tannin.

Sound Healthy, well balanced, clean-tasting.

Sour Like vinegar; wine that is spoiled and unfit to drink.

Spicy Definite aroma and flavor of spice arising from certain grape varieties (Gewürztraminer). The aroma is richer and more pronounced than what we call "fruity."

Spritzig A pleasant, lively acidity and effervescence noticeable only to the tongue and not to the eye and mostly found in young wines. *Frizante* in Italian.

Sulphury Disagreeable odor reminiscent of rotten eggs. If the smell does not disappear after the wine is poured, it is an indication that the wine is faulty.

Sweet Having a high content of residual sugar either from the grape itself or as the product of arrested fermentation.

Tannic The mouth-puckering taste of young red wines, particularly from Bordeaux. Too much tannin makes the wine hard and unyielding but also preserves it longer. Aging in the bottle diminishes the tannin and softens the wine.

Tart Sharp, with excessive acidity and tannin. In the case of a young red wine, this may be an element necessary for its development.

Thin Lacking body and alcohol. It is too watery to be called light and will not improve with age.

Tonne 1,000 kilograms

Velvety A mellow red wine and a smooth, silky texture that will leave no acidity on the palate.

Vigorous Healthy, lively, firm, and youthful. Opposite of insipid and flabby.

Watery Thin and small without body or character.

Woody Odor and flavor of oak due to long storage in the cask. Often found in Spanish and Australian wines.

Yeasty Smelling of yeast in fresh bread. Sign that the wine is undergoing a second fermentation, possibly because it was bottled too early and is therefore faulty.

Notes

PREFACE

1. Katherine Bitting, *The Wine-Drinker's Manual* (London: Marsh and Miller, 1830), 6.

INTRODUCTION: THE WORLD OF WINE

1. Origins.wine. Origins.wine/about. Accessed August 1, 2016.

2. Teresa Bulman, "The Joy of Geography." Presidential Address delivered to the Association of Pacific Coast Geographers, 66th Annual Meeting, Portland, Oregon. September 20, 2003, 5.

3. John Dickenson and John Salt, "In Vino Veritas: An Introduction to the Geography of Wine," *Progress in Human Geography* 6 (1982): 159.

4. Warren Moran, "Rural Space as Intellectual Property," *Political Geography* 12, no. 3 (1993): 263–277.

5. Justin Scheck, Tripp Mickle, and Saabira Chaudhuri, "With Moderate Drinking under Fire, Alcohol Companies Go on Offensive," *Wall Street Journal*, http://www.wsj.com/articles/with-moderate-drinking-under-fire-alcohol-companies-go-on-offensive-1471889160, August 22, 2016.

6. Donald Horton, "The Function of Alcohol in Primitive Societies: A Cross Cultural Study," *Quarterly Journal for the Study of Alcohol* 4 (1943): 199–320.

7. Richard Alleyne, "Red Wine Can Protect Against Radiation," *Telegraph*, http://www.telegraph.co.uk/news/3067317/Red-wine-can-protect-against-radiation.html, September 23, 2008.

CHAPTER 1: THE HISTORICAL GEOGRAPHY OF WINE

1. Edward Hyams, *Dionysus: A Social History of the Wine Vine* (New York: Macmillan, 1965), 5.

2. Clifton Fadiman and Sam Aaron, *The Joys of Wine* (New York: Harry N. Abrams, 1975).

3. Ronald Irvine and Walter J. Clore, *The Wine Project: Washington State's Winemaking History* (Vashon, WA: Sketch Publications, 1997).

4. National Geographic. http://video.nationalgeographic.com/video/botswana_okavangodelta. N.d.

5. Patrick McGovern, *Uncorking the Past* (Berkeley: University of California Press, 2009), 17.

6. Henry J. Bruman, *Alcohol in Ancient Mexico* (Salt Lake City UT: University of Utah Press, 2000).

7. Patrick McGovern, "Prehistoric China" (Philadelphia: University of Pennsylvania, 2016), http://www.penn.museum/sites/biomoleculararchaeology/?page_id=247. Accessed April 6, 2016.

8. Patrick McGovern, *The Origins and Ancient History of Wine* (Singapore: Overseas Publishing Associates, 1996).

9. Thomas H. Maugh, "Ancient Winery Found in Armenia," *Los Angeles Times*, January 11, 2011.

10. P. E. McGovern, J. Zhang, J. Tang, Z. Zhang, G. R. Hall, R. A. Moreau, A. Nuñez, E. D. Butrym, M. P. Richards, C.-S. Wang, G. Cheng, Z. Zhao, and C. Wang, "Fermented Beverages of Pre- and Proto-Historic China," *Proceedings of the National Academy of Sciences USA* 101, no. 51 (2004): 17593–17598. http://www.penn.museum/sites/biomoleculararchaeology/wp-content/uploads/2010/04/PNASChina.pdf

11. Hugh Johnson, *Vintage: A History of Wine*. Video series, Episode 1, "The Origins of Wine."

12. Patrick McGovern, *Ancient Wine* (Princeton, NJ: Princeton University Press, 2003).

13. Marvin Powell, "Wine and the Vine in Ancient Mesopotamia: The Cuneiform Evidence," in *The Origins and Ancient History of Wine*, ed. Patrick E. McGovern, Stuart J. Fleming, and Solomon H. Katz (Amsterdam, The Netherlands: Gordon and Breach, 1996), 101.

14. Jeffrey Munsie, "A Brief History of the International Regulation of Wine Production" (Unpublished manuscript, 2002).

15. Jean-Louis Flandrin and Massimo Montana, eds., *Food: A Culinary History* (New York: Columbia University Press, 2012), 57.

16. Anon. *The Rapiuma: A Phoenician Text*. Phoenician Wines and Vines, http://phoenicia.org/wine.html#null#ixzz1SfyRS8kJ. Accessed July 20, 2011. El was the consort of Ba'al, the dominant female deity of the time and place. El, it is thought, later evolved into Dionysus.

17. Samuel Morewood, *A Philosophical and Statistical History of the Inventions and Customs of Ancient and Modern Nations in the Manufacture and use of*

Inebriating Liquors; with the Present Practice of Distillation in all its Varieties: Together with an extensive illustration of the Consumption and Effects of Opium and other Stimulants used in the East, as Substitutes for Wine and Spirits (Dublin, Ireland: William Curry Jr., 1838), 18.

18. George Rawlinson, *History of Phoenicia* (1889). Project Gutenberg.

19. Hyams, *Dionysus,* 205.

20. Sanford Holst, *Phoenician Secrets: Exploring the Ancient Mediterranean* (Los Angeles, CA: Santorini Books, 2011), 213–24.

21. Rawlinson, *History of Phoenicia.* This topic is further discussed in the Old Testament book of Kings.

22. R. Arroyo-Garcia, L. Ruiz-Garcia, L. Bolling, et al., "Multiple Origins of Cultivated Grapevine (Vitis vinifera L. ssp. sativa) Based on Chloroplast DNA Polymorphisms," *Mol. Ecol.* 15, no. 12 (2006): 2707–14. PubMed PMID: 17032268.

23. Flandrin and Montana, *Food,* 57.

24. Thomas Pellechia, *Garlic, Wine and Olive Oil: Historical Anecdotes and Recipes* (Santa Barbara, CA: Capra Press, 2000), 38.

25. Leonard L. Lesko, "King Tut's Wine Cellar," cited in Thomas Pellechia, *Wine* (New York: Thunder's Mouth Press, 2006), 30.

26. McGovern, *Uncorking the Past.* McGovern maintains that the ancients added whatever fruits were available to their grain brew, resulting in constantly changing products.

27. Holst, *Phoenician Secrets,* 108–126.

28. John William Donaldson, *The Theatre of the Greeks* (London: George Bell and Sons, 1875), 12.

29. Daniel Stanislawski, "Dionysus Westward: Early Religion and Economic Geography of Wine," *Geographical Review* 65, no. 4 (1975): 435.

30. Hyams, *Dionysus,* 88.

31. Morewood, *A Philosophical and Statistical History,* 17.

32. Hugh Johnson, *Vintage: The Story of Wine* (New York: Simon & Schuster, 1989), 144.

33. William Davis, *A Day in Old Athens* (1914). Project Gutenberg, 167.

34. Morewood, *A Philosophical and Statistical History,* 11.

35. Evan T. Sage (trans.), Livy, *A History of Rome.* Books XXXVIII–XXXIX with an English translation (Cambridge, MA: Harvard University Press; William Heinemann, 1936).

36. Stuart Fleming, *Vinum: The Story of Roman Wine* (Glen Mills, PA: Art Flair, 2001), 81.

37. Rod Phillips, *A Short History of Wine* (New York: HarperCollins, 2000), 52.

38. Hyams, *Dionysus,* 104–5.

39. Didore the Sicilian, ca. 50 BCE.

40. "History of Inebriating Liquors," *Dublin University Magazine* 14 (1839): 590.

41. Martin Levey, "Babylonian Chemistry: A Study of Arabic and Second Millenium B.C. Perfumery," *Osiris* 23 (1956): 376–89. http://www.jstor.org/stable/301716. Accessed August 10, 2016.

42. Arthur Black, "Agave Intense, No Really," Society of Wine Educators Annual Conference, New Orleans, August 13, 2015.

43. Gordon Brown, *Classic Spirits of the World: A Comprehensive Guide* (New York: Abbeville Press, 1996), 12.

44. R. J. Forbes, *A Short History of the Art of Distillation* (Leiden, Netherlands: E. J. Brill, 1970), 95.

45. Conrad Gesner, *The Newe Iewell of Health* (London: Da Capo Press, 1971 [1576]).

46. Phillips, *Short History*, 124–5.

CHAPTER 2: THE BIOGEOGRAPHY OF THE VINE

1. Jack Keller, "Winemaking: Native North American Grapes and Wines." http://winemaking.jackkeller.net/natives.asp, 2008.

2. A. Smith, *An Inquiry into the Nature and Causes of the Wealth of Nations*, Vols. I & II, R. H. Campbell and A. S. Skinner, eds. (Indianapolis, IN: Liberty Fund, 1981), 170.

3. George Gale, *Dying on the Vine: How Phylloxera Transformed Wine* (Berkeley: University of California Press, 2011), 253.

4. Jancis Robinson, Julia Harding, and José Vouillamoz, *Wine Grapes* (New York: HarperCollins, 2012), xv.

5. Patrice This, Thierry Lacombe, and Mark R. Thomas, "Historical Origins and Genetic Diversity of Wine Grapes," *Trends in Genetics* 22, no. 9 (2006): 511–19.

6. Sean Myles et al., "Genetic Structure and Domestication History of the Grape," *PNAS* 108 (2011): 3530–3535.

7. Robinson, Harding, and Vouillamoz. *Wine Grapes*, 806.

8. A. C. F. Hui and S. M. Wong "Deafness and Liver Disease in a 57-Year-Old Man: A Medical History of Beethoven?" *Hong Kong Medical Journal* 6, no. 4 (2000): 433. PDF file accessed through Google Scholar, February 13, 2010.

9. Christy Campbell, *The Botanist and the Vintner* (Chapel Hill, NC: Algonquin Books, 2005), 7.

10. Eduardo Pastrana-Bonilla, et al., "Phenolic Content and Antioxidant Capacity of Muscadine Grapes," *Journal of Agricultural and Food Chemistry* 51 (2003): 5497–5503.

11. Minnesota Agricultural Experiment Station, "The Marquette Grape." St. Paul, MN, 2009. http://www.grapes.umn.edu/marquette/index.html. Accessed March 13, 2011.

12. Campbell, *The Botanist and the Vintner*, 43.

13. W. M. Davidson and R. L. Nougaret, *The Grape Phylloxera in California*. United States Department of Agriculture, Bulletin No. 903 (Washington, DC: Government Printing Office, 1921).

14. A. E. Bateman, "Wine Production in France," *J. Statistical Society of London* 47, no. 4 (Dec. 1884): 609.

CHAPTER 3: TERROIR

1. M. A. Amerine and R. M. Wagner, "The Vine and Its Environment," in D. Muscatine, M. A. Amerine, and B. Thompson, eds., *The Book of California Wine* (Berkeley: University of California Press, 1984), 97, 118.

2. A. Smith, *An Inquiry into the Nature and Causes of the Wealth of Nations*, Vols. I and II, R. H. Campbell and A. S. Skinner, eds. (Indianapolis, IN: Liberty Fund, 1981), 170.

3. Thomas Jefferson, *Journal* (1797). Quoted in Alexis Bespaloff, *The Fireside Book of Wine* (New York: Simon & Schuster, 1977), 269.

4. Hugh Johnson, *Vintage: The Story of Wine* (New York: Simon & Schuster, 1989), 148.

5. Bill Pregler, "Do You Need a Weather Station in Your Vineyard?" https://kestrelmeters.com/blogs/news/3457062-do-you-need-a-weather-station-in-your-vineyard. Kestrel Corp., 2016. Accessed April 18, 2016.

6. Vladimir Koppen, *Versuch einer Klassifikation der Klimate. Vorzugsweise nach iheren Beziehungen zur Pflanzenwelt* (Leipzig: B. G. Teubner, 1901); *Die Klimate der Erde* (Berlin and Leipzig: Walter de Gruyter, 1923).

7. M. Claussen, "On Coupling Global Biome Models with Climate Models," *Climate Research* 4 (1994): 203–221. http://hdl.handle.net/11858/00-001M-0000-0013-AF4F-E

8. Scott Burns, "The Importance of Soil and Geology in Tasting Terroir with a Case History from the Willamette Valley, Oregon," in *The Geography of Wine: Regions, Terroir, and Techniques*, ed. Percy Dougherty (Dordrecht: Springer, 2012), 98.

9. USGS, 2008. http://esp.cr.usgs.gov/info/eolian/14.gif

CHAPTER 4: HOW WINE IS MADE

1. Martin Luther, quoted in AZQuotes.com, http://www.azquotes.com/quote/365483. Accessed October 9, 2016.

2. Galileo, quoted in Helen Exley, *Wine Quotations* (New York: Exely, 1994), 6. Accessed October 9, 2016.

3. Wine Standards Bureau. *Guide to EC Wine Legislation* (Joint Publication of the UKVA and Wine Standards Board, 2008). http://www.food.gov.uk/multimedia/pdfs/euwineregs.pdf

4. K. E. Adams and T. S. Rans, "Adverse Reactions to Alcohol and Alcoholic Beverages," *Annals of Allergy, Asthma & Immunology* 111, no. 6 (2013): 439–45. http://dx.doi.org/10.1016/j.anai.2013.09.016. PMID 24267355.

CHAPTER 5: TASTING WINE

1. Stefan K. Estreicher, *Wine: From Neolithic Times to the 21st Century* (New York: Algora, 2006), 2.

2. Serafin Alvarado, "Interpreting, Analyzing, and Evaluating Wine Quality." Paper presented at the Society of Wine Educators Annual Conference, Providence, Rhode Island, August 4, 2011.

3. Tim Hanni, *Why You Like the Wines You Like* (Napa, CA: Hannico.com, 2013).

4. Stuart Fleming, *Vinum: The Story of Roman Wine* (Glen Mills, PA: Art Flair, 2001), 39.

5. Ann C. Noble, *The Wine Aroma Wheel*, http://winearomawheel.com/. Accessed October 9, 2016.

6. Rachel Eng, "Wine Down Wednesday: An Amateur's Guide to Reviewing Wine," October 22, 2014. http://blog.bazaarvoice.com/2014/10/22/wine-wednesday-amateurs-guide-reviewing-wine/. Accessed January 14, 2016.

CHAPTER 6: WINE IN THE GLOBAL ECONOMY

1. Adam Smith, quoted in http://www.azquotes.com/quote/602319. Accessed October 9, 2016.

2. Edward Hyams, *Dionysus: A Social History of the Wine Vine* (New York: Macmillan, 1965), 104.

3. Kym Anderson, David Norman, and Glyn Wittwer, "Globalization and the World's Wine Markets: Overview," CIES Discussion Paper 0143, University of Adelaide, 2001, 1. cies@adelaide.edu.au

4. Commission Europeenne, Direction Generale de l'agriculture et du Developpement Rural, "Common Market Organization WINE-R (CD) 555/2008 Grubbing-up scheme," 2012.

5. Anderson, Norman, and Wittwer, "Globalization and the World's Wine Markets."

6. Juliet Cox and Larry Bridwell, "Australian Companies Using Globalization to Disrupt the Ancient Wine Industry," *Competitiveness Review* 17, no. 4 (2007): 209–221.

7. Mitch Frank, "Italian Police Uncover Counterfeit Champagne Scheme," *Wine Spectator*, http://www.winespectator.com/webfeature/show/id/52684, February 2, 2016. Accessed June 11, 2016.

8. Michael Porter, "Location, Competition, and Economic Development: Local Clusters in a Global Economy," *Economic Development Quarterly* 14 (2000): 15–34.

9. Wine Institute, www.wineinstitute.org/communications/statistics.htm. 2014.

10. Wine Institute, www.wineinstitute.org/communications/statistics.htm. 2006.

11. Li Xiang, "China's EU Wine Probe Sparks Worries," ChinaDaily.com.cn. June 13, 2013. http://www.chinadaily.com.cn/business/2013-06/13/content_16612553.htm

12. Frank Schoonmaker and Tom Marvel, *American Wines* (New York: Duell, Sloan, and Pearce, 1941), 6.

13. T. M. Nephew, G. D. Williams, F. S. Stinson, K. Nguyen, and M. C. Dufour, "Apparent Per Capita Alcohol Consumption: National, State, and Regional Trends,

1977–98: Surveillance Report #55." National Institute on Alcohol Abuse and Alcoholism, Division of Biometry and Epidemiology, Alcohol Epidemiologic Data System, 2000; and Robin LaVallee, Trinh Kim, and Hsiao-ye Yi, "Apparent Per Capita Alcohol Consumption: National, State, and Regional Trends, 1977–2012: Surveillance Report #98." National Institute on Alcohol Abuse and Alcoholism, Division of Epidemiology and Prevention Research, Alcohol Epidemiologic Data System, http://pubs.niaaa.nih.gov/publications/surveillance98/CONS12.htm. Accessed July 22, 2014.

14. Nephew et al., ibid.; LaVallee et al., ibid.

15. The ethanol volume must be multiplied by eight (assuming 25 proof alcohol content) to determine the volume of wine consumed. Spirits value must be multiplied by 2.5 (assuming 80 proof alcohol content) to obtain true volume. The ethanol volume of beer must be multiplied by 20, assuming 5 percent alcohol content.

16. Donald A. Hodgen, Public Statement, 2003. U.S. Department of Commerce, donald_a_hodgen@ita.doc.gov, October 16, 2003.

17. Lewis Perdue, *The Wrath of Grapes* (New York: Avon Books, 1999), 23.

18. Philip Howard, "Big Wine Tightened Its Grip on the U.S. Market in 2013." http://winecurmudgeon.com/big-wine-tightened-its-grip-on-the-u-s-market-in-2013, March 13, 2014. Accessed April 24, 2016.

19. E. & J. Gallo, "Gallo Winery, Press Room: Fact Sheet," http://gallo.com/press-room/fact-sheet/CompanyFactSheet.html, 2016. Accessed April 24, 2016.

20. Philip Howard, "Concentration in the U.S. Wine Industry." Quoted in "Big Wine: 5 companies, 60 percent of sales, 200 brands," 2013. http://winecurmudgeon.com/big-wine-5-companies-60-percent-of-sales-200-brands/2013. Accessed April 25, 2016.

21. Simon Black, "China's Wine Production to Overtake Australia's in the Next Three Years," *Herald Sun*, Jan. 25, 2011.

22. Glyn Wittwer and Kym Anderson, *Global Wine Markets, 1961 to 2003: A Statistical Compendium* (Adelaide, Australia: University of Adelaide Press. 2004).

23. Wine Institute, "U.S. Wine Exports, 95% from California, Jump 30% to $876 Million in 2006," 2006. http://www.wineinstitute.org/resources/exports/article58

24. Liv-ex. "Fine Wine 100 Index." http://www.liv-ex.com/staticPageContent.do?pageKey=Fine_Wine_100

25. United Nations FAO, http://faostate.fao.org, 2011. Accessed May 28, 2011.

CHAPTER 7: WINES OF THE EUROPEAN UNION

1. European Union, Eurostat, http://ec.europa.eu/agriculture/wine/. Accessed March 11, 2012.

2. European Union, Eurostat, ibid.

3. Robert Sechrist, "Vine Grubbing in the European Union 1990–2010." Paper presented at the Association of American Geographers Annual Conference. Chicago, April 23, 2015.

4. Joanna Rothkopf, "170-Year-Old Champagne Tastes Like Wet, Cheesy Hair," www.Salon.com, April 21, 2015. http://www.salon.com/2015/04/21/170_year_old_champagne_tastes_like_wet_cheesy_hair/. Accessed April 29, 2016.

5. Jeffrey Munsie, "A Brief History of the International Regulation of Wine Production," Harvard Law School Unpublished manuscript (2002): 15. https://dash.harvard.edu/bitstream/handle/1/8944668/Munsie.html?sequence=2. Accessed January 31, 2017.

6. Rudolf Weinhold, *Vivat Bacchus: A History of the Vine and Its Wine* (Watford, UK: Argus Books, 1978), 65.

7. André Simon, *The History of Champagne* (London: Octopus Books, 1971), 54.

8. Simon, *History of Champagne*, 86.

9. Ibid., 81.

10. Comité Interprofessionnel du Vin de Champagne, "2016 Key Market Statistics," http://www.champagne.fr/en/champagne-economy/key-market-statistics, 2016. Accessed April 27, 2016.

11. André Simon, *In Vino Veritas: A Book about Wine* (London: Grant Richards, 1913), 60.

12. André Dominé, *Wine* (Cologne, Germany: Konemann, 2001): 181.

13. Wine Institute, "World Vineyard Acreage by Country," http://www.wineinstitute.org/files/2012_World_Acreage_by_Country_California_Wine_Institute.pdf. Accessed April 30, 2016.

14. Hubrecht Duijker, *Wine Atlas of Spain and Traveller's Guide to the Vineyards* (New York: Simon & Schuster, 1992), 112.

15. Dpto. Contabilidad y Registros, "Produccion de uva en las serie historica," http://www.dorueda.com/en/facts-figures/advances/309/2013-harvest-data-99828394-kilos-with-do/, 2013. Accessed May 1, 2016.

16. Carmen Valverde, "2015 Was a Record Year for Spanish Wine Exports," *USDA FAS GAIN Report* Number SP1607, April 1, 2016, http://gain.fas.usda.gov/Recent%20GAIN%20Publications/2015%20was%20a%20record%20year%20for%20Spanish%20wine%20exports_Madrid_Spain_4-1-2016.pdf. Accessed May 1, 2016.

17. Hugh Johnson, *Vintage: The Story of Wine* (New York: Simon & Schuster, 1989), 426.

18. Duijker, *Wine Atlas of Spain*, 107.

19. Daniel Stanislawski, *The Landscapes of Bacchus* (Austin: University of Texas Press, 1970), 7.

20. Charles Dickens, *Household Words* (Leipzig: Bernhard Tauchnitz, 1854), 328.

21. Godfrey Spence, *The Port Wine Companion: A Connoisseur's Guide* (New York: Macmillan, 1997), 19.

22. George Robertson, *Port* (London: Faber & Faber, 1978), 168.

23. Charles Tovey, *Wit, Wisdom, and Morals, Distilled from Bacchus* (London: Whittaker, 1878), 204.

24. Edward Hyams, *Dionysus* (New York: Macmillan, 1965), 191.

25. English Wine Producers, http://www.englishwineproducers.co.uk/. Accessed May 5, 2016.

26. Thomas Jefferson, *Journals*, 1797. Quoted in Alexis Bespaloff, *The Fireside Book of Wine* (New York: Simon & Schuster, 1977), 102.

27. Istituto Statistica Mercati Agroalimentari (ISMEA), Rome, Italy, 2005.

28. John Schreiner, *Icewine: The Complete Story* (Toronto, CA: Warwick, 2001).

29. Karl Bauer, Josef Pleil, and Victoria Loimer, "Austrian Wine—A Taste of Culture," www.austrianwine.com, 2013.

CHAPTER 8: WINES OF THE SOUTHERN HEMISPHERE

1. Anita Poddar, "Accolade Wines Submission to the Senate Standing Committee on Rural and Regional Affairs and Transport References Inquiry into the Australian Grape and Wine Industry. Australian Grape and Wine Industry Submission 26" (Reynella, South Australia: Accolande Wines, May 2015).

2. Australia Wine and Brandy Corp.

3. John Beeston, *The Wine Regions of Australia* (Crow's Nest, Australia: Allen & Unwin, 2002), 224.

4. New Zealand Winegrowers Association, *New Zealand Winegrowers Annual Report: 2015* (Auckland, NZ, 2016).

5. Harm De Blij, *Wine Regions of the Southern Hemisphere* (Totowa, NJ: Rowman and Allenheld, 1985), 29–30.

6. De Blij, *Wine Regions,* 36.

7. Evan Goldstein, *Wines of South America: The Essential Guide* (Berkeley: University of California Press, 2014), 112.

8. Wines of Chile http://www.winesofchile.org/content/1266. 2008. Accessed October 5, 2009.

9. Goldstein, *Wines of South America,* 110.

10. Goldstein, *Wines of South America,* 42.

11. Andrea Yankelevich, "Argentina Wine Annual 2015 Gain Report." *USDA FAS*, March 31, 2015.

12. Brazilian Ministry of Development, Industry, and Foreign Trade. August 7, 2013. http://www.winesofbrasil.com/CentralArquivos/Exporta%C3%A7%C3%A3o%20Brasileira%20GERAL%20FINAL%202012%20-%20English.pdf

13. Jan van Riebeeck, *Journal of Jan van Riebeeck. Volume II, III, 1656–1662*, ed. H. B. Thom and trans. J. Smuts (Cape Town: A. A. Balkema, 1954).

14. Joanna Breslin, *South African Wine History* (Southern Stars Inc. 2011). http://www.southernstarz.com/images/edu_region_art/sa_wine_history.pdf. Accessed May 14, 2016.

15. SAWIS, *2015—South Africa Wine Industry Statistics NR 39* (Paarl, South Africa: South African Wine Industry Statistics). www.sawis.co.za

CHAPTER 9: WINE IN NORTH AMERICA

1. John Lawson, *The History of Carolina* (Raleigh, NC: Strother & Marcon, 1860), 135. http://books.google.com. Accessed August 15, 2016.

2. Robert Louis Stevenson, *The Silverado Squatters*. Transcribed from the 1906 Chatto and Windus edition. Project Gutenberg, ebook #516. www.gutenberg.org /files/516. Accessed August 15, 2016.

3. Richard G. Leahy, *Beyond Jefferson's Vines: The Evolution of Quality Wine in Virginia,* 2nd ed. (CreateSpace, 2014), 230.

4. Rick Hendricks, "Viticulture in El Paso Del Norte during the Colonial Period," *Agricultural History* 78, no. 2 (2004): 192. http://www.jstor.org/stable/3744900.

5. Albert Cook Myers, *Narratives of Early Pennsylvania West New Jersey and Delaware* (New York: Charles Scribner's Sons, 1912), 269.

6. United States House of Representatives. *Abstract of the Seventh Census, 1850* (Washington, DC: Robert Armstrong, 1853), 74.

7. Arthur Henry Hirsch, *The Huguenots of Colonial South Carolina* (Durham, NC: Duke University Press, 1928), 207–11.

8. Thomas Pinney, *A History of Wine in America from the Beginnings to Prohibition* (Berkeley: University of California Press, 1989), 96.

9. Robert Sechrist, "A Good Place for Vines." Paper presented at the Association of American Geographers Annual Conference, Boston, MA. March 2008.

10. Robert Fuller, *Religion and Wine* (Knoxville: University of Tennessee Press, 1996).

11. Samuel Purchas, *Hakluytus Posthumus or Purchas His Pilgrimes*. Vol. 19 (Glasgow, Scotland: James MacLehose & Sons, 1906), 145.

12. Marie Kimball, "Some Genial Old Drinking Customs," *William and Mary Quarterly*, 3rd Ser, 2, no. 4 (1945): 355.

13. Alice Morse Earle, *Stage-Coach and Tavern Days* (New York: Macmillan, 1900), 88.

14. John Adlum, *A Memoir on the Cultivation of the Vine in America and the Best Mode of Making Wine* (Washington, DC: Davis and Force, 1823).

15. Alan Rocke, "Buckeyes, Corncrackers, and Suckers: Culinary Episodes in Ohio History," http://www.case.edu/artsci/wrss/documents/2002rocke_001.pdf, 2002. Accessed Feb. 13, 2010.

16. Henry Wadsworth Longfellow, "Catawba Wine," from *The Poetical Works of Henry Wadsworth Longfellow* (London: Henry G. Bohn, 1861), 290–92.

17. Robert Fuller, *Religion and Wine* (Knoxville: University of Tennessee Press, 1996), 80.

18. Albert Sidney Bolles, *Industrial History of the United States, from the Earliest Settlement to the Present Time* (Norwich, CT: Henry Bill, 1878), 171.

19. Christy Campbell, *The Botanist and the Vintner: How Wine Was Saved for the World* (Chapel Hill, NC: Algonquin Books, 2005), 7.

20. Carl Sauer, "The Geography of the Ozark Highland of Missouri," *Geographic Society of Chicago Bulletin No. 7* (Chicago: University of Chicago Press, 1920), 170.

21. Missouri State Board of Agriculture, *2nd Annual Report of the Missouri State Board of Agriculture* (Jefferson City, MO: Emory Foster, 1866), 505.

22. L. U. Reavis, *St. Louis, the Future Great City* (St. Louis, MO: E. F. Hobart, 1853), 266.

23. William Godfrey Bek, *The German Settlement Society of Philadelphia and Its Colony Hermann, Missouri* (Philadelphia: American Germanica Press, 1907), 151.

24. William Dana, "Production of Wine in Missouri," *Merchants' Magazine and Commercial Review* (New York: Freeman Hunt, 1853), 123.

25. Daniel Sumner, Helene Bombrun, Julian Alston, and Dale Heien, "An Economic Survey of the Wine and Wine Grape Industry in the United States and Canada" (Unpublished. UC Davis, 2001), 3.

26. Robert Louis Stevenson, *The Silverado Squatters* (London: Chatto & Windus, 1906), 34.

27. Frank Schoonmaker and Tom Marvel, *American Wines* (New York: Duell, Sloan, & Pearce, 1941), 181.

28. USDA, 2014, "Washington Wine Grape Release, 2013." https://www.nass.usda.gov/Statistics_by_State/Washington/Publications/Fruit/winegrape14.pdf. Accessed May 21, 2016.

29. Oregon Wine, "Discover Oregon Wines Fact Sheet," http://industry.oregonwine.org/wp-content/uploads/2010/08/2010-Industry-Facts-w_background.pdf. 2011. Accessed July 6, 2011.

30. Southern Oregon University Research Center, "2014 Oregon Vineyard and Winery Census Report," Ashland, Oregon, http://industry.oregonwine.org/wp-content/uploads/Final-2014-Oregon-Vineyard-and-Winery-Report.pdf, Accessed Aug. 2015.

31. MKF Research, "The Economic Impact of Wine and Grapes on the State of Texas 2007," MKF Research LLC, https://www.depts.ttu.edu/hs/texaswine/docs/FINAL_Economic_Impact_TX_2008.pdf, 2008. Accessed June 9, 2011.

32. Wines & Vines, 2016, "Wine Industry Metrics," http://www.winesandvines.com/template.cfm?section=widc&widcDomain=wineries. Accessed May 22, 2016.

33. Andrew Adams, "2014 Virginia Grape Harvest Tops 8,000 Tons," *Wines & Vines*, May 9, 2015. http://www.winesandvines.com/template.cfm?section=news&content=148479. Accessed May 22, 2016.

34. Virginia Wine Marketing Office, 2008, www.virginiawine.org.

35. Pennsylvania Winery Association, "About Pennsylvania Wine" (Harrisburg, PA: Pennsylvania Winery Association, 2016). http://www.pennsylvaniawine.com/about-pennsylvania-wine. Accessed May 22, 2016.

36. Frank, Rimerman + Co., 2013. "The Economic Impact of Pennsylvania Wine, Wine Grapes and Juice Grapes 2011." http://pennsylvaniawine.com/sites/default/files/Pennsylvania%202011%20EI%20Report_FINAL.pdf (St. Helena, CA). Accessed May 22, 2016.

37. James Dombrosky and Shailendra Gajanan, "Pennsylvania Wine Industry—An Assessment" (Harrisburg, PA: Center for Rural Pennsylvania, 2013), 4.

38. Michigan Grape and Wine Industry Council, "Michigan's Grape and Wine Industry Fast Facts" (Lansing, MI: Michigan Grape and Wine Industry Council, 2015).

39. Tax and Trade Bureau, "Tax and Trade Bureau List of Permittees," https://www.ttb.gov/foia/frl.shtml. Accessed May 22, 2016.

40. Samuel de Champlain, *The Voyages and Explorations of Samuel de Champlain (1604–1616), Narrated by Himself*, trans. Annie Nettleton Bourne (Toronto: Courier Press, 1911), 180.

41. Vintners Quality Alliance Ontario, "2015 Annual Report" (Toronto, ON: Vintners Quality Alliance Ontario, 2016). http://www.vqaontario.ca/Resources/Library.

CHAPTER 10: FRONTIERS

1. Austin Ramzy, "China's Craft Breweries Find They May Have a 5,000 Year Old Relative," *New York Times*, May 25, 2016, http://www.nytimes.com/2016/05/26/world/asia/china-beer-history.html?_r=2.

2. Ibis World, "Wine Production in China: Market Research Report," IBISWorld, Inc. http://www.ibisworld.com/industry/china/wine-production.html. 2013. Accessed June 12, 2014.

3. Aaron Kingsbury, "Constructed Heritage and Co-Produced Meaning: The Re-Branding of Wines from the Koshu Grape," *Contemporary Japan* 26 (2014): 29–48. De Gruyter, http://dx.doi.org/10.1515/cj-2014-000, 2014, p. 36.

4. Skyscanner.net, "Extreme Wines: 5 Unusual Wine Regions," https://www.skyscanner.net/news/extreme-wines-5-unusual-wine-regions, May 30, 2011. Accessed October 16, 2016.

5. Dhruv Sood and Adam Branson, "India: Wine Production and Trade Update," USDA-FAS Gain Report, #IN4096, October, 22, 2014. http://gain.fas.usda.gov/Recent%20GAIN%20Publications/Wine%20Production%20and%20Trade%20Update_New%20Delhi_India_10-22-2014.pdf. Accessed October 16, 2016.

6. Florence Fabricant, "Red, White, Sultry: The Wines of India," *New York Times*, June 4, 2008, http://www.nytimes.com/2008/06/04/dining/04india.html.

7. Giulia Meloni and Johan Swinnen, "The Rise and Fall of the World's Largest Wine Exporter—and Its Institutional Legacy," AAWE Working Paper No. 134, 8. American Association of Wine Economists. www.wine-economics.org.

8. Giulia Meloni and Johan Swinnen, "The Rise and Fall of the World's Largest Wine Exporter—and Its Institutional Legacy," *Journal of Wine Economics* 9 (2014): 3.

9. UN FAO 2010. United Nations FAOSTAT. http://faostat.fao.org/site/342/default.aspx, accessed May 28, 2011.

CHAPTER 11: MODERN WINE CONTAINERS

1. Virginia R. Grace, *Amphoras and the Ancient Wine Trade* (Princeton, NJ: American School of Classical Studies at Athens, 1979).

2. Joachim Radkau, *Wood*, transl. Patrick Camiller (Malden, MA: Polity Press, 2012), 125.

3. Amhurst, “The Bottlescrew,” in *The Muse in Good Humour: A Collection of Comic Tales, Vol. 2* (London: F. Noble, 1766).

4. Ronald MacLean and Bob Nugent, *William Rockwell Clough: Inventor and Manufacturer of Over a Billion Corkscrews* (Wirtz, VA: Bullworks, 2004).

5. *My Little Chickadee*. Directed by Edward F. Cline. 1940.

6. Andrew Adams, “Strong Barrel Sales Herald Strong Industry: American Oak Supply Trained as U.S. Becomes Largest Barrel Market in the World,” *Wines and Vines*, February 2015. Accessed June 3, 2016.

7. Bob Walker, vice president for marketing operations, The Wine Group Inc. of Ripon, California, which uses Scholle bags for its Franzia Winetap® products. 2002.

CHAPTER 12: CULTURAL GEOGRAPHY OF WINE

1. Tyler Coleman, quoted in Evan Drake, “The Spirits of Politics,” *Daily Northwestern*, May 13, 2003.

2. Donald Horton, “The Function of Alcohol in Primitive Societies: A Cross Cultural Study,” *Quarterly Journal for the Study of Alcohol* 4 (1943): 199–320.

3. David Hanson, “Alcohol Problems and Solutions.” http://www2.potsdam.edu/hansondj/Controversies/20070604112246.html. 2008.

4. Anonymous, Epic of Gilgamesh. Project Gutenberg edition.

5. Edward Hyams, *Dionysus: A Social History of the Wine Vine* (London: Thames and Hudson, 1965), 260.

6. Ibid.

7. Polish American Center, “Sharing the Bread, Salt, and Wine at a Polish Wedding.” http://www.polishamericancenter.org/Wedding.htm. Accessed June 22, 2011.

8. Daniel Stanislawski, “Dionysus Westward: Early Religion and Economic Geography of Wine,” *Geographical Review* 65, no. 4 (1975): 429.

9. Hyams, *Dionysus*, 38–39.

10. Robert Fuller, *Religion and Wine* (Knoxville: University of Tennessee Press, 1996), 4.

11. Stanislawski, “Dionysus Westward,” 428.

12. Thucididyes, *The Peleponnesian War*, ca. 400 BCE (London: Penguin Books).

13. Bertrand Russell, *A History of Western Philosophy* (New York: Simon & Schuster, 1945), 16.

14. Eubulus, “Semele or Dionysus,” in John Maxwell Edmonds, *The Fragments of Attic Comedy* (Leiden: E. J. Brill, 1959), 125.

15. Colin Flint, *Introduction to Geopolitics* (New York: Routledge, 2006), 205.

16. National Council on Alcoholism and Drug Dependence, https://www.ncadd.org/about-addiction/alcohol-drugs-and-crime. Accessed October 11, 2016.

17. Amethyst Initiative, http://www.theamethystinitiative.org/. Accessed October 11, 2016.

18. Jonathan Swift, “On the Irish Club,” in *The Works of Jonathan Swift Volume One* (London: Henry G. Bohn, 1856), 729.

19. Ben Blanchard and Francesco Guarascio, "EU, China End Wine Dispute Ahead of Xi's European Tour," Reuters, March 21, 2014. http://www.reuters.com/article/us-china-europe-wine-idUSBREA2K0QE20140321

20. Tim Worstall, "Welcome to the Trade War—China Imposes Tariffs on EU, US, and Japanese Steel Imports," *Forbes*, June 24, 2016. http://www.forbes.com/sites/timworstall/2016/07/24/welcome-to-the-trade-war-china-imposes-tariffs-on-eu-us-and-japanese-steel-imports/#3066547c618e

21. Sandy Block, "Politics and Wine," www.beveragebusiness.com (Boston, MA: Massachusetts Beverage Business, 2002).

22. Lewis Perdue, *The Wrath of Grapes* (New York: Avon Books, 1999), 71–72.

23. Lawrence Gostin, *Public Health Law and Ethics: A Reader*, http://www.publichealthlaw.net/reader/docs/44liquormart.pdf, 2002.

24. National Institute on Alcohol Abuse and Alcoholism, "Drinking Levels Defined," https://www.niaaa.nih.gov/alcohol-health/overview-alcohol-consumption/alcohol-facts-and-statistics. Accessed June 13, 2016.

25. National Council on Alcoholism and Drug Dependence. "Facts About Alcohol," https://www.ncadd.org/about-addiction/alcohol/facts-about-alcohol. Accessed July 31, 2016.

26. Richard Mendelson, *From Demon to Darling: A Legal History of Wine in America* (Berkeley: University of California Press, 2009), 32.

27. David Hanson, "Alcohol Problems and Solutions." http://www2.potsdam.edu/hansondj/Controversies/20070604112246.html. 2008.

28. Ibid.

29. United Nations World Health Organization, "World Health Report 2002: Reducing Risks, Promoting Healthy Life" (New York: United Nations, 2002).

30. C. J. Chivers and Kevin Flynn, "35 Scary Minutes: Women Tell Police of Assaults in Park," *New York Times*, June 13, 2000, http://www.nytimes.com/2000/06/13/nyregion/35-scary-minutes-women-tell-police-of-assaults-in-park.html?_r=0

31. Wine Business.com, "Wine Market Council Unveils New Stats on Consumer Wine Consumption Habits," https://www.winebusiness.com/news/?go=getArticle&dataid=164519, Feb. 9, 2016. Accessed October 12, 2016.

32. Wine Market Council, "Wine Market Council Releases Latest Research on the Online Wine Shopping Behaviors of Wine Consumers," http://winemarketcouncil.com/wp-content/uploads/2016/01/FinalWineMarketCouncilOct2015MediaAlert.pdf, October 19, 2015. Accessed October 13, 2016.

33. Barrett Ludy, "Confusion: A Quick Summary of the EU Wine Reforms," https://www.guildsomm.com/stay_current/features/b/guest_blog/posts/confeusion-a-quick-summary-of-the-eu-wine-reforms, October 5, 2012. Accessed October 13, 2016.

34. Craig Silver, "Wine Names Are Getting Wackier and Wackier, and at No Extra Price," *Forbes*, http://www.forbes.com/sites/craigsilver/2013/08/16/wine-names-are-getting-wackier-and-wackier-and-at-no-extra-price/#1b658f5e610d, August 16, 2013. Accessed October 13, 2016.

35. Nancy Friedman, "The World's Dirtiest Wine Names," http://www.slate.com/blogs/lexicon_valley/2015/04/03/read_the_dirty_profane_obscene_names_that_companies_give_their_wines.html. Accessed October 11, 2016.

CONCLUSION

1. André Simon, *Wines of the World*, 2nd ed. (New York: McGraw-Hill, 1981), 152.

2. Robert Louis Stevenson, *The Silverado Squatters*. Transcribed from the 1906 edition (London: Chatto and Windus). Project Gutenberg, ebook #516, Chapter 3. www.gutenberg.org/files/516. Accessed August 15, 2016.

Bibliography

Adams, Andrew. "Strong Barrel Sales Herald Strong Industry: American Oak Supply Trained as U.S. Becomes Largest Barrel Market in the World." *Wines and Vines*. February 2015. http://www.winesandvines.com/template.cfm?section=features&content=145533&ftitle=Strong%20Barrel%20Sales%20Herald%20Strong%20Industry. Accessed June 3, 2016.

Adams, Andrew. "2014 Virginia Grape Harvest Tops 8,000 Ton." *Wines & Vines*. Last Modified May 9, 2015. http://www.winesandvines.com/template.cfm?section=news&content=148479. Accessed May 22, 2016.

Adams, K. E., and T. S. Rans. "Adverse Reactions to Alcohol and Alcoholic Beverages." *Annals of Allergy, Asthma & Immunology* 111, no. 6 (December 2013): 439–45. http://dx.doi.org/10.1016/j.anai.2013.09.016. PMID 24267355.

Adlum, John. *A Memoir on the Cultivation of the Vine in America and the Best Mode of Making Wine*. Washington, DC: Davis and Force, 1823.

Allen, H. Warner. *The Romance of Wine*. New York: E. P. Dutton, 1932.

Alleyne, Richard. "Red Wine Can Protect Against Radiation." *Telegraph*, http://www.telegraph.co.uk/news/3067317/Red-wine-can-protect-against-radiation.html, September 23, 2008.

Alvarado, Serafin. "Interpreting, Analyzing, and Evaluating Wine Quality." Paper presented at Society of Wine Educators Annual Conference. Providence, RI. August 4, 2011.

Amerine, M. A., and R. M. Wagner. "The Vine and Its Environment." In D. Muscatine, M. A. Amerine, and B. Thompson, eds., *The Book of California Wine*. Berkeley: University of California Press, 1984.

Amethyst Initiative. http://www.theamethystinitiative.org/. Accessed October 11, 2016.

Amhurst. "The Bottlescrew." In *The Muse in Good Humour: A Collection of Comic Tales by the Most Eminent Poets*, Vol. 2. Edited by F. Noble. London, 1766.

Anderson, Kym. *Growth and Cycles in Australia's Wine Industry: A Statistical Compendium, 1843 to 2013*. Adelaide: University of Adelaide Press, 2013. www.adelaide.edu.au/press.

Anderson, Kym. *Which Winegrape Varieties Are Grown Where? A Global Empirical Picture*. Adelaide: University of Adelaide Press, 2013.

Anderson, Kym, David Norman, and Glyn Wittwer. "Globalization and the World's Wine Markets: Overview." Adelaide: University of Adelaide. Centre for International Economic Studies (CIES). Discussion Paper 0143, 2001. cies@adelaide.edu.au.

Anonymous. *Epic of Gilgamesh*. New Haven, CT: Yale University Press, Yale Oriental Series Vol. IV, Part III, trans. Morris Jastrow and Albert Clay, *An Old Babylonian Version of the Gilgamesh Epic, On the Basis of Recently Discovered Texts.* Project Gutenberg edition. EBook #11000. Released July 4, 2006.

Arroyo-Garcia, R., L. Ruiz-Garcia, L. Bolling, R. Ocete, M. A. Lopez, C. Arnold, A. Ergul, G. Soylemezo'lu, H. I. Uzun, F. Babello, J. Ibanex, M. K. Aradhya, A. Atanassov, I. Atanassov, S. Balint, J. L. Cenis, L. Costantini, S. Gorislavets, M. S. Grando, B. Y. Klein, P. E. McGovern, D. Merdinoglu, I. Pejic, F. Pelsy, N. Primikirios, V. Risovannaya, K. A. Roubelakis-Angelaiks, H. Snoussi, P. Sotiri, S. Tamhankar, P. This, L. Troshin, L. M. Malpica, F. Lefort, and J. M. Martinex-Zapater. "Multiple Origins of Cultivated Grapevine (Vitis vinifera L. spp. Sativa) Based on Chloroplast DNA Polymorphisms." *Molecular Ecology* 15, no. 12 (2006): 3707–14. http://dx.doi.org/10.1111/j.1365-294X.2006.03049.x.

Asbury, Herbert. *The Great Illusion: An Informal History of Prohibition*. Garden City, New York: Doubleday, 1950.

Atkins, Tim. *Chardonnay.* New York: Penguin Books, 1991.

Barr, Andrew. *Pinot Noir.* New York: Penguin Books, 1992.

Bateman. A. E. "Wine Production in France." *J. Statistical Society of London* 47, no. 4 (Dec. 1884): 609.

Bauer, Karl, Josef Pleil, and Victoria Loimer. "Austrian Wine—A Taste of Culture." www.austrianwine.com. 2013. Accessed July 2, 2014.

Baxevanis, John. *The Wines of Bordeaux and Western France*. Totowa, NJ: Rowman and Littlefield, 1987.

Baxevanis, John. *The Wines of Champagne, Burgundy, Eastern and Southern France.* Totowa, NJ: Rowman and Littlefield, 1987.

Beeston, John. *The Wine Regions of Australia*. Crow's Nest, Australia: Allen & Unwin, 2002.

Bek, William Godfrey. *The German Settlement Society of Philadelphia and Its Colony Hermann, Missouri*. Philadelphia: American Germanica Press, 1907.

Benton, A. L. *The Century of Temperance Reform,* 1885, as quoted in Mark Lender and James Martin, *Drinking in America: A History.* Revised and Expanded Edition. New York: Free Press, 1987.

Bespaloff, Alexis. *The Fireside Book of Wine*. New York: Simon & Schuster, 1977.

Bierce, Ambrose. *The Devil's Dictionary.* New York: Neale, 1911.

Bitting, Katherine. *The Wine Drinker's Manual.* London: Marsh and Miller, 1830.

Black, Arthur. "Agave Intense, No Really." Paper presented at the Society of Wine Educators Annual Conference, New Orleans, August 13, 2015.

Black, Simon. "China's Wine Production to Overtake Australia's in the Next Three Years." *Herald Sun,* Jan. 25, 2011.

Blanchard, Ben, and Francesco Guarascio. "EU, China End Wine Dispute Ahead of Xi's European Tour." Reuters, March 21, 2014. http://www.reuters.com/article/us-china-europe-wine-idUSBREA2K0QE20140321.

Block, Sandy. "Politics and Wine." *Massachusetts Beverage Business*, 2002. www.beveragebusiness.com. Accessed October 14, 2003.

Bolles, Albert Sidney. *Industrial History of the United States, from the Earliest Settlement to the Present Time.* Norwich, CT: Henry Bill, 1878.

Brazilian Ministry of Development, Industry, and Foreign Trade. http://www.winesofbrasil.com/CentralArquivos/Exporta%C3%A7%C3%A3o%20Brasileira%20GERAL%20FINAL%202012%20-%20English.pdf. Last updated August 7, 2013.

Breslin, Joanna. *South African Wine History.* Cape Town: Southern Stars, 2011. http://www.southernstarz.com/images/edu_region_art/sa_wine_history.pdf. Accessed May 14, 2016.

Bresnahan, S. "MADD Struggles to Remain Relevant." *Washington Times*, August 6, 2002.

Brook, Stephen. *Sauvignon Blanc and Semillon.* New York: Penguin Books, 1992.

Brown, Gordon. *Classic Spirits of the World: A Comprehensive Guide.* New York: Abbeville Press, 1996.

Bruman, Henry J. *Alcohol in Ancient Mexico.* Salt Lake City: University of Utah Press, 2000.

Bulman, Teresa. "The Joy of Geography." Presidential Address delivered to the Association of Pacific Coast Geographers, 66th Annual Meeting. Portland, Oregon. September 20, 2003.

Burns, Scott. "The Importance of Soil and Geology in Tasting Terroir with a Case History from the Willamette Valley, Oregon." In *The Geography of Wine: Regions, Terroir, and Techniques*, Percy Dougherty, ed. Dordrecht: Springer, 2012.

Butler & Butler, *Indiana Wine: A History.* Bloomington: Indiana University Press, 2001.

California Department of Food and Agriculture. "California Grape Acreage Report 2009" and "California Grape Acreage Report 2014." Sacramento, CA. www.nass.usda.gov/ca. Accessed September 16, 2015.

Campbell, Christy. *The Botanist and the Vintner: How Wine Was Saved for the World.* Chapel Hill, NC: Algonquin Books, 2005.

Champlain, Samuel de. *The Voyages and Explorations of Samuel de Champlain (1604–1616), narrated by Himself,* Translated by Annie Nettleton Bourne. Toronto: Courier Press, 1911.

Chivers, C. J., and Kevin Flynn. "35 Scary Minutes: Women Tell Police of Assaults in Park." *New York Times*, June 13, 2000. http://www.nytimes.com/2000/06/13/nyregion/35-scary-minutes-women-tell-police-of-assaults-in-park.html?_r=0.

Clarion Associates. "Executive Summary." *Study on the Costs of Sprawl in Pennsylvania*. Harrisburg, PA: 10,000 Friends of Pennsylvania, 2002.

Claussen, M. "On Coupling Global Biome Models with Climate Models." *Climate Research* 4 (1994): 203–221. http://hdl.handle.net/11858/00-001M-0000-0013-AF4F-E.

Coleman, Tyler, quoted in Evan Drake, "The Spirits of Politics." *Daily Northwestern*, May 13, 2003.

Comite Champagne. "Key Market Statistics." http://www.champagne.fr/en/champagne-economy/key-market-statistics, 2016. Accessed April 27, 2016.

Commission Europeenne, Direction Generale de L'agriculture et du Developpement Rural. "Common Market Organization WINE-R(CD)555/2008 Grubbing-up Scheme," 2012. Accessed June 2, 2013.

Cox, Juliet, and Larry Bridwell. "Australian Companies Using Globalization to Disrupt the Ancient Wine Industry." *Competitiveness Review* 17, no. 4 (2007): 209–221.

Dana, William. *Merchants' Magazine and Commercial Review* 28. New York: Freeman Hunt, 1853.

Davidson, W. M., and R. L. Nougaret. *The Grape Phylloxera in California*. United States Department of Agriculture, Bulletin No. 903. Washington, DC: Government Printing Office, 1921.

Davis, William Stearns. *A Day in Old Athens*. Project Gutenberg Edition EBook #4167, 1914. Posted March 6, 2002.

De Blij, Harm. *Wine: A Geographic Appreciation*. Totowa, NJ: Rowman and Allanheld, 1983.

De Blij, Harm. *Wine Regions of the Southern Hemisphere*. Totowa, NJ: Roman and Allenheld, 1985.

Detroit Publishing Co., Copyright Claimant, and Publisher Detroit Publishing Co. *[Original Concord grape vine, Concord, Mass]*. [Between 1910 and 1920] Image. Retrieved from the Library of Congress, https://www.loc.gov/item/det1994024654/PP/. Accessed July 10, 2016.

Deutsches Weininstitut GmbH. *Deutscher Wein Statistik 2014/2015*. Mainz, Germany: Deutscher Wein Institut, 2015. PDF accessed June 9, 2016.

Deutsches Weininstitut GmbH. *A Short Guide to German Wines*. Mainz: Deutsches Weininstitut, 1997.

Dewey, Suzette. *Wines for Those Who Have Forgotten and Those Who Want to Know*. Chicago: Lakeside Press, 1934.

Dickens, Charles. *Household Words*. Leipzig: Bernhard Tauchnitz, 1854.

Dickenson, John, and John Salt. "In Vino Veritas: An Introduction to the Geography of Wine." *Progress in Human Geography* 6 (1982).

Diodorus of Sicily, ca. 50 BCE.

Dombrosky, James, and Shailendra Gajanan. "Pennsylvania Wine Industry—An Assessment." Harrisburg, PA: Center for Rural Pennsylvania, 2013.

Dominé, André. *Wine*. Cologne, Germany: Konemann, 2001.

Donaldson, John William. *The Theatre of the Greeks*. London: George Bell and Sons, 1875.

Dpto. Contabilidad y Registros. "Produccion de uva en las serie historica" (2013-harvest-data-99828394-kilos-with-do.pdf). http://www.dorueda.com/en/facts-figures/advances. Posted November 29, 2013. Accessed May 1, 2016.

Duijker, Hubrecht. *Wine Atlas of Spain and Traveler's Guide to the Vineyards*. New York: Simon & Schuster, 1992.

E. & J. Gallo Co. "Gallo Winery, Press Room: Fact Sheet." http://gallo.com/press-room/fact-sheet/CompanyFactSheet.html, 2016. Accessed April 24, 2016.

Earle, Alice Morse. *Stage-Coach and Tavern Days*. New York: Macmillan, 1900.

Eng, Rachel. "Wine Down Wednesday: An Amateur's Guide to Reviewing Wine." http://blog.bazaarvoice.com/2014/10/22/wine-wednesday-amateurs-guide-reviewing-wine/. Posted October 22, 2014. Accessed Jan. 14, 2016.

Engelmann, Larry. *Intemperance: The Lost War against Liquor*. New York: Free Press, 1979.

English Wine Producers. http://www.englishwineproducers.co.uk/. Accessed May 5, 2016.

Estreicher, Stefan K. *Wine: From Neolithic Times to the 21st Century*. New York: Algora, 2006, p. 2.

Estreicher, Stefan K. *Wine: The Past 7,400 Years*. http://www1.mpi-halle.mpg.de/~md_simul/data/special-data/wine-history.pdf. 2004. Retrieved July 30, 2013.

Eubulus. "Semele or Dionysus," in John Maxwell Edmonds, *The Fragments of Attic Comedy*. Leiden: E. J. Brill, 1959.

European Union, Eurostat. http://ec.europa.eu/agriculture/wine/. First Accessed March 11, 2012.

Eyres, Harry. *Cabernet Sauvignon*. New York: Penguin Books, 1991.

Fabricant, Florence. "Red, White, Sultry: The Wines of India." *New York Times*, June 4, 2008. http://www.nytimes.com/2008/06/04/dining/04india.html.

Fadiman, Clifton. *The Joys of Wine*. New York: Galahad Books, 1981.

Fadiman, Clifton. "Origins of Wine." *New York Times*, March 8, 1987.

Flandrin, Jean-Louis, and Massimo Montana, eds. *Food: A Culinary History*. New York: Columbia University, 2012.

Fleming, Stuart. *Vinum: The Story of Roman Wine*. Glen Mills, PA: Art Flair, 2001.

Flint, Colin. *Introduction to Geopolitics*. New York: Routledge, 2006.

Forbes, R. J. *A Short History of the Art of Distillation*. Leiden, Netherlands: E. J. Brill, 1970.

Frank, Mitch "Italian Police Uncover Counterfeit Champagne Scheme." *Wine Spectator*. http://www.winespectator.com/webfeature/show/id/52684. February 2, 2016. Accessed June 11, 2016.

Frank, Rimerman + Co., "The Economic Impact of Pennsylvania Wine, Wine Grapes and Juice Grapes 2011." http://pennsylvaniawine.com/sites/default/files/Pennsylvania%202011%20EI%20Report_FINAL.pdf (St. Helena Ca), 2013. Accessed May 22, 2016.

Franklin, William Temple. *The Posthumous and Other Writings of Benjamin Franklin*. London: Henry Colburn, 1819.

Friedman, Nancy. "The World's Dirtiest Wine Names." http://www.slate.com/blogs/lexicon_valley/2015/04/03/read_the_dirty_profane_obscene_names_that_companies_give_their_wines.html, accessed October 11, 2016.

Fuller, Robert. *Religion and Wine*. Knoxville: University of Tennessee Press, 1996.

Gabler, James. *Wine into Words*. Baltimore, MD: Bacchus Press, 1985.

Gale, George. *Dying on the Vine: How Phylloxera Transformed Wine*. Berkeley: University of California Press, 2011.

Galileo, quoted in Helen Exley, *Wine Quotations*. New York: Exley, 1994, p. 6. Accessed October 9, 2016.

Gawel, Richard. "Common Wine Faults and Their Causes." www.aromadictionary.com, 2005.

Gesner, Conrad. *The Newe Iewell of Health*, 1576. London: Da Capo Press, 1971.

Goldstein, Evan. *Wines of South America: The Essential Guide*. Berkeley: University of California Press, 2014.

Gostin, Lawrence. *Public Health Law and Ethics: A Reader.* http://www.publichealthlaw.net/reader/docs/44liquormart.pdf. 2002.

Grace, Virginia R. *Amphoras and the Ancient Wine Trade*. Princeton, NJ: American School of Classical Studies at Athens, 1979.

Hanni, Tim. *Why You Like the Wines You Like*. Napa, CA: Hannico.com, 2013.

Hanson, David. "Alcohol Problems and Solutions." 2008. http://www2.potsdam.edu/hansondj/Controversies/20070604112246.html.

Hanson, David. "Crash Course on MADD." 2008. http://www.alcoholfacts.org/CrashCourseOnMADD.html#111. 2002–2014.

Harrell, Alfred. *Wine Cask*. October 1980, Image. Retrieved from the Library of Congress. https://www.loc.gov/item/ncr002171/. Accessed July 10, 2016.

Heinlein, Robert. *Time Enough for Love*. New York: Ace Books, 1973.

Hendricks, Rick. "Viticulture in El Paso Del Norte during the Colonial Period." *Agricultural History* 78, no. 2 (2004):191–200.

Hirsch, Arthur Henry. *The Huguenots of Colonial South Carolina*. Durham, NC: Duke University Press, 1928.

Hodgen, Donald A. Public Statement, U.S. Department of Commerce. Oct. 16, 2003. donald_a_hodgen@ita.doc.gov, Oct. 16, 2003.

Holst, Sanford. *Phoenician Secrets: Exploring the Ancient Mediterranean*. Los Angeles, CA: Santorini Books, 2011.

Horton, Donald. "The Function of Alcohol in Primitive Societies: A Cross Cultural Study." *Quarterly Journal for the Study of Alcohol* 4 (1943): 199–320.

HortResearch. "Fresh Facts: New Zealand Horticulture." http://www.hortresearch.co.nz/files/aboutus/factsandfigs/ff2006.pdf. 2006. Accessed November 11, 2008.

Howard, Philip. "Big Wine Tightened Its Grip on the U.S. Market in 2013." http://winecurmudgeon.com/big-wine-tightened-its-grip-on-the-u-s-market-in-2013/. March 13, 2014, accessed April 24, 2016.

Hui, A. C. F., and S. M. Wong. "Deafness and Liver Disease in a 57-Year-Old Man: A Medical History of Beethoven?" *Hong Kong Medical Journal* 6, no. 4 (December 2000): 433–38. PDF file accessed through Google Scholar, February 13, 2010.

Hyams, Edward. *Dionysus: A Social History of the Wine Vine*. New York: Macmillan, 1965.

IbisWorld. "Wine Production in China: Market Research Report." *IBISWorld, Inc.* http://www.ibisworld.com/industry/china/wine-production.html. 2013. Accessed June 12, 2014.

Inter Rhone. "Key Figures 2011 of Rhone Valley AOCs." http://www.vins-rhone.com/en/5100-Wine Statistics.htm. 2011. Accessed May 30, 2011.

Irvine, Ronald, and Walter J. Clore. *The Wine Project: Washington State's Winemaking History*. Vashon, WA: Sketch Publications, 1997.

Istituto Statisctica Mercati Agro-Alimentari. Rome, Italy: ISMEA, 2005.

Jefferson, Thomas. *Journal*, 1797. Quoted in Alexis Bespaloff, *The Fireside Book of Wine*. New York: Simon & Schuster, 1977, p. 269.

Johnson, Hugh. "The Origins of Wine." *Vintage: A History of Wine*. Video series, Episode 1, 1986.

Johnson, Hugh. *Vintage: The Story of Wine*. New York: Simon & Schuster, 1989.

Johnson, Hugh, and James Halliday. *The Vintner's Art*. New York: Simon & Schuster, 1992.

Johnson, Hugh, and Jancis Robinson. *The World Atlas of Wine*. London: Octopus Publishing Group, 2001.

Johnson, Samuel. *The Life of Samuel Johnson*. London: Penguin Classics, 1791.

Jones, Gregory V. "Climate and Terroir: Impacts of Climate Variability and Change on Wine." In *Fine Wine and Terroir—The Geoscience Perspective,* in R. W. Macqueen and L. D. Meinert, eds., *Geoscience Canada* Reprint Series Number 9, Geological Association of Canada. St. John's, Newfoundland, 2005.

Jones, Gregory V., R. Reid, and A. Vilks, "Climate, Grapes, and Wine: Structure and Suitability in a Variable and Changing Climate." In *The Geography of Wine: Regions, Terroir, and Techniques,* edited by P. Dougherty. Leipzig: Springer Press, 2012.

Jones, Gregory, Nicholas Snead, and Peder Nelson. "Geology and Wine 8. Modeling Viticultural Landscapes: A GIS Analysis of the Terroir Potential in the Umpqua Valley of Oregon." *Geoscience Canada* 31, no. 167 (2004).

Keller, Jack. "Winemaking: Native North American Grapes and Wines." http://winemaking.jackkeller.net/natives.asp, 2008. Accessed September 19, 2011.

Kimball, Marie. "Some Genial Old Drinking Customs." *William and Mary Quarterly*, 3rd Series, 2, no. 4 (1945): 349–358.

Kingsbury, Aaron. "Constructed Heritage and Co-Produced Meaning: The Re-Branding of Wines from the Koshu Grape." *Contemporary Japan* 26 (2014): 29–48. De Gruyter.

Koppen, Vladimir. *Versuch einer Klassifikation der Klimate. Vorzugsweise nach iheren Beziehungen zur Pflanzenwelt*. Leipzig: B. G. Teubner, 1901. *Die Klimate der Erde*. Berlin and Leipzig: Walter de Gruyter, 1923.

Korshin, Nathanie. *Better Wines for Less Money*. New York: Stein & Day, 1974.

Lambert-Gocs, Miles. *Tokaji Wine*. Williamsburg, VA: Ambeli Press, 2010.

LaRaia, Anita. *Pick a Perfect Wine . . . in No Time*. Indianapolis, IN: Que Books, 2006.

Lawson, John. *The History of Carolina*. Raleigh, NC: Strother & Marcon, 1860. http://books.google.com. Accessed August 15, 2016.

Leahy, Richard G. *Beyond Jefferson's Vines: The Evolution of Quality Wine in Virginia,* 2nd ed. CreateSpace, 2014.

Lender, Mark, and James Martin. *Drinking in America: A History.* New York: Free Press, 1987.

Lesko, Leonard L. "King Tut's Wine Cellar," Cited in Thomas Pellechia, *Wine: The 8,000-Year-Old History of the Wine Trade* New York: Thunder's Mouth Press, 2006.

Liem, Peter. "Tasting Notes." *Riesling Report*, Nov./Dec. 2000. Portland, OR. www.reislingreport.com.

Liv-ex. "Fine Wine 100 Index." http://www.liv-ex.com/staticPageContent.do?pageKey=Fine_Wine_100.

Longfellow, Henry Wadsworth. "Catawba Wine." *The Poetical Works of Henry Wadsworth Longfellow*. London: Henry G. Bohn, 1861. PDF downloaded from books.google.com.

Luther, Martin. Quoted in AZQuotes.com, http://www.azquotes.com/quote/365483. Accessed October 9, 2016.

MacDonogh, Giles. *Syrah, Grenache, and Mourvedre*. New York: Penguin Books, 1992.

MacLean, Ronald, and Bob Nugent. *William Rockwell Clough: Inventor and Manufacturer of Over a Billion Corkscrews*. Wirtz, VA: Bullworks, 2004.

Mann, Windsor. "The Ignorance of Neo-Prohibitionists." American Spectator. http://spectator.org/58435_ignorance-neo-prohibitionists/. Accessed May 26, 2016.

Maugh, Thomas H. "Ancient Winery Found in Armenia." *Los Angeles Times*, January 11, 2011.

McGovern, Patrick. *The Origins and Ancient History of Wine*. Singapore: Overseas Publishing Associates, 1996.

McGovern, Patrick. "Prehistoric China." http://www.penn.museum/sites/biomolecul ararchaeology/?page_id=247. 2016. Accessed 4/6/2016.

McGovern, Patrick. *Uncorking the Past*. Berkeley: University of California Press, 2009.

Meloni, Giulia, and Johan Swinnen. "The Rise and Fall of the World's Largest Wine Exporter—and Its Institutional Legacy." *Journal of Wine Economics* 9 (2014): 3–33.

Mendelson, Richard. *From Demon to Darling: A Legal History of Wine in America*. Berkeley: University of California Press, 2009.

Michigan Grape and Wine Industry Council. "Michigan's Grape and Wine Industry Fast Facts." Lansing, MI: Michigan Grape and Wine Industry Council, 2015.

Milner, Duncan. *Lincoln and Liquor* New York: Neale, 1920.

Minnesota Agricultural Experiment Station. "The Marquette Grape." St. Paul, MN, 2009. http://www.grapes.umn.edu/marquette/index.html. Accessed March 13, 2011.

Mishkin, David. "The American Colonial Vineyard: An Economic Interpretation." *Journal Econ. History* 25, no. 4 (1965): 683–85.

Missouri State Board of Agriculture. *2nd Annual Report of the Missouri State Board of Agriculture*. Jefferson City, MO: Emory Foster, 1866.

MKF Research. "The Economic Impact of Wine and Grapes on the State of Texas 2007." *MKF Research LLC*. https://www.depts.ttu.edu/hs/texaswine/docs/FINAL_Economic_Impact_TX_2008.pdf, 2008. Accessed June 9, 2011.

Moran, Warren. "Rural Space as Intellectual Property." *Political Geography* 12, no. 3 (1993): 263–277.

Moran, Warren. "The Wine Appellation as Territory in France and California." *Annals, AAG* 83 (1993): 694–717.

Morewood, Samuel. *A Philosophical and Statistical History of the Inventions and Customs of Ancient and Modern Nations in the Manufacture and use of Inebriating Liquors; with the Present Practice of Distillation in all its Varieties: Together with an extensive illustration of the Consumption and Effects of Opium and other Stimulants used in the East, as Substitutes for Wine and Spirits*. Dublin, Ireland: William Curry Jr., 1838.

Munsie, Jeffrey. "A Brief History of the International Regulation of Wine Production." Harvard Law School (2002). https://dash.harvard.edu/bitstream/handle/1/8944668/Munsie.html?sequence=2. Accessed January 31, 2017.

My Little Chickadee. Directed by Edward F. Cline. 1940.

Myers, Albert Cook. *Narratives of Early Pennsylvania West New Jersey and Delaware*. New York: Charles Scribner's Sons, 1912.

Myles, Sean, Adam Boyko, Christopher Owens, Patrick Brown, et al. "Genetic Structure and Domestication History of the Grape." *PNAS* 108, no. 9 (2011): 3530–3535.

National Council on Alcoholism and Drug Dependence. "Facts About Alcohol." https://www.ncadd.org/about-addiction/alcohol/facts-about-alcohol. Accessed July 31, 2016.

National Institute on Alcohol Abuse and Alcoholism. "Drinking Levels Defined." https://www.niaaa.nih.gov/alcohol-health/overview-alcohol-consumption/alcohol-facts-and-statistics. Accessed June 13, 2016.

Nephew, T. M., G. D. Williams, F. S. Stinson, K. Nguyen, and M. C. Dufour. "Apparent Per Capita Alcohol Consumption: National, State, and Regional Trends, 1977–98: Surveillance Report #55." *National Institute on Alcohol Abuse and Alcoholism, Division of Biometry and Epidemiology*, Alcohol Epidemiologic Data System, 2000.

New Zealand Wine Growers Association. *2005 New Zealand Wine Growers Association Annual Report*. Auckland, NZ: Wine Institute of New Zealand, 2006.

New Zealand Wine Growers Association. *New Zealand Winegrowers Annual Report: 2015*. Auckland, NZ: Wine Institute of New Zealand, 2016.

Noble, Ann C. *The Wine Aroma Wheel*. http://winearomawheel.com/. Accessed October 9, 2016.

Oregon Wine. "Discover Oregon Wines Fact Sheet." http://industry.oregonwine.org/wp-content/uploads/2010/08/2010-Industry-Facts-w_background.pdf. 2011. Accessed July 6, 2011.

Origins.wine. Origins.wine/about. Accessed August 1, 2016.

Pastrana-Bonilla, Eduardo, et al. "Phenolic Content and Antioxidant Capacity of Muscadine Grapes." *Journal of Agricultural and Food Chemistry* 51 (2003): 5497–5503.

Pellechia, Thomas. *Garlic, Wine and Olive Oil: Historical Anecdotes and Recipes.* Santa Barbara, CA: Capra Press, 2000.

Pellechia, Thomas. *Wine.* New York: Thunder's Mouth Press, 2006.

Penington, John. *An Examination of Beauchamp Plantagenet's Description of the Province of New Albion.* Philadelphia, 1840.

Pennsylvania Winery Association. "About Pennsylvania Wine." Harrisburg, PA: Pennsylvania Winery Association, 2016. http://www.pennsylvaniawine.com/about-pennsylvania-wine. Accessed May 22, 2016.

Perdue, Lewis. *The Wrath of Grapes.* New York: Avon Books, 1999.

Philip, Howard. "Concentration in the U.S. Wine Industry." Quoted in "Big Wine: 5 companies, 60 Percent of Sales, 200 Brands." http://winecurmudgeon.com/big-wine-5-companies-60-percent-of-sales-200-brands/. 2013. Accessed April 25, 2016.

Phillips, Rod. *A Short History of Wine.* New York: HarperCollins, 2000.

Pigott, Stuart. *Riesling.* New York: Penguin, 1991.

Pinney, Thomas. *A History of Wine in America from the Beginnings to Prohibition.* Berkeley: University of California Press, 1989.

Poddar, Anita. "Accolade Wines Submission to the Senate Standing Committee on Rural and Regional Affairs and Transport References Inquiry into the Australian Grape and Wine Industry. Australian Grape and Wine Industry Submission 26." Reynella, South Australia: Accolade Wines, May 2015.

Poletti, Peter. "An Interdisciplinary Study of the Missouri Grape and Wine Industry, 1650 to 1989." Doctoral dissertation, St. Louis University, 1989.

Polish American Center. "Sharing the Bread, Salt and Wine at a Polish Wedding." http://www.polishamericancenter.org/Wedding.htm. Accessed 6/22/2011.

Porter, Michael. "Location, Competition, and Economic Development: Local Clusters in a Global Economy." *Economic Development Quarterly* 14 (2000): 15–34.

Powell, Marvin. "Wine and the Vine in Ancient Mesopotamia: The Cuneiform Evidence." In *The Origins and Ancient History of Wine*, Patrick E. McGovern, Stuart J. Fleming, and Solomon H. Katz, eds. Amsterdam, Netherlands: Gordon and Breach, 1996.

Pregler, Bill. "Do You Need a Weather Station in Your Vineyard?" https://kestrelmeters.com/blogs/news/3457062-do-you-need-a-weather-station-in-your-vineyard. Kestral Corp., 2016. Accessed April 18, 2016.

Purchas, Samuel. *Hakluytus Posthumus or Purchas His Pilgrimes.* Glasgow, Scotland: James MacLehose & Sons, 1906. Vol. 19: 145.

Radkau, Joachim. *Wood.* Translated by Patrick Camiller. Malden, MA: Polity Press, 2007, 2012.

Ramzy, Austin. "China's Craft Breweries Find They May Have a 5,000 Year Old Relative." *New York Times*, May 25, 2016. http://www.nytimes.com/2016/05/26/world/asia/china-beer-history.html?_r=2.

Rawlinson, George, *History of Phoenicia*. Oxford: Longmans, Green. Project Gutenberg edition. 1889.

Reavis, L. U. *St. Louis, the Future Great City*. St. Louis, MO: E. F. Hobart, 1853.

Riebeeck, Jan van. *Journal of Jan van Riebeeck. Volumes II, III, 1656–1662*. Edited by H. B. Thom and translated by J. Smuts. Cape Town: A. A. Balkema, 1954.

Robertson, George. *Port*. London: Faber and Faber, 1978.

Robinson, Jancis, Julia Harding, and José Vouillamoz. *Wine Grapes*. New York: HarperCollins, 2012.

Rocke, Alan. "Buckeyes, Corncrackers, and Suckers: Culinary Episodes in Ohio History." Cleveland, OH: Case Western Reserve University, 2002. http://www.case.edu/artsci/wrss/documents/2002rocke_001.pdf. Accessed Feb. 13, 2010.

Rockefeller, John D. Quoted in S. Michael Craven, "Changing Culture: A Study in Cultural Engagement—Part 2." Battlefortruth.org. Last updated October 14, 2010.

Roger, J-R. *The Wines of Bordeaux*. New York: E. P. Dutton, 1960.

Rothkopf, Joanna. "170-Year-Old Champagne Tastes Like Wet, Cheesy Hair." www.Salon.com. April 21, 2015. http://www.salon.com/2015/04/21/170_year_old_champagne_tastes_like_wet_cheesy_hair/. Accessed April 29, 2016.

Russell, Bertrand. *A History of Western Philosophy*. New York: Simon & Schuster, 1945.

Sage, Evan T., trans. Livy. *A History of Rome*. Books XXXVIII–XXXIX with an English Translation. Cambridge, MA: Harvard University Press; London: William Heinemann, 1936: published without copyright notice. Perseus Edition. Medford, MA: Tufts University. https://www.google.com/search?q=tufts+univ&ie=utf-8&oe=utf-8.

Sauer, Carl. "The Geography of the Ozark Highland of Missouri." *Geographic Society of Chicago Bulletin No. 7*. Chicago, IL: University of Chicago Press, 1920.

SAWIS. "2015—SA Wine Industry Statistics NR 39." Paarl, South Africa. www.sawis.co.za. 2015.

Scheck, Justin, Tripp Mickle, and Saabira Chaudhuri. "With Moderate Drinking under Fire, Alcohol Companies Go on Offensive." *Wall Street Journal*. http://www.wsj.com/articles/with-moderate-drinking-under-fire-alcohol-companies-go-on-offensive-1471889160. August 22, 2016.

Schoonmaker, Frank, and Tom Marvel. *American Wines*. New York: Duell, Sloan, and Pearce, 1941.

Schreiner, John. *Icewine: The Complete Story*. Toronto, Canada: Warwick, 2001.

Sechrist, Robert. "The Expanding International and Varietal Composition of the Wine Spectator Top 100, 1988–2010." Paper presented at the AAG Annual Conference, Seattle, WA. 2011.

Sechrist, Robert. "A Good Place for Vines." Paper presented at the AAG Conference, Boston, MA. March 2008.

Sechrist, Robert. "The Origin, Diffusion and Globalization of Riesling." In *Geography of Wine*, Percy Dougherty, ed. New York: Springer Verlag, 2012.

Sechrist, Robert. "Pennsylvania Liquor Control Board Wine Sales: Sept. 2000 to Aug. 2002." *Pennsylvania Geographer* (2004) 42: 39–59.

Sechrist, Robert. *The Pennsylvania Wine Monopoly.* American Association of Wine Economists. Working Paper 109, 2012. http://www.wine-economics.org/aawe/wp-content/uploads/2012/10/AAWE_WP109.pdf.

Sechrist, Robert. "Vine Grubbing in the European Union 1990–2010." Paper presented at the AAG Annual Conference. Chicago, April 23, 2015.

Shrady, Nicholas. *The Last Day*. New York: Penguin Books, 2008.

Simon, André L. *Bottlescrew Days: Wine Drinking in England during the Eighteenth Century*. London: Small, Maynard, 1927.

Simon, André L. *The History of Champagne*. London: Octopus Books, 1971.

Simon, André L. *In Vino Veritas: A Book about Wine*. London: Grant Richards, 1913.

Simon, André L., and S. F. Hallgarten. *The Great Wines of Germany*. New York: McGraw-Hill, 1963.

Smith, Adam. *An Inquiry into the Nature and Causes of the Wealth of Nations* (Vols. I & II), R. H. Campbell & A. S. Skinner, eds. Indianapolis, IN: Liberty Fund, 1981.

Smith, Lloyd, and Peter Whigham. "Spatial Aspects of Vineyard Management and Wine Grape Production." Paper presented at the 11th Colloquium of the Spatial Information Research Center, New Zealand. 1999.

Sonoma County Grape Growers Association. *Exploring the Appellations of Sonoma County.* Sebastopol, CA: Sonoma County Grape Growers Association, 2006.

Sood, Dhruv, and Adam Branson. "India: Wine Production and Trade Update." USDA-FAS GAIN Report, #IN4096, October 22, 2014. http://gain.fas.usda.gov/Recent%20GAIN%20Publications/Wine%20Production%20and%20Trade%20Update_New%20Delhi_India_10-22-2014.pdf. Accessed October 16, 2016.

Sournia, Jean-Charles. *A History of Alcoholism*. New York: Blackwell, 1990.

Southern Oregon University Research Center. "2014 Oregon Vineyard and Winery Census Report." Ashland, OR. 2015. http://industry.oregonwine.org/wp-content/uploads/Final-2014-Oregon-Vineyard-and-Winery-Report.pdf.

Spence, Godfrey. *The Port Wine Companion: A Connoisseur's Guide*. New York: Macmillan, 1997.

Stanislawski, Daniel. "Dionysus Westward: Early Religion and Economic Geography of Wine." *Geographical Review* 65, no. 4 (1975): 427–444.

Stanislawski, Daniel. *The Individuality of Portugal*. Austin: University of Texas Press, 1959.

Stanislawski, Daniel. *The Landscapes of Bacchus*. Austin: University of Texas Press, 1970.

Stevenson, Robert Louis. *The Silverado Squatters*. London: Chatto and Windus, 1906. Project Gutenberg, ebook #516. www.gutenberg.org/files/516. Accessed August 15, 2016.

Stewart, Kathryn. "How Alcohol Outlets Affect Neighborhood Violence." Prevention Research Center. Pacific Institute for Research and Evaluation. www.resources .prev.org. n.d.

Sumner, Daniel, Helene Bombrun, Julian Alston, and Dale Heien. "An Economic Survey of the Wine and Wine Grape Industry in the United States and Canada." Unpublished. UC Davis, 2001.

Swift, Jonathan. "On the Irish Club." In *The Works of Jonathan Swift,* Vol. 1. London: Henry G. Bohn, 1856.

Tax and Trade Bureau. "Tax and Trade Bureau List of Permittees." 2016. https://www.ttb.gov/foia/frl.shtml. Accessed May 22, 2016.

Telegraph. "SA Wines for India, Thank Rand & Scam." Calcutta, India: Telegraph, Dec. 14, 2004.

Theiss, Lewis. *A Journey through Pennsylvania Farmlands.* Harrisburg, PA: Pennsylvania Book Service, 1936.

This, Patrice, Thierry Lacombe, and Mark R. Thomas. "Historical Origins and Genetic Diversity of Wine Grapes." *Trends in Genetics* 22, no. 9 (2006): 511–19.

Thucydides. *The Peloponnesian War,* ca. 400 BCE. London: Penguin Books.

Tovey, Charles. *Wine and Wine Countries: A Record and Manual for Wine Merchants and Wine Consumers*. London: Hamilton, Adams, 1862.

Tovey, Charles. *Wit, Wisdom, and Morals, Distilled from Bacchus*. London: Whittaker, 1878.

United Nations Food and Agricultural Organization. *United Nations FAOSTAT.* http://faostat.fao.org/site/342/default.aspx, 2010. Accessed May 28, 2011.

United Nations, World Health Organization. "World Health Report 2002: Reducing Risks, Promoting Healthy Life." New York: United Nations, 2002.

United States House of Representatives. Abstract of the Seventh Census, 1850. Washington, DC: Robert Armstrong, 1853.

Unwin, Tim. *Wine and the Vine: An Historical Geography of Viticulture and the Wine Trade*. New York: Routledge, 1991.

USDA. "Washington Wine Grape Release, 2013," 2014. https://www.nass.usda.gov /Statistics_by_State/Washington/Publications/Fruit/winegrape14.pdf. Accessed May 21, 2016.

USGS. http://esp.cr.usgs.gov/info/eolian/14.gif, 2008.

Utah Alcoholic Beverage Control, *65th Annual Report, Summary of Operations July 1, 1999 to June 30, 2000*. Salt Lake City, UT: Utah Alcoholic Beverage Control. Accessed June 19, 2003.

Valverde, Carmen. "2015 Was a Record Year for Spanish Wine Exports." USDA FAS GAIN Report No. SP1607. April 1, 2016. http://gain.fas.usda.gov/Recent%20 GAIN%20Publications/2015%20was%20a%20record%20year%20for%20 Spanish%20wine%20exports_Madrid_Spain_4-1-2016.pdf. Accessed May 1, 2016.

Vintners Quality Alliance Ontario. "2015 Annual Report." Toronto, ON: Vintners Quality Alliance Ontario, 2016. http://www.vqaontario.ca/Resources/Library.

Virginia Wine Marketing Office. 2008. www.virginiawine.org.

Walker, Bob. The Wine Group Inc., Ripon, CA, 2002.

Washington State Liquor Control Board. *Annual Report, FY 2002*. 2003.

Washington Wine Commission. "Washington State Wine History." http://www.washingtonwine.org/wine-101/history/. Accessed July 6, 2011.

Weinhold, Rudolf. *Vivat Bacchus: A History of the Vine and Its Wine*. Watford, England: Argus Books, 1978.

Welles, Orson. "Lunch Conversations with Orson Welles." 1983. http://www.vulture.com/2013/06/orson-welles-lunch-with-henry-jaglom.html?mid=imdb. Accessed Nov. 15, 2013.

Wille, Kirk. "Ice Wine Is Hot." *Riesling Report*, March/April 2002. *www.rieslingreport.com*.

Wille, Kirk. "New York's Finger Lakes." *Riesling Report*, July/August 2002. *www.rieslingreport.com*.

Wine Business Monthly. "The Top 30 U.S. Wine Companies of 2003—Profiles." *Wine Business Monthly*, Feb. 2004.

Wine Business Monthly. "The Top 30 U.S. Wine Companies of 2004—Profiles," *Wine Business Monthly*, Feb. 2005.

Wine Business Monthly. "The Top 30 U.S. Wine Companies of 2005—Profiles." *Wine Business Monthly*, Feb. 2006.

Wine Business Monthly. "The Top 30 U.S. Wine Companies of 2006—Profiles." *Wine Business Monthly*, Feb. 2007.

Wine Business Monthly. "The Top 30 U.S. Wine Companies of 2007—Profiles." *Wine Business Monthly*, Feb. 2008.

Wine Business Monthly. "The Top 30 U.S. Wine Companies of 2008—Profiles." *Wine Business Monthly*, Feb. 2009.

Wine Business Monthly. "The Top 30 U.S. Wine Companies of 2009—Profiles." *Wine Business Monthly*, Feb. 2010.

Wine Business Monthly. "The Top 30 U.S. Wine Companies of 2010—Profiles." *Wine Business Monthly*, Feb. 2011.

Wine Business Monthly. "The Top 30 U.S. Wine Companies of 2011—Profiles." *Wine Business Monthly*, Feb. 2012.

Wine Business Monthly "The Top 30 U.S. Wine Companies of 2012—Profiles." *Wine Business Monthly*, Feb. 2013.

Wine Business Monthly "The Top 30 U.S. Wine Companies of 2013—Profiles." *Wine Business Monthly*, Feb. 2014.

Wine Curmudgeon. "Big Wine Tightened Its Grip on the U.S. Market in 2013." http://winecurmudgeon.com/big-wine-tightened-its-grip-on-the-u-s-market-in-2013/. March 13, 2014. Accessed April 24, 2016.

Wine Institute. "World Vineyard Acreage by Country." http://www.wineinstitute.org/files/2012_World_Acreage_by_Country_California_Wine_Institute.pdf. Accessed April 30, 2016.

Wine Institute. "U.S. Wine Exports, 95% from California, Jump 30% to $876 million in 2006." http://www.wineinstitute.org/resources/exports/article58. Accessed September 14, 2008.

Wine Institute. www.wineinstitute.org/communications/statistics.htm. 2006.

Wine Institute, www.wineinstitute.org/communications/statistics.htm. 2014.

Wine Standards Bureau. Guide to EC Wine Legislation. UKVA & Wine Standards Board. http://www.food.gov.uk/multimedia/pdfs/euwineregs.pdf. 2008.

Wines & Vines. "Wine Industry Metrics." http://www.winesandvines.com/template.cfm?section=widc&widcDomain=wineries. First update January 2016. Accessed May 22, 2016.

Wines of Chile. http://www.winesofchile.org/content/1266. 2008. Accessed October 5, 2009.

Winkler, A. J., J. A. Cook, M. W. Kliewer, and L. A. Lider. *General Viticulture*. Berkeley: University of California Press, 1962.

Wittwer, Glyn, and Kym Anderson. *Global Wine Markets, 1961 to 2003: A Statistical Compendium*. Adelaide, Australia: University of Adelaide Press, 2004.

Worstall, Tim, "Welcome to the Trade War—China Imposes Tariffs on EU, US, and Japanese Steel Imports." *Forbes*, June 24, 2016. http://www.forbes.com/sites/timworstall/2016/07/24/welcome-to-the-trade-war-china-imposes-tariffs-on-eu-us-and-japanese-steel-imports/#3066547c618e

www.oregonwine.org. "Discover Oregon Wines Fact Sheet." http://industry.oregonwine.org/wp-content/uploads/2010/08/2010-Industry-Facts-w_background.pdf. Accessed 7/6/2011.

Xiang, Li. "China's EU Wine Probe Sparks Worries." *ChinaDaily.com.cn*. Last update 2013-06-13-03:05. http://www.chinadaily.com.cn/business/2013-06/13/content_16612553.htm.

Yankelevich, Andrea. "Argentina Wine Annual 2015 Gain Report." USDA FAS, issued March 31, 2015.

Zelenko, Sergei. "Underwater Archaeology of the Black Sea: Crimean Coastal Survey 1997." Last update April 9, 1997. http://nautarch.tamu.edu/projects/crimea/final.htm.

Index

About the Author

Robert Sechrist was born in Lewistown, Pennsylvania and grew up in Downingtown, Pennsylvania. He graduated from the University of Pittsburgh in 1977 where he majored in Geography and Anthropology. After obtaining a master's degree in Geography from SUNY–Binghamton, he earned his doctorate in Geography from Louisiana State University in 1986. While at LSU he met his wife Gail; together they have two daughters, Emma and Katherine.

Robert joined the faculty at Indiana University of Pennsylvania in 1986. While there he has taught a number of courses, including Geographic Information Systems, Quantitative Methods, Computer Programming, and Wine. He has provided GIS implementation consulting services to numerous utility companies. He earned the Geographic Information Science Professional post nominal in 2008 and the Certified Specialist in Wine post nominal from the Society of Wine Educators in 2009. Past chair of the Association of American Geographers' Wine, Beer, and Spirits specialty group, he continues to play a role in that group and in the Society of Wine Educators.

His interest in wine goes way back. Bob started making wine at 15 and beer at 23. In 2012 he began distilling at his newly formed company, Disobedient Spirits. Bob is also an avid fisherman and kayak paddler.

For the past eighteen years Sechrist has taught and continues to teach the Geography of Wine course every fall and spring semester to about seventy undergraduate students. This book was written to accompany that course. In the summer and winter students from around the world take the course online for college credit.